OCT 2007

W9-BVA-686

The Deer
of North America

The Deer of North America

Leonard Lee Rue III

*All photography by the author unless
otherwise indicated*

THE LYONS PRESS
Guilford, Connecticut
An imprint of The Globe Pequot Press

This book is dedicated to four men who
enriched my life through their friendship—

George Homer Hicks
Bill Shipley
Jess Staugaard
Joe Taylor

Copyright © 1997 by Leonard Lee Rue III

First Lyons Press paperback edition, 2004

ALL RIGHTS RESERVED. No part of this book may be reproduced or transmitted in any form by any means, electronic or mechanical, including photocopying and recording, or by any information storage and retrieval system, except as may be expressly permitted in writing from the publisher. Requests for permission should be addressed to The Lyons Press, Attn: Rights and Permissions Department, P.O. Box 480, Guilford, CT 06437.

The Lyons Press is an imprint of The Globe Pequot Press.

10 9 8 7 6 5 4 3 2 1

Printed in the United States of America

ISBN 1-59228-465-5

The Library of Congress has previously cataloged an earlier (hardcover) edition as follows:

Rue, Leonard Lee.
    The deer of North America / Leonard Lee Rue III.
        p. cm.
    Originally published: Danbury, Conn.: Grolier Book Clubs, 1989.
    Includes bibliographical references and index.
    ISBN 1-55821-577-8
    1. Deer—North America. I. Title.
QL737.U55R83   1997
599.65'097—dc21   97-8276
                                                         CIP

CONTENTS

Part II: How the Year Goes

Part III: Toward Sound Deer Management

FOREWORD
to the 1997 Edition

I t was back in 1939 that I saw my first deer tracks, back on the family farm. Mind you, I hadn't seen a deer yet; they were so scarce that their same numbers today would make them candidates for the endangered-species list. I don't know how many deer were in New Jersey back in 1939, but I do know there were less than two hundred of them in the late 1800s. All of that has changed, dramatically. Today we have more deer in New Jersey, about 15,000, than we actually have food and space for.

The whitetail deer is one of conservation's great success stories. The total population of the whitetail north of the Mexican border is estimated at being over twenty-one million. We have more whitetail deer on our continent now than we have ever had. So far as whitetail deer are concerned, the "good old days" are today. Those who would debate that point fail to take into account the fact that, despite the destruction of habitat by civilization and urbanization that has diminished many species, the same opening up of the virgin forests and the elimination of most of the major deer predators in many of the areas has created optimum conditions for deer. In fact, in many areas, the whitetail deer is own worst enemy.

Whitetails are not cyclic in the same manner as are lemmings, grouse, varying hares, gray squirrels, etc. They are not subject to the same disease population controls as are raccoons, foxes, skunks, muskrats, etc. They do have a population controlling "boom-or-bust" cycle that is brought about by overpopulation, range destruction, and then starvation. With the elimination of the deer's major predators over most of their range, man has to fill the void. With the increasing deer population, most of the states have had to liberalize their hunting laws and seasons. In my own state of New Jersey, we have gone from one buck in one short season to as many as forty to fifty deer over a period of four months. Some states allow even more deer to be taken. And we still have problems.

Whereas in the past we used to be concerned with just the physical carrying capacity of the land, today we must also be concerned about the social carrying capacity of the land. We must now take into consideration the wishes and concerns of the population as a whole—hunters, landowners, and the general public. After all, the deer are state-owned, but 93 percent or more of the deer live and feed on private land. Hunters want to have the largest population possible to hunt, whereas many of the suburban home owners want the deer herd drastically reduced to prevent the destruction of their expensive shrubbery and the possibility of a collision with their even more expensive automobiles.

Because of the increased urbanization of what was once prime deer habitat, many areas are now off limits to gun hunting. In some of those areas, some of the states have tried to reduce the deer population by giving extra incentives to bowhunters to hunt there. In some areas no hunting of any kind is allowed.

The off-limit areas present a dilemma, offering both pluses and minuses. On the debit side are increased possibility of auto/deer collisions, the destruction of shrubbery and everything edible so that the deer will be weakened by starvation, which opens them up to disease. On the credit side, urban areas bordered by open areas act as a reservoir for the open-hunted areas as the young deer move out to fill the niche left vacant by the deer that have been harvested. The restricted areas are also allowing many of the bucks to live longer and thus offer the hunters more mature trophy animals when they do venture out into the open areas.

The ever-increasing deer population has created a problem for which there is no simple condition. On one hand you have the hunters, who want to hunt more deer. On the other hand you have antihunters, who don't want any animal, especially deer, to be hunted. In the middle you have the bulk of our human population, who understand that the deer are a renewable resource that must be controlled to prevent the destruction of the deer's habitat and the resultant starvation. These folks are neither for nor against hunting, they just want to do what is best for the deer.

It is realized that live trapping deer is too labor-intensive and too costly and there are few places in which to relocate the live-trapped deer. The use of sterilants is being considered but this, too, is labor-intensive, costly, and has not proven to work on the large scale that is needed to remedy the problem. At present, regulated sport hunting is still the most effective deer-management tool available. What is needed is new research, effort, and dialogues between all interested parties. That's what it will take to ensure the future of the whitetail deer. And that is what we are all interested in.

Dr. Leonard Lee Rue, III
Blairstown, New Jersey

PREFACE

When I finished writing the first edition of this book in 1978, I knew the book would never be really finished. After 12 printings of the original edition, the Outdoor Life Book Club and I decided that the book just had to be updated to incorporate all of the newest data from the field, and the tremendous amount of information I have learned in the intervening years.

I have now taken over 200,000 photographs of deer. I have no idea how many thousands of hours with deer that figure represents. Lest anybody get the idea that I think that I know all there is to know about deer, I don't. Despite the tremendous numbers of photographs I have taken, the hundreds of research reports I have read, and the countless thousands of hours I have spent doing both, I seldom observe deer without learning something new. As I said in one of my books, *How I Photograph Wildlife and Nature*, I learn more each day so I can teach more tomorrow. And I expect to be learning more and sharing more with you who read my books and listen to my seminars until the day I die. The acquisition of knowledge makes life so exciting.

My thanks again go to all who helped me to do the original book, and I extend it to those whose works have been added to the bibliography. One of my greatest regrets is that I have not met, personally, more of the researchers listed. I regret not having gone to many of the colleges; I regret not having been able to ask questions that needed answering. I wish I could have studied under many of these people. In particular, I wish I could have been a student of Dr. Richard Goss of Brown University. His scope, depth, insights, range, and application of knowledge is mind-boggling. His ability to communicate that knowledge is appreciated. Among many, many other good researchers, I would also like to have studied under are George Bubenik of the University of Toronto, John Ozoga and Louis J. Verme at the Cusino Research Station in Michigan, and Larry Marchinton at the University of Georgia.

Joe Taylor continues to be an unending source of knowledge to me and acts as a sounding board for the discoveries, ideas, and hypotheses that I come up with. I want to extend special thanks to Gary Knepp and his folks, Thelma and Richard. Don Edwards, of Louisiana, helped me to extend my deer knowledge there, and in Texas; thanks, Don. My dear friend, or deer friend, Helen Whittemore has passed away, but, through her will, I am able to continue to photograph the deer on her estate for the rest of my life.

My oldest son, Leonard Lee Rue IV, who goes by the professional name of Len Rue, Jr., has come back into partnership with me. He has taken over the running of the business, freeing me to do more writing, photography, and seminars. He has also taken some of the photos depicted in this book. It is one of the greatest satisfactions of my life to be able to work with him each day.

Marilyn Maring, who is a superb teacher, is my friend, my assistant. Putting these additions in the book had to be done here at home. At times the manuscript resembled a puzzle more than it did a book. It was through Marilyn's unstinting efforts that you will be able to read this updated and expanded edition. I cannot thank her enough.

A substantial foundation for this book was the work done by the people listed in my bibliography. I hope you will browse through the list and perhaps have occasion to read the original books or papers for further information. My thanks to all of those authorities.

The following people are personal friends and family members who have helped me immeasurably: Elaine and Manny Barrone, Denise Hendershot, Irene Vandermolen, my sons Tim and Jim Rue, Bill Shipley, Fred Space, Danita and Glen Wampler. Also thanks to Bob La Rose, who prints my thousands of photos; Susan Dulaney, who dries them; Mary Ann Johnston, who copies the captions for them; and Barbara Dalton, who files and refiles them.

It is always a pleasure to do a book for the Outdoor Life Book Club. My thanks to my good friends, retired publisher John W. Sill, editor Jerry Hoffnagle, and editorial director Neil Soderstrom.

God be with all of you and them.

Leonard Lee Rue, III
Blairstown, New Jersey

I
THE ANIMAL AND
ITS BEHAVIOR

1

Man's Historical Impact on Deer

Biologically as well as in their behavior, deer are exquisitely adapted creatures. Like a number of other herbivorous mammals, they are ruminants—in other words, cud-chewers—and this digestive adaptation is one of many that helps them avoid predators, as will be shown in later chapters. They are ungulates—or hoofed quadrupeds—and their particular type of cloven hoof structure is another aid to survival that will be explored in later pages. Members of the order Artiodactyla, family Cervidae, genus *Odocoileus,* they evolved during the Miocene, 15 to 20 million years ago, and they have outlasted many of the related and unrelated species and genera that evolved during that period.

They have also outlasted some forms of wildlife that thrived until the advent of civilization on this continent. The history of the deer population in North America is one of the greatest success stories in conservation literature. But it cannot be credited entirely to the efforts of conservationists, because today's population level is the result primarily of the tremendous adaptability of the deer themselves.

It is often said that there are more deer in North America today than there have ever been. I have made that statement myself, but my present research forces me to question its accuracy. I cannot prove the point one way or the other, however, and don't think anyone else can.

One of my earliest heroes was Ernest Thompson Seton, an excellent naturalist who really knew the wildlife he wrote about. I've read everything he wrote that I could get my hands on. In his memorable book *Lives of Game Animals,* Seton went out on a limb by trying to estimate the population of various game animals before Europeans came to this continent. He based his work on the writings of the earliest explorers, naturalists, and hunters. Seton then compared this research with reports from the various game departments of all the states and provinces. He considered all these findings in the light of his own remarkable expertise, and his calculations gave us the first major wildlife census figures. Many of his estimates have stood the test of time, having been backed up by research in recent years. But I must risk charges of sacrilege by refusing to accept his figures on the population of the deer in North America.

Seton calculated that 2 million square miles of whitetail country originally existed, with an average of 20 deer per square mile. That added up to 40 million whitetails. He estimated that mule deer occupied 2.5 million square miles, with 4 or 5 to the square mile, for a total of about 10 million mule deer. He figured that blacktail deer had about 300,000 square miles of territory, and allowed 10 deer to the square mile—or 3 million blacktails.

Deer were frequent subjects in the Currier and Ives lithographs, Audubon prints, and other nineteenth-century art. This is "Startled Deer—A Prairie Scene" (ca. 1848) painted by Victor Gifford Audubon and John Woodhouse Audubon. courtesy of the Brooklyn Museum

The greatest discrepancy is in the allotment of deer range. The continental United States, including Alaska, has 3,669,209 square miles of territory. Most of Alaska should be discounted because its deer (blacktails) inhabit only a portion of the southeastern Alaskan panhandle. If we take off most of Alaska, we have about 3,089,200 square miles. Of this area, Seton allowed 2,000,000 square miles for the whitetail and 2,500,000 square miles for the mule deer. The whitetail has always been found in every one of the lower 48 states, although the population is minimal in such western states as California, Oregon, Nevada, and Utah. The whitetail's main range today is the eastern two-thirds of the United States.

Seton's estimate of 20 whitetails per square mile is now considered to be about the optimum number of deer *on the best of ranges,* although we do have areas with 50 deer per square mile.

With regard to whitetails, then, Seton evidently underestimated the range but drastically overestimated the population density throughout that range. The result was much too high a figure for whitetails. His figures for muleys and blacktails were not as far off the mark though they, too, were probably inflated. If there is any truth in the claim that we now have more deer than ever—or even as many—all of Seton's estimates were fantastically magnified. It is unlikely that the continent ever had enough habitat or just the right kind to support 40 million whitetails, though it might possibly have supported slightly more than the present 19.5 million. The latest census of mule deer puts their population at just a bit over a third of Seton's 10-million estimate, and the census of blacktails is a little more than half of his 3-million estimate.

All three groups now total about 24,270,250. That is a high population for any single kind of wildlife. Whether or not there were more deer before the coming of the Europeans, deer were always abundant on this continent, and they were important to man from the time of the earliest human presence here.

The whitetail was as important to the eastern woodland Indians as the bison was to the Indians of the plains. Venison was a dietary staple, their "bread of life." Deer hide was made into clothing, the sinew was used to sew the skins, the bones were fashioned into splinter awls to make the holes in the leather so that the sinew could be used, and the hoofs provided glue, ornaments, and rattles. The deer's hair was stuffed inside moccasins in the winter to act as insulation. The bones and antlers were made into tools, weapons, decorations, and religious implements. The mound-building Indians of Ohio deified the deer and made sacred head-dresses adorned with antlers.

Deer were equally important to the earliest European settlers, providing them with almost as many of the basic necessities of life as they did for the Indians. The new settlers, in fact, hunted deer far more intensively than did the Indians. There were soon more settlers than there were Indians. Moreover, deerskins and venison became major items of trade with Europe—a situation that gave rise to professional hunting. Thomas Meacham of Hopkinton, in St. Lawrence County, New York, kept an exact record of the game he killed as a professional hunter. When he died in 1850, he had killed 214 wolves, 77 cougars, 219 bears, and 2,550 deer. Nathaniel Foster of Herkimer, New York, killed 76 deer in a single season. Meshach Browning, who died in 1859, was a professional hunter from Garrett County, Maryland. In 44 years he killed between 1,800 and 2,000 whitetails. There were many such hunters.

The records of His Majesty's Custom Service show that 2,601,152 pounds of deerskins from 600,000 whitetails were shipped to England from Savannah, Georgia, in the years 1755 to 1773.

A traveler named Captain Marcy told of passing through southern Texas in 1846 and seeing thousands of deer daily—as many as 100 to 200 in a single herd. If Captain Marcy were to travel now in the Llano Basin of Texas, he would find the deer there as plentiful as they were in his day. Texas currently has over 3 million whitetails, and there are as many as 121 deer to the square mile in the Llano Basin.

Today, ill-informed people tend to condemn the professional hunters of yesterday. These men were a product of their time, a time when it was believed that the supply of game was endless. The professional hunters were heroes to the boys of that era and the envy of most of the men.

The citizens of the short-lived state of Franklin (Tennessee today) even paid their officials' salaries in deerskins. After all, they had no money but they did have lots of deerskins.

According to the records of the railroads in St. Paul, Minnesota, 7,409 venison saddles and carcasses, 4,000 venison hams in boxes, and 750 pounds (340.2 kg) of venison in barrels were shipped to markets in November and December of 1877. It is estimated that about 2,000 deer were killed within 15 miles (24.1 km) of Osakis, Minnesota, in October, November, and December of

1880 by professional hunters. The venison was shipped to markets as far away as Boston.

Records such as these abound—and would seem to support Seton's statistics. However, there are also many records such as those of Major Robert Rogers, who found deer exceedingly scarce in the Northeast. During the French and Indian Wars (1689–1763), Rogers led his famous company of rangers north from Massachusetts through Maine to attack the Abnaki Indians on their home ground. His men, although skilled hunters and frontiersmen, saw no game and were actually starving by the time they returned.

Most of my home state of New Jersey was originally covered with mature forests, and we know that deer are not creatures of the mature forests. Most of the eastern half of the United States was also covered with such forests and undoubtedly those woods did not have high deer populations.

The "good old days" for deer in most of North America are right now: there were 110 whitetails in these fields and the fencerow adjoining them, that I could count. In many parts of the country, whitetail populations have expanded to fill the carrying capacity of the range.

In 1646, Rhode Island became the first colony to pass a law protecting the deer from hunting for a part of the year. Other colonies and then the states followed Rhode Island's lead in protecting the deer in some manner. Often the laws were considered just a nuisance to be ignored. After all, no law has ever prevented anyone from doing anything; laws merely punish those who are caught breaking them.

The late 1800s were the blackest period for all kinds of wildlife throughout the continent. As the deer herds were reduced, conservationists labored to alert the populace to the fact that this splendid game species was on the verge of being annihilated. The northeastern states were the hardest hit. By the end of the 19th century, New Jersey was down to fewer than 200 deer. In Massachusetts, New Hampshire, and Vermont, deer were so scarce that just sighting the tracks of one made headlines in local newspapers.

Finally, because of the increased awareness of the general public, more rigid game laws, better enforcement of the laws, and the importation of deer from states that still had them to the areas where they had been extirpated, the pendulum started to swing back the other way. Today, that pendulum is still swinging in the direction of a population boom for whitetails, although the populations of both mule deer and blacktails have declined slightly in recent years. Much of the decline in the mule and blacktail deer populations stems from the competition of domestic live-stock— that is, the practice of overgrazing on federal lands by stockmen, particularly those who own sheep.

Despite that situation, the future of deer of North America is bright. Deer thrived before the coming of man and they still thrive. Barring some new calamity, they should continue to inhabit this continent farther into the future than we can speculate.

2

Varieties and Distribution

M embers of the deer family—17 genera, 40 distinct species, and more than 190 subspecies—range throughout most of the world, although they are not found in Antarctica and have an exceedingly limited range in northern Africa. They have been introduced into Australia and New Zealand. The North American deer family includes the elk, moose, and caribou, as well as the animals of the genus *Odocoileus,* more commonly known as deer.

Scientifically, a deer is assigned to the phylum Chordata because it has a backbone. It belongs to the class Mammalia because the females have mammary glands and suckle their young, which are born alive rather than as egg-enclosed embryos. A mammal is also characterized by a four-chambered heart, hair covering at least a part of its body, and a homoiothermal system—meaning the animal is warm-blooded, with a fairly constant body temperature regardless of the temperature of its surroundings. The deer is in the order Artiodactyla, meaning "even-toed," because it has four toes (the two hoof lobes plus the dewclaws) on each foot. It is in the suborder Ruminantia because it has a four-compartmented stomach and chews a cud. The family grouping is Cervidae.

The exclusively American genus of deer, *Odocoileus,* is generally considered to include only two species—the whitetail *(O. virginianus)* and the mule deer, or mule-deer group *(O. hemionus).* The Columbian blacktail and the Sitka blacktail are ranked as subspecies of the mule deer. For that matter, there are a good many other subspecies of both the whitetail and the muley. As will be seen, these classifications have engendered some controversy among biologists and some confusion among the rest of us.

A species is an organism that is genetically linked so that its members are sexually compatible, making reproduction possible. Occasionally, matings take place between different species, but only rarely are the resulting crosses fertile. Most members of most species look alike, having similar characteristics. Variations that

Varieties and Distribution

The Florida Key deer, O.v. clavium, *is the smallest subspecies of the white-tail deer north of Mexico. It illustrates Bergmann's Rule, which states that body mass of a geographic race is in direct relation to the distance from the equator. The Key deer's body mass is about one-fourth that of our northernmost subspecies,* O.v. borealis.

evolve within a species, creating subspecies, are usually the result of geographic and physiographic conditions causing spatial isolation of one or more populations. The response by the organism to a given environment affects its size, weight, body conformation, and color as the species adapts to differences in temperature, light, moisture, altitude, regional vegetation, and so on.

The different subspecies of deer conform to four biological laws of natural selection. One of these, called Bergmann's Rule, states that the farther a geographic race is found north or south of the equator, the larger the mass of its body will be. The larger the body, the smaller is its relative surface area, resulting in a reduced loss of body heat. Conversely, the hotter the habitat is, the smaller the body and the larger its relative surface will be, allowing for greater heat dissipation. Thus, the Key deer *(O. v. clavium)*, our most southern deer, is the smallest subspecies, and the northern woodland whitetail *(O. v. borealis)* and Dakota whitetail *(O. v. dacotensis)* are our most northern deer and our largest.

I have written "Rue's Rule" to amend Bergmann's Rule to read, "the larger the members of the same species will be, so long as their preferred food is abundant and meets their nutritional needs; body size decreases in direct proportion to decreasing food supplies.

The third biological law, Allen's Rule, states that among warm-blooded creatures the physical extremities—ears, tail, and legs—are shorter in the cooler part of their range than in the warmer part. This rule is borne out by the Coues whitetail *(O. v. couesi)* of southern Arizona, which has larger ears and tail relative to body size than the northern deer. When hunters or other observers see the relatively mule-eared Coues deer—also known as the Arizona whitetail—at a distance, they occasionally mistake it for a muley. Nevertheless, the color and shape of the tail and, in a mature buck, the conformation of the antlers will usually distinguish a subspecies of whitetail from a muley where both species are encountered. In subsequent pages I will give detailed descriptions to facilitate field identification.

The fourth law, Gloger's Rule, states that among warm-blooded animals dark pigments are most prevalent in warm, humid habitat. I must add that dark coloration also tends to prevail in forested regions. Note the darkness depends on humidity as well as temperature. Hot, *dry* habitat does not produce darker coloration, for pigment is a survival factor and a dark deer (or other prey species) would invite predation by standing out conspicuously against a background of desert rock and sand. Red and yellow tints tend to dominate in arid regions, paler tones—reduced pigmentation—in colder climates. You will see the operation of Gloger's rule in the descriptions of the various subspecies.

There are 30 recognized subspecies of the whitetail deer. Seventeen are found north of the Mexican border, and these are the ones I will be discussing. There are 11 subspecies of the mule deer, and I will discuss the 8 that are found north of the Mexican border. Although the blacktail deer is scientifically classified as a subspecies of the mule deer because of close biological links, most wildlife biologists believe that the blacktail is a nascent species—an emerging species which, in time, will evolve into a separate species. This process would be speeded if barriers such as impassable mountain ranges, barren deserts, or great bodies of water caused spatial isolation. At present, however, the blacktails are widening

Varieties and Distribution

their range and frequently hybridizing with other mule deer sub-species. All the same, the two varieties of blacktails differ mark-edly from their close relatives in some respects, and at times I will discuss them separately, as if they already constituted a distinct species.

The deer of our continent are native Americans, descended from a common ancestor of the Pliocene period of 10 million years ago. This ancestor was itself evolved from Miocene types that ex-isted perhaps twice as long ago. Deer as we know them today de-veloped during the Pleistocene period, about one million years ago, when they spread throughout the continent and diversified into the two separate whitetail and mule deer species. Geographical and physical barriers tended to isolate them while continued evo-lution created further differences, the bases for the subspecifi-cation.

The genus name of *Odocoileus* was bestowed on all of them by the French-American naturalist Constantine Samuel Rafinesque, a gifted but slightly eccentric and glory-hungry scientist with a mania for discovering and naming species—sometimes erron-eously. For example, between 1818 and 1820 John James Audu-bon, who enjoyed practical jokes, tricked him into publishing de-scriptions of almost a dozen purely fictitious creatures. Rafinesque added substantially to 19th-century knowledge of American wild-life, but in his rush to claim credit, he occasionally grew careless. In 1832, while exploring caves in Virginia, he found a fossilized deer tooth, the last remains of some ancestor of the whitetails browsing in Virginia's woods. From this prototype he named the genus and species. Evidently he meant to call the genus *Odonto-coelus,* meaning "hollow tooth" or "concave tooth" (why, no one is certain), but Greek was not his strong suit and the result was *Odo-coileus.* The error has been perpetuated because taxonomists honor the earliest recorded name for a creature. *Virginianus* was added because the fossil was discovered in Virginia, and so the whitetail deer of that region became the "type species."

It was also Rafinesque who christened the mule deer *O. hemio-nus.* He named it—in this instance more accurately—from a speci-men taken at the mouth of the Sioux River in what is now South Dakota. *Hemionus* is another Greek word, meaning "mule" or "part-ass," a reference to the animal's very large ears.

All living things are scientifically classified according to a sys-tem originated by the great Swedish botanist and naturalist Caro-lus Linnaeus, in the mid-1700s. Thanks to this system, organisms are cataloged and given a distinctive, often descriptive, universal Latin or Greek name. The chief advantage is that scientists throughout the world, regardless of their own language, know by an organism's scientific name that they are all talking about the same genus, species, or whatever.

The classification of living things, known as *taxonomy*, is a complex but not always exact science. Among taxonomists there are two warring factions, the "lumpers" and the "splitters." The lumpers are those biologists who want to simplify (and sometimes oversimplify) the divisions and differences found within a single genus, species, or other classification. The splitters are those who emphasize slight (and sometimes imagined) differences. Quite often the arguments advanced by both groups seem beyond any reasoning, but they have brought order out of what would otherwise be chaos. As one side or the other prevails in a controversy, and as new discoveries are made, reclassification continues.

The Major Musculature of the Whitetail

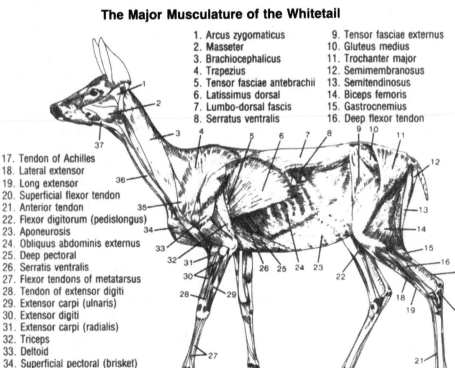

1. Arcus zygomaticus
2. Masseter
3. Brachiocephalicus
4. Trapezius
5. Tensor fasciae antebrachii
6. Latissimus dorsal
7. Lumbo-dorsal fascis
8. Serratus ventralis
9. Tensor fasciae externus
10. Gluteus medius
11. Trochanter major
12. Semimembranosus
13. Semitendinosus
14. Biceps femoris
15. Gastrocnemius
16. Deep flexor tendon

17. Tendon of Achilles
18. Lateral extensor
19. Long extensor
20. Superficial flexor tendon
21. Anterior tendon
22. Flexor digitorum (pedislongus)
23. Aponeurosis
24. Obliquus abdominis externus
25. Deep pectoral
26. Serratis ventralis
27. Flexor tendons of metatarsus
28. Tendon of extensor digiti
29. Extensor carpi (ulnaris)
30. Extensor digiti
31. Extensor carpi (radialis)
32. Triceps
33. Deltoid
34. Superficial pectoral (brisket)
35. Shoulder-transverse process muscle
36. Sternocephalicus
37. Sternomandibularis

Illustration by Robert Pratt and Wayne Trimm from "White-tailed Deer Ecology and Management." Used by permission of the Wildlife Management Institute.

Varieties and Distribution

The original division of subspecies was based on physical differences in skull characteristics, dentition, body size, and geographical locations. Such differences are sometimes clear-cut, sometimes subtle. The little Key deer of Florida, *Odocoileus virginianus clavium,* is the smallest deer north of the Mexican border, and no one would disagree that this whitetail is a distinct subspecies. The largest whitetail, *Odocoileus virginianus borealis,* is found in the northeastern United States and Canada. Most people, including me, could not tell it from any of the other large subspecies.

The deer of New Jersey are classified as *Odocoileus virginianus borealis.* I defy any expert in taxonomy to prove this classification is correct and I'll tell you why. According to Dr. Witmer Stone, an eminent naturalist whose writings appeared at about the turn of the century, the whitetail in New Jersey had been reduced to less than 200 individuals by the end of the 1800s. In an effort to

The Major Skeletal Structure of the Whitetail

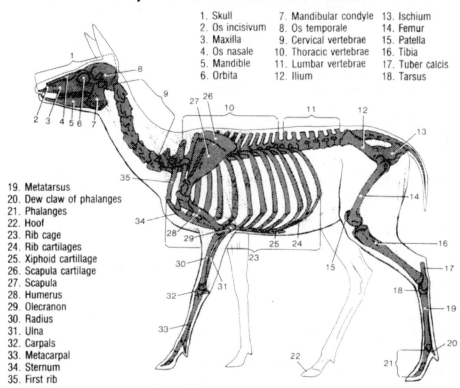

1. Skull
2. Os incisivum
3. Maxilla
4. Os nasale
5. Mandible
6. Orbita
7. Mandibular condyle
8. Os temporale
9. Cervical vertebrae
10. Thoracic vertebrae
11. Lumbar vertebrae
12. Ilium
13. Ischium
14. Femur
15. Patella
16. Tibia
17. Tuber calcis
18. Tarsus

19. Metatarsus
20. Dew claw of phalanges
21. Phalanges
22. Hoof
23. Rib cage
24. Rib cartilages
25. Xiphoid cartillage
26. Scapula cartilage
27. Scapula
28. Humerus
29. Olecranon
30. Radius
31. Ulna
32. Carpals
33. Metacarpal
34. Sternum
35. First rib

Illustration by Wayne Trimm from "White-tailed Deer Ecology and Management." Used by permission of the Wildlife Management Institute.

save the deer, hunting was prohibited for 21 years. New and stronger game laws were passed, and they were more strictly enforced. Many private individuals, notably Charles C. Worthington of Pahaquarry Township, and the State of New Jersey purchased hundreds of deer from Virginia, Maine, Michigan, Wisconsin, and perhaps other states. There were no centrally kept records of the private purchases, and current information is sketchy. The deer were released or escaped into various parts of the state, breeding with the remnants of the native deer. Since the original importation, many more deer have been transplanted to many different locations throughout the state. Those imported from Maine, Michigan, and Wisconsin were the *borealis* subspecies, as were the original New Jersey deer, but those from Virginia were not. Yet the subspecies in New Jersey is still classified as *O. v. borealis*.

This has been the situation in most eastern states. Even within a single state (and within a subspecies) there are considerable variations in body size and conformation, depending on habitat and the amount, quality, and types of food available. To further complicate the issue, unless there are definite geographic features dividing the ranges of subspecies, there is always an overlapping and interbreeding, making positive identification of many races arbitrary.

Before describing the species and subspecies in more detail, it may be helpful to discuss the marks of identification that enable a naturalist, hunter, or other observer in the field to recognize whitetails, mule deer, and blacktails. The three simplest keys to field identification are geography, tails, and antlers.

East of the Mississippi River, there are no mule deer or blacktails, except for small transplanted populations such as Oregon blacktails that have been stocked in Tennessee. These introduced herds are usually well publicized. In the East, therefore, you can generally assume that a wild deer is a whitetail. The range of the blacktail is restricted to coastal areas in California, Oregon, Washington, British Columbia, and Alaska. So, in most of the West, if a deer is not a whitetail it must be a mule deer.

Muleys tend to be larger than blacktails or whitetails but, as the chapter on weights will show, size is not a reliable criterion. The tail is. A mule deer has a relatively short, narrow, pendant-shaped tail, mostly white but tipped with black. Some subspecies also have a bit of black at the base of the tail. The blacktail has a somewhat fuller tail, but it is considerably smaller than a whitetail's. Its dorsal (outer, or rear) surface is black or blackish from base to tip. A blacktail-muley hybrid has a black stripe, with white showing on each side, from tail base to tip. The whitetail has a longer, fuller tail, brown or with gradations of black on the dorsal surface. It is white underneath, and when the tail is raised, as in alarm, this white is very conspicuous—hence, the name of the species.

Bucks with sizable racks can usually be identified by their antler structure as well. The points, or *tines*, of a whitetail's antler branch up from the two main beams. On a mule deer or blacktail, each main beam is bifurcated—that is, it forks, and then forks again. Generally, there are two such Y-forks, or four points (though there may be more), on each side, plus a pair of small brow tines, on a fully grown set of muley antlers. But again, antler conformation is less reliable in some regions than the size, shape, and color of the tail. There are occasional blacktails with antlers that look like a whitetail's, and a few whitetails grow bifurcated antlers that look like a mule deer's.

In addition to these marks of identification, the size and placement of the glands on the hind legs furnish a key to species recognition. But you have to get a close look to make such use of these glands. The tails and antlers are discussed in more detail in Chapter 4, the glands in Chapter 5.

The whitetail deer, also called *white-tailed* deer and sometimes *flagtail*, is found in all of the contiguous states and in eight of the Canadian provinces. The latest estimates put the total whitetail population at over 19,316,350. The following list enumerates some general physical characteristics and the distribution of the seventeen North American whitetail subspecies, or races.

1. The Virginia whitetail, *Odocoileus virginianus virginianus*, is the prototype of all whitetail deer. Its range includes Virginia, West Virginia, Kentucky, Tennessee, North Carolina, South Carolina, Georgia, Alabama, and Mississippi. This is a moderately large deer with fairly heavy antlers. It is hunted in all of the states it inhabits, and each state has a good deer population. It has a widely diversified habitat, varying from woodlands, coastal marshes, swamplands, and pinelands to the "balds" atop the Great Smoky Mountains.

2. The northern woodland whitetail, *O. v. borealis*, is the largest and generally the darkest in coloration. It also has the largest range, being found in Maryland, Delaware, New Jersey, Pennsylvania, Ohio, Indiana, Illinois, Minnesota, Wisconsin, Michigan, New York, Connecticut, Rhode Island, Massachusetts, New Hamp-

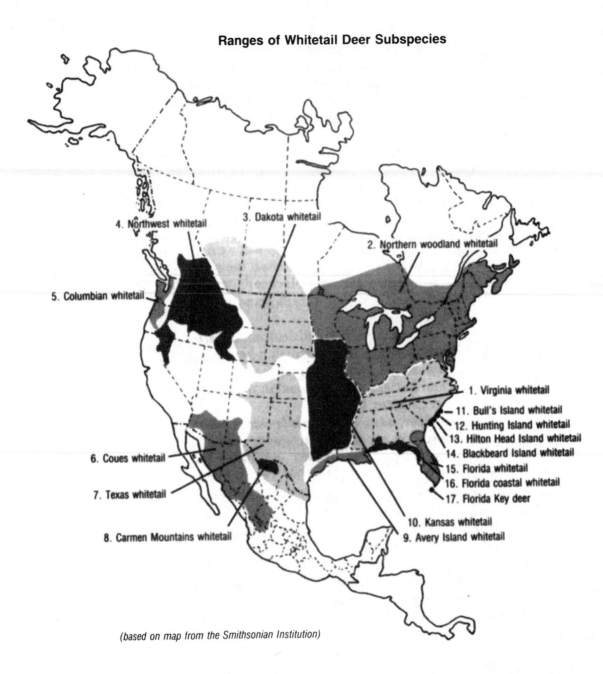

(based on map from the Smithsonian Institution)

shire, Vermont, Maine, and the Canadian provinces of New Brunswick, Nova Scotia, Quebec, Ontario, and a portion of Manitoba. More whitetails of this subspecies are hunted than of any other.

This area has also produced 8 of the top 20 record whitetail heads listed in the Boone and Crockett Club's official records book,

Records of North American Big Game. Founded by Theodore Roosevelt, the Boone and Crockett Club is dedicated to big-game hunting and to the study and conservation of wildlife. Since 1932, the organization has been publishing and updating a series of record books in which the most outstanding trophies are listed for 32 species of North American game animals. Official scoring charts and a very careful scoring system govern the measurement of these trophies. A Big Game Awards Program is administered, and the records book is sponsored by the Boone and Crockett Club.

The ninth edition of *Records of North American Big Game* was published in 1988. For the sake of dispensing information, it lists not only the outstanding animals harvested by hunters but also trophy animals of unknown origin—some killed by anonymous hunters long before the record-keeping began, some simply found and submitted because of impressive size. Unless otherwise specified, the record trophies I mention in my book will be those officially recognized by the Boone and Crockett Club.

As might be expected, the current world-record whitetail head is from the northern woodland region of *O. v. borealis.* The deer was killed near Danbury, Wisconsin, in 1914 by James Jordan. Its antlers score 206⅛ in the Boone and Crockett system. Both main beams measure 30 inches (762 mm) and each has 5 *points*—tines more than an inch (25.4 mm) long. Each main beam has a circum-

An eight-point Virginia whitetail buck, O.v. virginianus, *the deer of our eastern and southern forests and lowlands. At right, the northern woodland whitetail buck,* O.v. borealis, *the largest and darkest of the North American whitetail deer, in prime condition in September.*

ference of more than 6 inches (152.4 mm) at the smallest place below the first point. A good many trophies have longer beams or more points, but scoring takes into account such other factors as the girth of the beams, their inside spread, and the lengths of the points. Incidentally, I am speaking here of typical heads; there is a separate classification for "nontypical antlers," which are generally asymmetric and sometimes studded with so many points that they look like stalagmite formations in a limestone cavern. Second place among typical whitetails is held by a 12-point buck shot in 1971 in Missouri. Third place is held by a 13-pointer shot in 1965 in Illinois—again, prime *borealis* country.

The range of *borealis* is expanding steadily northward. In the 17 years when I guided canoe trips into the wilderness areas of Quebec, I witnessed a northward expansion of more than 100 miles (160.9 km). As a lumber company cut the virgin forests of spruce for paper pulp, the network of logging roads opened up vast areas for hunters, fishermen, and tourists. Before such an onslaught of disturbances, the moose and wolves retreated northward. The cut-over land sprang back with newly sprouting aspen, birch, berry bushes, and conifers. With the increased food and decreased competition and predation, this whitetail is enlarging both its range and its numbers.

3. The Dakota whitetail, *O. v. dacotensis,* is another very large deer, about equaling the northern woodland whitetail in weight (see Chapter 9 for tabulations of body weights). This subspecies has produced even more of the high-ranking trophy heads than the *borealis* race. The range covers North Dakota, South Dakota, and parts of Nebraska, Kansas, Wyoming, Montana, and the Canadian provinces of Manitoba, Saskatchewan, and Alberta. Dakota bucks have heavy, fairly widespread antlers. The winter coat is a little paler than that of *borealis*. This is a deer of the breaks. Its home is in the timbered coulees, gullies, draws, and river and stream bottoms that cut through the prairies.

4. The Northwest whitetail, *O. v. ochrourus,* is also a large deer. It inhabits parts of Montana, Idaho, Washington, Oregon, California, Nevada, Utah, and the Canadian provinces of British Columbia and Alberta. The biggest-bodied whitetail I've ever seen was in Glacier National Park. It could have been either this subspecies or a Dakota. The two races intergrade in that area. This subspecies has very widespread antlers and a winter coat of relatively pale cinnamon-brown.

5. The range of the Columbia whitetail, *O. v. leucurus,* has

The Dakota whitetail, O.v. dakotensis, *is a deer of the high plains in the United States and Canada; in this northern range it has evolved as a large-bodied subspecies often equalling* borealis *in body size and antler production. At right, the Northwestern whitetail,* O.v. ochourus, *in mid-summer velvet. These two subspecies intergrade along the continental divide, which marks their range boundary.*

been so greatly reduced that most of these deer are now found only on the Federal Columbian White-tailed Deer Refuge, on the Columbia River near Cathlamet, Washington. The subspecies formerly ranged along the Pacific coast in Washington and Oregon, spreading eastward to intergrade with the Northwest whitetail. The Columbian whitetail is not hunted: it is now on the endangered-animal list.

The population is estimated at 500 animals or fewer. Unfortunately, of the 13 Columbian whitetails I've seen, 9 wore large numbered collars and ear tags. I fully realize the importance of research to improve our management of wildlife, but I was terribly disappointed to find most of the Columbians wearing tags. They are large deer with high but narrow-spreading antlers.

6. The Coues (pronounced "cows"), or Arizona whitetail, *O. v. couesi,* is a small variety. At one time it was thought to be a distinct species, but more recent research has relegated it to the status of subspecies. It has larger ears and tail relative to its body size than most whitetails, but I found that the ears and tail were not as large as I had been led to believe. This deer is found in the dry, desert regions of southeastern California, southern Arizona, southwestern New Mexico, and on down into Old Mexico. The Coues is apparently isolated from areas where it could intergrade

Left: O.v. couesi, *the Coues deer, is a smaller whitetail subspecies that ranges from desert Arizona through Texas down into Mexico. Right:* O.v. texanus, *the Texas whitetail, shares range borders with Coues deer from New Mexico to central Texas.*

with the Texas whitetail, but in the southern part of its range it probably intergrades with several Mexican subspecies. Even in Arizona the Coues whitetails are more or less isolated in the mountainous areas that rise above the desert, such as the Chiricahua and Huachuca Mountains. Arizona estimates it has about 25,000 Coues deer but does not give any harvest figures. New Mexico has a hunting season for this deer but gives neither a population estimate nor the hunters' take. The Coues deer has its own classification in the Boone and Crockett Club, dating back to when it was considered a distinct species. From the hunter's point of view, the separate classification remains legitimate, since the little Coues deer has a light rack. A trophy that is outstanding by Coues standards could hardly compete with a northern woodland or Dakota whitetail trophy.

7. The Texas whitetail, *O. v. texanus,* is found in western Texas, Oklahoma, Kansas, southeastern Colorado, eastern New Mexico, and the northern portion of Old Mexico. Everything about Texas is big, even its population of whitetail deer. It has an estimated 4²⁄₁₀ million of them, more than any other state. The annual harvest usually goes over 350,000. Texas has four whitetail sub-

Varieties and Distribution

The Carmen Mountain whitetail, O.v. carminis, *remains a distinct subspecies isolated in the middle of* O.v. texanus *range by mountain barriers. At right is another of the four subspecies found in Texas,* O.v. mcilhennyi, *the Avery Island whitetail, also found in the Mississippi delta.*

species, of which the most abundant is the Texas whitetail. Its body is much smaller than that of the more northerly deer, but it is the largest of the southern forms. The antlers are slender but widespread, and there are several record heads among the top 25. I have photographed a world-record whitetail in south Texas.

8. The Carmen Mountains whitetail, *O. v. carminis,* is a small deer found in the Big Bend region of Texas. Its range is limited to the Carmen Mountains on both sides of the Rio Grande. Not many of these deer are hunted because most of their range falls within the boundaries of Big Bend National Park, where hunting is prohibited. Here is a good example of isolation. A buffer strip of semi-desert, inhabited by mule deer, separates this subspecies from the Texas whitetail and prevents intergrading.

9. The range of the Avery Island whitetail, *O. v. mcilhennyi,* stretches along the Gulf Coast in Texas and Louisiana. This is the deer of the Texas Big Thicket Country. It is a large one with a dark, brownish winter coat, and it intergrades with the whitetail subspecies found to the west, north, and east. Some of the whitetails I have photographed in Louisiana's bayous had the largest antlers I have ever seen.

10. The Kansas whitetail, *O. v. macrourus,* is the fourth subspecies occurring in Texas. Found in eastern Texas, Oklahoma, Kansas, Nebraska, Iowa, Missouri, Arkansas, and Louisiana, it is a large deer with heavy main antler beams and short tines. Several deer of this type are listed among the top 25 heads.

11. The Bull's Island whitetail, *O. v. taurinsulae,* is an isolated and very limited race of whitetail deer, found only on Bull's Island, South Carolina.

12. The Hunting Island whitetail, *O. v. venatorius,* is another of South Carolina's minor variations, found only on Hunting Island.

13. The Hilton Head Island whitetail, *O. v. hiltonensis,* is still another South Carolinian variation, limited to Hilton Head Island.

14. The Blackbeard Island whitetail, *O. v nigribarbis,* is found only on the Georgia islands of Blackbeard and Sapelo.

All of the last four subspecies named are medium-size deer with fairly small antlers that are heavily ridged or wrinkled at the base. The islands they inhabit are far enough out in the ocean to prevent intergrading with mainland subspecies or one another.

15. The Florida whitetail, *O. v. seminolus,* is a good-size deer with a good rack. Some of the Florida deer that I have photographed in Okefenokee Swamp in Georgia were every bit as large as those photographed back home in New Jersey. Some had antlers as impressive as those of *borealis,* though the spread was not as wide. The Florida whitetail is the deer of the Everglades.

16. The Florida coastal whitetail, *O. v. osceola,* is found in the Florida panhandle, southern Alabama, and Mississippi. It is not as large as the Florida or the Virginia whitetail, but it intergrades with both.

17. The diminutive Florida Key deer, *O. v. clavium,* is the smallest of our native deer. No hunting is allowed for this subspecies, which is on the endangered-animal list. By 1949, the Key deer population had plummeted to an all-time low of 30 individuals. This reduction was brought about by habitat destruction, fires, hurricanes, automobile kills, and hunting. The Key Deer National Wildlife Refuge was established in 1953. With the protection thus provided, the deer population has crept back up to about 300. Today the automobile is the number-one killer; the highway linking the Florida Keys passes through the center of the range. Commercial and residential land development in the Keys is a constant threat. Additional land should be added to the refuge.

A Kansas whitetail doe, O.v. macrourus, *which is also found in parts of Texas. The bucks are heavy-antlered and several specimens are near the top of the whitetail trophy records.*

The stately Florida whitetail deer, O.v. seminolus, *is a good-size whitetail deer found mainly in the everglades, northern Florida, and the lowlands of Georgia.* (photo by Len Rue, Jr.) *The range of the smaller* O.v. osceola, *includes the Gulf coastal areas of Florida, Alabama, and Mississippi.*

Ranges of Mule Deer Subspecies

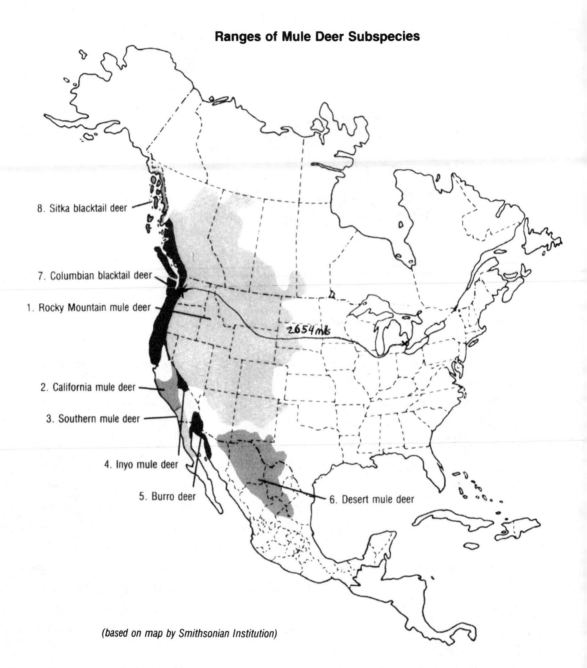

8. Sitka blacktail deer

7. Columbian blacktail deer

1. Rocky Mountain mule deer

2654 mls

2. California mule deer

3. Southern mule deer

4. Inyo mule deer

5. Burro deer

6. Desert mule deer

(based on map by Smithsonian Institution)

The mule deer, also known as the *muley*, inhabits the mountains, deserts, coast, and high plains of the western half of the United States and Canada. In 1984 the North American mule deer population was estimated to be about 3,443,900. The blacktail deer is generally regarded as a coastal variety, but in California black-

Varieties and Distribution

tails are found as far east as the mountains of Yosemite National Park. The total blacktail population was estimated in 1984 to be 1.5 million. The following list enumerates the eight subspecies of mule deer, including the two races of blacktails.

1. Rocky Mountain mule deer, *Odocoileus hemionus hemionus,* is the prototype. It has a larger range than any other subspecies of mule deer or whitetail. This is the northernmost of the mule deer, the largest, heaviest, and darkest in coloration. (In view of its northerly distribution, you might expect it to be paler, but much of its habitat is forested, whereas some other subspecies—most notably the desert muley—have a paler coloration that blends best with their surroundings.) The Rocky Mountain subspecies is found in the Canadian provinces of British Columbia, Alberta, Saskatchewan, Manitoba, and in the Northwest Territories, as well as in Washington, Idaho, Montana, North Dakota, Minnesota, South Dakota, Nebraska, Wyoming, Oregon, California, Nevada, Utah, Colorado, Oklahoma, New Mexico, and Arizona. And I have received reports indicating it has spread its range to Iowa.

The average mule deer is larger than the average whitetail, and all of the top 25 mule deer trophies in the Boone and Crockett records are of this subspecies. At the western fringe of its range, the Rocky Mountain muley intergrades with the Columbian blacktail.

2. The California mule deer, *O. h. californicus,* is found only in California, from the Sierras to the Pacific. It intergrades with the Columbian blacktail along the northern border of its range. It is smaller than the Rocky Mountain mule deer and also has a smaller white rump patch.

3. The southern mule deer, *O. h. fuliginatus,* is found along California's coast, from the vicinity of Los Angeles south into Mexico's Baja California Peninsula. It is a very dark subspecies, of about the same size as the California mule deer.

4. The Inyo mule deer, *O. h. inyoensis,* has only a small range in and around Inyo County, California. In size and color, it is intermediate between the Rocky Mountain mule deer to the east and the California mule deer to the west. Its white rump patch is also intermediate. The black on its tail goes farther up the dorsal surface than does that of the Rocky Mountain deer, but not quite as far as that of the California mule deer.

5. The burro deer, *O. h. eremicus,* is found in the extreme southeastern portion of California and from southwestern Arizona down into Mexico. This subspecies is smaller than the Rocky

Mountain mule deer and much lighter in coloration.

6. The desert mule deer, *O. h. crooki*, has the second largest distribution of any of the mule deer. Its range includes southeastern Arizona, southern New Mexico, and western Texas, and it extends hundreds of miles south into Old Mexico. This is a large-boned, good-size deer, and it would probably be heavier if more forage were available. The scantiness of desert vegetation, however, may be only one of two factors limiting this muley's weight. In the hotter parts of its range, Bergmann's Rule may also apply; that is, if the animal's mass were greater relative to its surface area, it could not dissipate heat as efficiently.

The desert mule deer copes with some of the harshest conditions imaginable for a deer. Much of its habitat is so arid that forage is relatively meager and water is scarce during much of the year. In much of its habitat there is intense heat—and intense cold, as well. And most of its range is open country, with relatively little cover in which to hide from enemies. Yet a desert muley can not only sustain itself in its seemingly impoverished habitat but can make itself almost invisible when it beds down in a little clump of vegetation or even a small depression in the terrain, blending with its background. The desert muley is the palest of mule deer—a very light tannish gray. It has a very small rump patch and a dark stripe running partway down the tail. There is some concern that expanding whitetail populations may cause declines in the desert mule deer.

The prototype of all mule deer, the Rocky Mountain "muley" O.h. hemionus, with the characteristic two forks on each antler and the white rump patch. The does' ears at right show how the mule deer won its name.

The California mule deer, O.h. californicus, *here in velvet, is a smaller subspecies found from the Sierras to the Pacific coast in California. The desert mule deer,* O.h. crooki, *at bottom, is actually the second-largest mule deer despite the forage limits and other hardships of its range.*

7. The Columbian blacktail deer, *O. h. columbianus,* is found in California, Washington, Oregon, and British Columbia. It manages to thrive in the mountains, in dry chaparral, and in the densest, wettest forests on the North American continent. Although both of the blacktail subspecies—the Columbian and the Sitka—occasionally exhibit very respectable weights, these two subspecies usually are smaller than the other varieties of mule deer. They are, in fact, smaller than the average whitetail. The Columbian blacktail has a redder coat than the mule deer, with

A ten-point Columbian blacktail deer, O.h. columbianus, *one of two North American blacktail subspecies. The other is the Sitka deer,* O.h. sitkensis.

which it interbreeds. Its tail top is all black and its ears are intermediate in length—shorter than the muley's, longer than the whitetail's.

8. The Sitka blacktail deer, *O. h. sitkensis,* has a cinnamon brown coat and is smaller on average than the Columbian. This would seem to be a refutation of Bergmann's Rule, but in this case the size is more directly connected to diet than to latitude, bringing Rue's Rule into effect. Forage becomes meager in the winter. When deep snows occur, this deer is forced down to the beaches to subsist on a marginal or starvation diet.

There are parts of the Sitka range where deep snows are seldom a problem, but in those more temperate pockets of habitat severe cold is seldom a problem, either. Therefore, a large body mass relative to the deer's surface area is not needed to conserve body heat, and the Sitka's size does not refute Bergmann's Rule after all.

The Sitka blacktail is an animal of the coastal rain forests. Originally, its range extended only from the Queen Charlotte Islands, off British Columbia, northward through the islands and mainland of the southeastern Alaska Panhandle. Beginning in 1916, herds were successfully introduced to several islands in Prince William Sound near Cordova, and later to Kodiak Island and the Yakutat area.

Harsh conditions are the limiting factor in this subspecies' northern range expansion. The Sitka blacktail is very important to the hunters of the areas it inhabits because no other members of the deer family live there.

3

In the Tracks
of Our Deer

We read almost daily about the depletion of our wildlife resources, as one creature or another is being relentlessly pushed down the road to extinction. Each time it happens, no matter what creature is involved, the world is poorer. The fact that most of us have never heard of some of the threatened species does not lessen the tragedy when existence ends for some variety of wildlife. However, our perspective may be a little clearer if we bear in mind that some 94 percent of all species known to have existed on Earth had already disappeared before the coming of man.

There are approximately 4,500 species of mammals alive today. Some, such as the opossum, are from very ancient orders. Some, such as the deer, are of rather recent origin. All of the mammals that are alive today share many characteristics, but perhaps the most significant one is that they have evolved to survive under specialized conditions. The geological features of the world have been shaped by the stresses of continental drift, disturbances of the earth's crust, climatic conditions, and erosion. The creatures of the world have also been shaped so that they can obtain the necessities to suppo.t life and can withstand their adversaries.

The deer we know today are beautiful, alert, graceful, and intelligent creatures because in the past all inferior deer were weeded out by predators, disease, competition for food, and all the rigors of their environment. I have always said that the one word that best describes deer is *adaptability*. Today, scientists such as Dr. Michael Smith at the Savannah River Ecology Laboratory in Aiken, South Carolina, are able to prove what outdoorsmen have believed all along.

All living creatures have different bands of protein in their blood and tissues. Using a process called *electrophoresis*, scientists can separate and distinguish the protein bands by means of an electrical charge. The protein bands allow positive differentiation of whitetail, blacktail, and mule deer just from a sample of tissue. Electrophoresis has proved crossbreeding between subspecies, shown by tissue that contains the protein bands of two subspecies. The process is also the nemesis of poachers, because it affords positive identification of any meat, even if it has been ground up into hamburger.

Electrophoresis allows the genetic variability shown in the protein bands of an individual animal to be compared not only with that of others of the same species but of other species as well. Dr. Smith has found that the deer family has twice as many genetic variables in its protein bands as has any other creature he has tested. Mature bucks have the greatest number of variables. It is these genetic variables that allow some deer to live in the tropics or dense rain forests and others in the dry deserts, lowlands, or high country. The variables allow deer to feed on the very varied foods found in these different habitats and to survive whatever adversity is encountered in them.

The whitetail, in particular, has the widest range and is found in the greatest variety of habitats. It is primarily a deer of the woodlands, frequenting the edges and openings that have been created by natural conditions, fire, or humans. It is also at home in the swamps of the South, the river bottoms of the prairies, and the forest edges of the high country. The mule deer is a deer of rough, broken, tumbled, and mountainous country, though it is also found

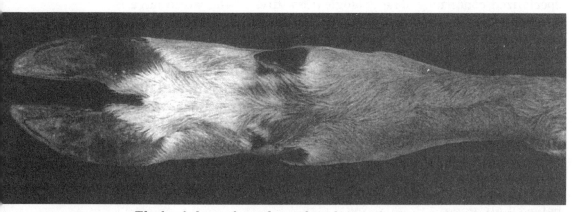

The hoof shown from the underside reveals clues to the phylogeny of the deer family. The ancestors of deer possessed five toes, two of which remain visible in modern deer as vestigial "dewclaws."

on deserts and prairies. Its primary habitat is more open than that of other deer. The blacktail is equally at home in the dense, dark, tangled rain forests of the coastal areas and the dry, hot, and brush-covered chaparral regions. Each of the deer has been strengthened by its environment.

Although the deer is not a long-distance runner, it depends primarily on speed to escape its enemies. The ancestors of our modern deer had five toes on each foot, an adaptation to living in the forested areas that dominated much of the world at that time. As the climate of the earth changed, many of the forested areas gave way to savannas and grasslands. The need for more sustained speed in these open spaces called for a sacrifice in the number of toes so that those remaining could be enlarged and strengthened. The deer lost its thumb or great toe completely. It raised up on its two center toes, and the digits that correspond to our "pinky" and index finger became vestigial bone splints that moved to the rear and are known as dewclaws. The toenails lengthened, hardened, and evolved into the deer's cloven hoofs.

Although extra digits are fairly common among the fleshy-toed animals, particularly six-toed cats, I had never known any of the *Artiodactyla* animals to have more than four.

In November 1984, Gary Beeland of Jackson, Mississippi, killed a six-point buck in Mississippi's Scott County that was perfectly normal in every respect except one. Gary's deer had three main toes on its left hind foot. This extra toe grew on the outside of the regular second and third digits that are the two functional toes that deer walk on. There was some speculation that the toe had grown where the thumb would have been. But it is more likely that this was just an extra toe, because the thumb would have had to grow on the outside of the right dewclaw. This third toe was functional because it showed as much wear as did the two normal toes. It will be interesting to see if this buck has passed on the extra-toe characteristic to future generations of deer in that area.

Hoofs are composed of keratin, the same kind of substance that makes up our fingernails. Keratin is actually a type of solidified hair. On a deer's hoofs, the outer rim is much harder than the central portion. Usually the central portion retracts somewhat in the winter so that the hoof has a concave appearance.

With this slight retraction, only the hard outer rim makes contact with the ground. This provides good traction on a rough surface but is treacherous on any smooth surface such as ice or an oiled blacktop road. When snow is on the ground, deer often run

The hard edges of the deer's hooves make for bad traction on smooth surfaces such as ice.

out on ice to escape from dogs or other predators. If these deer slip, they are doomed, because they seldom can regain their footing. Usually they dislocate their hind legs or strain the muscles in the front legs. I have seen a number of deer that fell on ice and had to be killed. They could not rise again, and their legs would not function when they were taken off the ice.

In the winter of 1981–82—a bad one for ice storms—Pennsylvania lost so many deer that a notice was put out through every newspaper in the state stating that the excessive number of ice-related deer injuries might mean a reduction in the number of antlerless deer permits offered the following year. Although some of the deer were injured on lakes and ponds, most were injured sliding off mountains.

Of course, every rule about deer seems to have its exceptions. My friend Joe Taylor, who owns a campground and deer preserve, has reported observing three whitetails walking on very smooth ice. They had learned to keep their front legs rigid and slide with their front hoofs flat on the ice, while they splayed their hind feet and pushed forward with the tips of their rear hoofs. Indeed, under most circumstances, the hoof structure helps a deer move about sure-footedly, find forage, and escape enemies. It is an evolutionary adaptation to general conditions—far more an advantage than a disadvantage.

Most of the year, the central portion of a deer's hoof is almost convex, swelling outward to provide the foot with a spongy surface similar to that of wild goats and sheep. The hoof substance grows continuously throughout the deer's life but is kept worn down by the abrasion of rock and hard soil. The hoofs of deer that live in swampy areas wear down at a much slower rate than those of deer

In the Tracks of Our Deer

that live on hard ground. Their hoofs often become elongated, providing a much greater support area that prevents their sinking into the mud.

Central Africa has marsh-dwelling antelopes called the sitatunga and lechwe. Their two center toes are twice as long as those of other species of antelope or deer of the same body weight. These toes can also be more widely spread to increase their support surface when walking in the muck and mud of the animals' habitat.

North American deer hoofs would also grow that long if they were not worn down by rock. I have seen injured deer that walked on their pasterns rather than the hoof surface. Such deer had hoofs that were at least 4 inches (101.6 mm) long.

A very old desert bighorn ram kept in captivity on a desert sheep range in Nevada also walked on his pasterns. His hoofs got to be 5 to 6 inches (127 to 152.4 mm) in length, resembling skateboards.

The front hoofs of most animals are larger than those on the hind feet. One whitetail buck I measured had a live weight of 151 pounds (68.5 kg). One of its front hoofs measured 3¼ inches (82.6 mm) in overall length and 2 inches (50.8 mm) in width. The hind-foot hoof measured 2¾ inches (69.9 mm) in overall length by 1¾ inches (44.51 mm) in width. On a hard surface, a deer does not

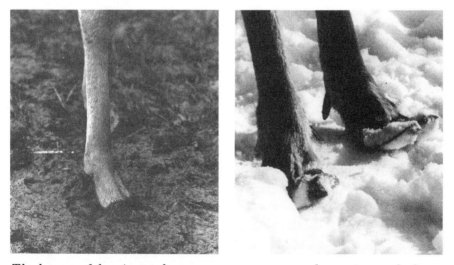

The hooves of deer in sandy or swampy areas wear down at a much slower rate, causing longer-looking hooves and dewclaws. The ski-like hooves at the right are the result of this deer walking on the rear of its hooves so the tips are not worn down.

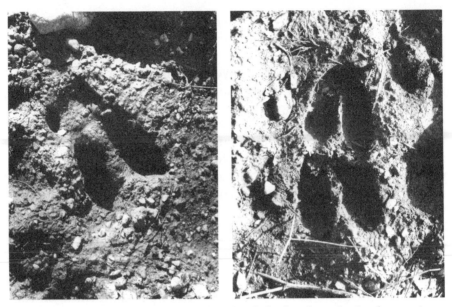

The imprint of the dewclaws show behind the hoof marks in these prints only because the animal was walking in soft earth or snow. Under ordinary conditions, the hind foot of the deer will come down on top and slightly in front of the track made by the front foot, as at right.

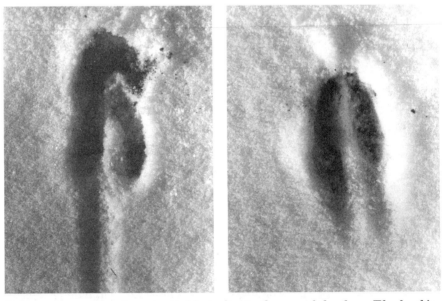

Only one type of track can reliably indicate the sex of the deer. The buck's characteristic "drag" marks in shallow snow. The doe's tracks, at right, are usually neater.

walk on the overall length of its hoofs because they slope upward at the rear. The front hoof of this deer had a *bearing surface* that measured 2¼ inches (57.2 mm) in length, then angled upward at 38° for another inch (25.4 mm). Rarely are the two toes on a hoof of equal length. Usually the outside toe is about ¼ inch (6.4 mm) longer.

When the ground is soft, the entire hoof surface comes into bearing. When the ground is very soft or muddy, the dewclaws are also used. On the buck I measured, the front foot's overall length, including the dewclaws, was 4½ inches (114.3 mm) by 2⅜ inches (60.3 mm) in width. The slightly greater width is because the dewclaws spread out a fraction wider than the hoof. The advantages of being able to splay the hoofs and to use the dewclaws are apparent. When this deer was walking on hard ground, it had about 5 square inches (32.3 sq cm) of bearing surface per hoof, or 20 square inches (129 sq cm) altogether. This works out to a body-weight ratio of only 7½ pounds (3.4 kg) per square inch of bearing surface. When the deer was on muddy ground, with hoofs and dewclaws splayed out, it effectively doubled its bearing surface, having about 40 square inches (258.1 sq cm) and a ratio of only 3¾ pounds (1.7 kg) of body weight per square inch. The hoofs of a blacktail deer are about the same size as a whitetail's. A mule deer's hoofs are slightly larger because in general it is a slightly larger animal.

Regardless of species, the feet are a marvel of adaptation. If you have ever watched a deer browse its way gracefully through the woods, you will agree that the feet look delicate—and delicate they must be to pick their way through tangles and brush—yet they provide an enormously greater bearing surface than their appearance suggests. And they provide the best possible traction under most common conditions. Moreover, when a deer must sprint for safety, the hoofs are shaped to dig in powerfully, forming perfect bases for a fast run or sudden bolt.

Many hunters claim they can often tell a deer's sex by looking at its tracks. They say a buck's hoofs are broader, longer, and have more rounded tips because a mature buck weighs more than a mature doe. The greater weight creates more wear on the toe tips, blunting them. This argument might hold water if the tracks were found in a national park where no hunting was permitted. In most regions, all of these supposed buck signs more often indicate a doe's track. In New Jersey—which is fairly typical—very few bucks ever live beyond 2½ years. They are usually harvested when they are 1½ or 2½ years old. Like many states, New Jersey now has an annual "antlerless" harvest, but it is much more limited than the buck season. Consequently, most does have the chance to become much older and larger than the bucks.

Biologists have proved that a buck's hoofs, measured a third of the way back from the tip, average 5 percent wider than those of a doe. Fine, but I defy anyone to detect the difference when looking at tracks in the wild.

The fact that the dewclaws show in the tracks does not indicate the deer's sex, either. What it does indicate is that the ground was soft when the animal left the tracks.

Some hunters claim that a buck's tracks angle out more from a central axis or, in other words, "toe out." There is a degree of truth in this, but the degree of splaying can be pretty subtle.

The most reliable sex-indicating track that I know can be found after a snow of one inch (25.4 mm) or less. A buck's feet produce a drag mark in the track because, like most males, a buck doesn't lift his feet any higher than he has to. The doe, being daintier, picks her feet clear and leaves just the track of the hoof imprint.

When there is more than a half-inch of snow on the ground, all deer leave drag marks. During the rutting season, bucks drag their feet even more than they ordinarily do, producing continuous drag marks in shallow snow. As to why, we can only speculate. Perhaps the bucks are conserving energy, or perhaps the shuffling walk is a means of leaving more glandular scent on the ground.

This test for drag marks also works on sand or any other surface where the deer sinks in about an inch. Under other conditions I cannot tell the sex of a deer by its tracks, though I have been studying and living with deer for more than 45 years. The so-called experts would stop claiming they could do so if they ever managed to see most of the deer that made the tracks.

The drag marks are made by the tips of the toes as they begin to straighten up so that their bearing surface is brought horizontal to the ground just before the next step is made. As the deer continues to walk, the front of the toes leave just a slight forward indentation in the snow as they leave the track to take the next step.

"Reading" tracks and other wildlife signs—that is, recognizing and interpreting them—is one of the most exciting and rewarding of outdoor activities, whether you do it for its own sake or in connection with hunting. Though deer tracks cannot often reveal sex, they can give you a great deal of other information. They can tell you the number of deer that went by and they do tell you which direction they took. And if you find more than one set of tracks, you may, after all, learn the sex of one of the deer even without snow or sand. When you find the tracks of one large deer accompanied by two smaller ones, it is safe to assume you are looking at the trail of a doe with her two fawns.

Depending on recent weather, you may even be able to date the tracks. A hard rain washes nature's blackboard clean, and even a moderate rain dims it somewhat. Sharp-edged tracks are fresh. They were made after the rainfall. Rounded edges mean the tracks were made prior to the rain.

If the tracks are in dirt or sand, the edge will tend to soften as wind brushes them. And if debris has blown into the tracks, you can date them to the last strong wind. Where cold keeps the snow powder-dry, tracks often can be dated to the last snowfall, or the last heavy wind. Powder snow drifts easily, and in a wind it will fill or partially fill the tracks. In many areas the wind blows all day but dies down at night, and this, too, may help in dating tracks.

The sequence of tracks in the snow here shows the deer at the right was trotting, while the one at the left was galloping. The galloping deer left only one hole in the snow for both hooves on each side, while the trotting deer left holes where each foot landed.

In my area, what begins as powder snow often takes on a glaze of ice as it melts and then freezes again with fluctuating temperatures. If the snow is powdery, I date the tracks back to the snowfall. If the tracks are glazed slightly (and they will glaze before the crust of the snow because they are punched into the surface and are protected from the wind), you can figure they were made the day following the snow. If they are punched through the glaze of the snow, they will have crumbly pieces of snow in them. In that event, they were made after the last warm day following the snow.

Moderating weather often fools an inexperienced tracker. The prints of small deer spread out in the melting snow, becoming larger as the trail grows old, and the prints of larger deer look as if they were left by monsters. The amount of melting can also help to date the tracks.

I love to go out after a fresh fall of slightly moist snow, a good tracking snow that is not too deep to obliterate what made the tracks. It's like reading a gossip column. You can tell who went where, when, with whom, and what they did there. That is, you can tell all this if you can read tracks.

This whitetail exhibits the deer's typical walking gait; at a walk, deer almost always have three feet on the ground, forming a tripod under the body.

This whitetail buck is "single footing," a graceful gait in which he lifts his front foot exceptionally high while he trots.

The average adult deer—whitetail, mule deer, or black-tail—has a hind foot about 19 inches (48.3 cm) long, the ankle being the portion we refer to as the *tarsal joint,* or *hock.* A deer's legs are long and heavily muscled to provide the power needed to run and jump.

Whitetail deer walk, trot, and gallop. Blacktails and muleys walk, trot, gallop, and bound. When a deer walks, it starts off by moving one of its hind feet forward. This pushes the weight of the body forward, allowing the other three legs to act as a weight-supporting tripod. After the left hind leg is halfway through the forward motion, the front leg on the same side starts forward. As the left hind leg touches the ground, the right hind leg starts forward, followed a half step afterward by the right front leg. This is the basic pattern of walking for most land mammals. A feeding deer may use its feet in other sequences, but it usually has three feet on the ground at the same time.

When a deer walks, the hind feet are placed in almost the exact spot just vacated by the front feet. With young deer, the hind foot usually overshoots the track made by the front foot by a little so that a portion of the front track is visible sticking out behind. Larger, heavier deer usually have a slightly shorter stride for their size, so the front foot toe tips show in the front part of the track

Gaits of Deer

A whitetail walks in this manner: the stride is about 18 inches, two or three inches longer for the mule deer species, at about 3½ to 4 mph.

A blacktail doe pictured in a very fast trot clearly illustrating the pattern of leg movement. The stride is increased to 30 or 36 inches at 10 to 12 mph.

The step sequence changes from a trot to a gallop; sometimes this becomes a single-footing pattern as the deer changes gear to a faster gallop.

made by the hind foot. This is particularly true during the rutting season, when the mature bucks walk with a much more stiff-legged gait. This gait causes the hind foot to step on or behind the rear of the track made by the front foot.

A whitetail deer's normal stride, the distance measured from toe tip to toe tip, is about 18 to 19 inches (45.7 to 48.3 cm). A blacktail's stride is comparable. A mule deer's stride is about 20 to 23 inches (50.8 to 58.4 cm).

Deer seldom run without cause. They usually walk or trot

In the Tracks of Our Deer

Now the deer is "swapping ends:" front and hind feet overlapping in powerful, long strides averaging 25 feet or more and achieving speeds of 35 to 40 mph for the whitetail, and 20 to 25 mph for the mule deer.

All four feet are off the ground twice in the eight-step rotatory gait of a full gallop; the deer almost seems to float between strides.

A young mule deer "stot-ting," a peculiar-looking but effective gait that allows them to change directions with each bound, since all four feet hit the ground at once and they can spring off again in any direction.

from one place to another because then they can use all of their senses to be alert for danger. They tend to walk rather slowly, perhaps feeding as they go. When walking steadily, they move a little faster than a man—about 3½ to 4 miles per hour (5.6 to 6.4 km per hour).

When a deer trots, the left hind foot is teamed with the right front foot. The left hind foot starts forward and, when it is halfway through the step, the right front foot starts forward. As the left hind foot touches the ground, the right hind foot starts forward.

Deer have no trouble clearing great heights on the run, as the deer at left is doing. The 5-foot high brushpile posed no obstacle. The difference between the fast gaits is clear in the sequence of photos at right: the deer on the top of the photos is trotting, while the one at the bottom is galloping.

When the right hind foot is halfway through the step, the left front foot starts forward. Deer trot if they have a clear destination, whether they are just in a hurry to get to their feeding area or, in the case of the blacktail and muley, they are migrating. They can trot hour after hour at a speed of about 10 to 12 miles per hour (16.1 to 19.3 km per hour). I have clocked deer trotting, and the speeds vary considerably, depending on the urgency that the animals feel.

When a deer trots, its stride lengthens to about 30 to 36 inches (76.2 to 91.4 cm). Because of the greater speed, the feet fall almost directly under the body, giving the track more of the appearance of a dotted line. You will see the track of each foot individually.

The whitetail has a variation of its trot that I call "single-footing." Usually done when the animal is nervous, it is an exceedingly graceful motion. It appears that the deer is about to stamp its foot but then changes its mind and starts to trot off. The raised front foot is held parallel to the ground while the other forefoot makes two steps. Then the first front foot makes two steps while the alternate front foot is held up. It is a beautiful, dainty motion that looks almost like a gesture of impatience or uncertainty—as if the animal were eager to trot off but could not quite make up its mind to do so.

Galloping deer use what is called a *rotatory* pattern. It entails an eight-step sequence in which all four feet are off the ground during two different steps. Some animals, such as the horse, use the transverse gallop and have all four feet off the ground during only one step. When deer gallop, the right front foot makes contact with the earth first. Then the left front foot hits the ground a little distance ahead of where the right foot touched down. Now the deer's body bunches up as the hind legs come forward and the right front foot leaves the ground. The momentum propels the body forward and all four feet are off the ground. The right hind foot touches the ground first, followed by the left hind foot. The hind feet land ahead of where the front feet touched down. Now the body lengthens out like an uncoiling strip of spring steel. The right hind foot and then the left hind foot thrust mightily and again the body is launched forward to complete and repeat the entire cycle.

A whitetail deer's top speed is about 35 to 40 miles per hour (56.3 to 64.4 km per hour). A blacktail's top speed is about 30 to 35 miles per hour (48.3 to 56.3 km per hour). A mule deer can match the whitetail for a short distance but it soon slows down. Ordinarily, whitetails run at speeds of 20 to 25 miles per hour (32.2

A Deer Galloping

Viewed left to right, these photos show the complete rotatory motion of a whitetail galloping.

The deer traveled just over 77½ (23.5m) feet during the elapsed time of these photos, which was less than 3½ seconds.

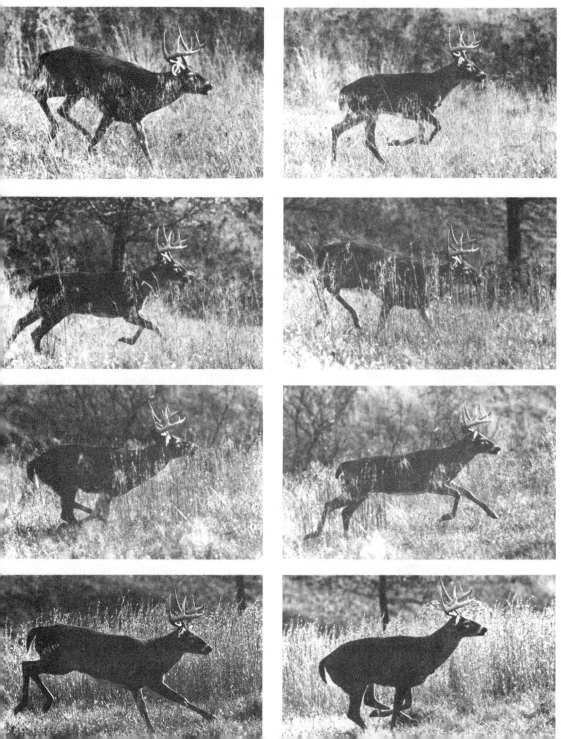

A mule deer running with his tail extended in the characteristic position. Mule deer seldom use their tails as "flags". photo courtesy Len Rue, Jr.

to 40.3 km per hour) and can maintain these speeds for a long period. Mule deer can do about 20 miles per hour for considerable distances.

Donald McLean tested the speed of mule deer by herding them with his car on Mosquito Lake Flats in Modoc County, California. The highest speed was 38 miles per hour (61.2 km per hour) and that only for a short distance. The deer then slowed down to about 23 miles per hour (37 km per hour) and were badly winded in less than a mile.

When a whitetail really gets into high gear, it appears almost to float over the terrain with its tremendous ground-eating gallop. The longest span between tracks that I have personally measured was a little over 26 feet (about 8 m). C. W. Severinghaus, of New York state, measured the distance between the tracks of a deer running down a slight grade. The deer had to leap over a 7½-foot (2.3-m) windfall, and it still cleared 29 feet (8.8 m) horizontally.

A number of years ago, I witnessed a very dramatic example of the jumping ability of the whitetail. At that time the Hercules Powder Company owned a large tract of land outside Belvidere, New Jersey, my hometown area. When the deer inside the enclosure became too numerous, the company decided to take down a wide section of the chain-link fence and drive the deer out through the opening to suitable habitat outside.

In the Tracks of Our Deer

The local hunting club and all interested local people, including me, turned out to participate in the deer drive. When the string of men began to encounter the deer, there was pandemonium. A very important lesson was learned that day. You don't really drive deer—even with a group of hunters experienced in conducting deer drives. All you do is stir them up, and they go where they want to. Most of the deer ran in front of the drivers, some between the drivers, and a couple ran right over them, though no one was seriously injured. I am sure that the deer, in their panic, never saw the drivers. Most of the deer probably never saw the hole in the fence, either; they simply jumped over the fence, which was 8 feet (2.4 m) high and topped by three strands of barbed wire, raising the total height to a little over 9 feet (2.7 m). I saw numerous deer jump the fence from both the running and standing positions. Many of the panicky deer also ran into it. The next day several of the deer were seen leaping over the fence to get back inside.

Although deer can easily jump such a fence, they don't exert themselves any more than they have to. Many times I have seen a doe, followed by her two fawns, come to a standard 4-foot (1.2-m), three-strand barbed wire fence. Sometimes the doe would jump over while the fawns jumped through the strands. Sometimes she, too, would go between the strands. Does and fawns can jump through strands of barbed wire that are 10 to 12 inches (25.4 to

This deer had no trouble clearing a 14-foot wide stream.

Some deer make an exceptionally high "observation leap" while they're galloping, to gain a view of what's ahead.

30.5 cm) apart at the speed of a gallop and not touch the wire. But frequently they will stop and crawl under the wire. Sometimes deer will jump over blowdowns, and at other times they will worm their way underneath. There is no predicting what wild creatures will do; they are all individuals.

Both the mule deer and the blacktail have another unique gait. They "*stott.*" All four legs are used simultaneously, thrusting down and backward, powerfully propelling the body up and forward. Although a stotting deer appears awkward, the gait is very functional. It must have evolved as an adaptation to the rough, broken, and steep country that these deer inhabit. The stott not only lets deer jump easily over rocks and brush, it allows them to gain elevation quickly when scrambling up a mountainside.

Naturally, the longest stotts are made when going downhill. Donald McLean, recording the jumps of two mule deer on level ground, found that they measured 19½ feet (5.9 m) and 23¼ feet (7.1 m). But on a 7° downward slope, one bound measured 28 feet 7 inches (8.7 m).

I have never seen a blacktail stott more than about 10 times before it settled down into a stiff-legged gallop. The muley can stott for a longer period of time but the gait is very tiring. After about 500 feet (152.4 m) the deer's mouth will be wide open and the tongue lolling out. A mule deer can stott about 8 feet (2.4 m) high and can cover 26 feet (7.9 m) horizontally. When winded by stotting, the muley then gallops like a whitetail.

Because a mule deer is usually a larger, heavier, more chunky-bodied animal than a whitetail, it does not move quite as gracefully. But a blacktail can slip through the most impenetrable tangles of underbrush as easily as the whitetail.

The oddest example of deer locomotion I ever heard of was related to me in a letter by Mrs. Paul Warnich of Williamsport, Pennsylvania. The incident was witnessed by five people. In July of 1967, the group had been out looking for deer when their trip was cut short by heavy rain. On the way home they saw two deer in a field of high weeds. Suddenly one of the deer raised both of his hind feet in the air, as if it were doing a handstand, balancing and walking on its forefeet. They thought the deer was injured and had to walk that way, but then the second deer did the same thing. Both deer walked on their front legs for a distance of about 75 feet (22.9 m) to where they could no longer be seen. I have never seen anything like this, but I received additional reports of it. I have included the incident here because I feel that it warrants recording. We are constantly learning new things about deer. Moreover, this incident demonstrates their marvelous coordination and balance. Perhaps it also demonstrates their adaptability to circumstances. Just as our own species engages in play, most of the highly evolved animals seem to indulge in some antics just for fun, but more often there is a serious reason for what they do. It may well be that walking on forelegs had a definite though undetected value to those deer.

When deer are startled, they dash off as if they were late for an appointment in the next county. This headlong flight is just long enough for the deer to reach protective cover. The mule deer almost always stops to look back just before it reaches heavy cover, and the whitetail does so quite often. Deer are skulkers; they prefer to sneak away from danger rather than expose themselves by flight. A running deer is at a tremendous disadvantage because it has no way of checking for danger ahead, and the deer knows it. A deer will not run when it can walk and will not walk away from danger if it believes it can escape detection by remaining motionless. I want to stress this because it explains why so few deer are actually seen even in areas where their population is high. Hunters walk past far more deer than they see. I always say that for every ten creatures we see in the woods, a hundred see us and either remain hidden or slip away undetected.

Deer often take to water to elude their enemies. In the summertime, especially in the north woods, they spend a great amount of time in the water feeding, cooling off, or just seeking relief from the stinging and biting insects that plague the woods in warm weather. Sometimes they enter the water just to play and splash about. I have seen fawns get as much enjoyment as a human child would get by splashing with one of its feet. Does frequently swim out to islands before giving birth to their fawns, thus gaining the protection of a water barrier.

Deer swim well, fast, and for long distances. Walter P. Taylor reports one man who was able to check the speed of a swimming whitetail doe by using the tachometer on his powerboat. The doe was frightened at first and swam about a quarter of a mile (402 m) at 13 miles per hour (20.9 km per hour). When the deer found she was in no real danger from the boat, she settled down to an easily maintained speed of 10 miles per hour (16.1 km per hour). Whitetails have been seen swimming out in the Atlantic Ocean 5 miles (8.1 km) from Cape Cod, Massachusetts.

In the Tracks of Our Deer

A Millbridge, Maine, lobsterman did more than clock the speed of a swimming deer—he gave it a ride. When hunters chased a spike buck into the ocean off Millbridge, the deer started to swim to an island a mile offshore. The lobsterman ran his boat alongside the deer, caught it by the antlers, and swung it aboard. He ran his boat over to the island and put the deer back in the water where it waded ashore to safety.

Sanford D. Schemnitz, in a deer study done in Maine, saw an adult doe swim from Isle au Haut to Vinalhaven Island, a distance of 6⅔ miles (10.9 km).

Blacktail deer often swim from island to island in the Pacific Ocean off the coast of British Columbia. Arthur Einarsen tells of one Sitka blacktail buck that swam 14 miles (22.5 km) during the breeding season from one island to another along Alaska's southeastern coast. What especially intrigued me about this record was that the deer could not have seen another island 14 miles away. What guided that deer? Could it have been scent?

For many years I lived along the Delaware River. The deer-hunting season in Pennsylvania always opened one week in advance of the New Jersey season. Every year when the guns started banging and the drives commenced in Pennsylvania, the deer

All deer possess excellent swimming ability; Florida Key does have been known to make daily round trips between islands to nurse their fawns.

Routine Rise

1. *A doe at rest decides to rise.*

2. *She heaves up onto her front knees.*

3. *She then raises her hindquarters.*

—sometimes more than a dozen at a time—would swim across the river to the Jersey side. A week later, when the New Jersey deer season opened, the Pennsylvania side of the river was comparatively quiet even though the season was still open, and the deer would then swim back to Pennsylvania.

Different types of animals have different ways of using their legs to lie down and to get up. When a deer or other cloven-hoofed animal lies down, it lowers itself to both knees of its front legs, then lowers its hindquarters to the ground. The deer generally lies

In the Tracks of Our Deer

4. Next she lifts a front leg forward.

5. She extends one leg.

6. Once up, she stretches all her muscles.

with one side of its body touching the ground, the legs on that side tucked underneath. Only rarely, as perhaps in cold weather, will a deer lie down with all four legs beneath the body. In fact, this position usually indicates that the deer suspects danger and is ready to jump up fast. Ordinarily, when the deer arises, it reverses the normal lowering procedure, throwing its weight forward and rising up on its front knees. In that same motion, the animal raises its hindquarters and extends its hind legs. Then the deer extends one front leg, which raises the forepart of the body so that the other

Startled Rise

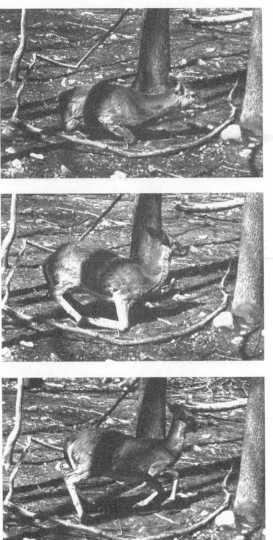

1. *When startled, a deer uses a faster rise to escape.*

2. *All four legs are flexed to catapult the body upward.*

3. *The deer pivots off the rear feet and is instantly running.*

foreleg can be extended. Usually the deer then takes a long, leisurely stretch before walking off. But when a deer is startled, it rolls its body off the ground so that all four legs are beneath the body and, using chiefly the hind legs, it literally explodes into action and is away.

In warm weather deer will often extend their leg or legs out sideways, and occasionally to the front, in an effort to cool off. In very cold weather the legs are folded up beneath the body as a heat-conservation measure.

In the Tracks of Our Deer

Bucks may shed their antlers simultaneously or they may carry one as much as one week longer.

A buck's growing antlers are covered with a nourishing network of blood vessels that we call velvet.

This newly scabbed, bloody pedicle shows that this whitetail buck shed its antlers just hours ago.

As the doe's newborn fawns nurse, the female licks their anal region, which stimulates them to pass feces.

The doe leads her four-hour-old fawn from the spot where it was born and will hide it in a safe spot.

A really fine whitetail buck in the 160 Boone & Crockett class.

While many Texas bucks have large antlers and small bodies, this buck was big in all respects.

Whitetail bucks fight for dominance because the boss buck does most of the breeding.

This buck is flehmening; he curls his upper lip, trapping the scent of the female's urine in his nostrils in searching for the pheromones that will indicate that she is in estrus and ready to be bred.

When a buck rubs a tree, he strengthens his neck muscles, deposits scent from his forehead glands, and makes a white visual marker by removing the bark.

Whitetail deer remain segregated throughout most of the year, with the bucks joining the does only during the breeding season.

This buck is unique in that it has palmated brow tines.

Deer in good condition can easily withstand the rigors of cold and snow, but they cannot tolerate strong winter winds and must seek shelter.

This one-day-old whitetail fawn is depending on its spotted coat to camouflage it so that it will not be detected.

This doe, like all deer, depends primarily on her nose, then her ears, and then her eyes to detect danger.

This mule deer buck approaches the doe in the "head-down" aggressive posture that will prompt her to urinate.

Whitetail deer are strong swimmers and do so readily.

He will immediately smell her urine and flehmen in the hopes that she will be receptive.

A mule deer buck can quickly be told from a whitetail buck by its forked antlers and white rump patch.

Whitetail bucks chew overhead branches, depositing their saliva, then paw a scrape and urinate in it.

This buck is working over a "lick stick." He will deposit forehead scent on the stick, lick it off, and repeat the process.

This whitetail buck would be considered nontypical because the antlers are not symmetrical.

A truly magnificent whitetail buck in his prime.

The Texas brush country provides such heavy protective cover that the bucks can only be seen and counted when flushed by a helicopter.

Whitetail bucks have been measured jumping over twenty-seven feet horizontally and over nine and one-half feet vertically.

Blacktail deer can be found in California's dry coastal regions or the Northwest's rain forests. This buck has an outstanding set of antlers.

4

Heads and Tails

F or purposes of identification, tails and antlers are the two deer characteristics that are easiest to distinguish. To many, the word *deer* summons an image of a whitetail with "flag" flying; others think of a whitetail or a muley with a fine rack.

TAILS

Tails alone would serve to distinguish the three types of deer. The whitetail deer's tail is the largest, performs the most functions, and is often conspicuously displayed. The size of the tail, of course, depends to a great extent on the size of the subspecies. Most of my work has been done on *borealis,* the largest of the whitetail deer. I have found that the average tail length from the rump to the bone tip is 9 to 10 inches (22.9 to 25.4 cm). From the rump to the hair tip is 11 to 13 inches (27.9 to 33 cm). The longest tail that I can find record of was 14 inches (35.6 cm). The hairs on the tail can be erected and flared out to the sides for a maximum width of 10 to 11 inches (25.4 to 27.9 cm).

The dorsal, or top, surface is brown, but the tips of the top hairs may be black. There is great variation in the amount of black; on some tails it is entirely lacking, on others it may be a third to half or even the entire dorsal surface. The deer gets its name from the white underside of the tail, which is seen when the tail is elevated. The white hairs always extend beyond the dark hairs of the top so that the tail is bordered with white.

In the Blue Ridge mountains of Virginia, I once photographed a whitetail doe with a normal tail that was entirely brown on its dorsal surface, while the entire dorsal surface of both of her fawns' tails was jet black. The amount of black coloration on the tail's dorsal surface in any whitetail is only a genetic characteristic. Some deer have it, some don't. If the dominant buck in the area has a black dorsal tail surface, many of his offspring will inherit this characteristic.

This tail is another marvel of evolutionary adaptation. Anyone who has stalked (or tried to stalk) whitetail deer, whether to hunt them, photograph them, or just get a close look at wildlife in its habitat, will have no doubt whatever that the tail is an efficient mechanism for communication. An alarmed whitetail deer raises, or flags, its tail to signal to other deer in the vicinity. When they see the danger signal (which operates simultaneously with other, more subtle means of communication) their own flags are raised and all the deer in the group make a hasty departure. There is also a theory that the suddenly raised tail conveys a message to a predator—that it has been discovered and its prey is too alert for a successful sneak attack. The chapter on communication will include further details regarding these functions.

The tail of the blacktail deer, from the rump to the bone tip, averages about 5½ to 6 inches (14 to 15.2 cm) long. From the rump to the hair tips, it averages about 9½ inches (24.1 cm). The dorsal hair is dark brownish black, and some of the dark hairs extend all

Tail size and coloration identify the three types of North American deer: from left, the tails in this photo typify whitetail, mule deer, and blacktail.

The whitetail's trademark: the waving "flag" of flared white hair is a conspicuous signal of alarm. photo courtesy Irene Vandermolen

the way to the tail tip. The hair on the underside is white, and there is a slight white bordering on the lower third of the tail. In blacktails, the hairs do not flare out, and the tail is not used for signaling. When the deer is running, the tail may droop or be carried almost horizontal to the ground, elevated almost vertically, or tilted forward toward the head.

The muley's tail is a pendant, wide at the top, tapering to a narrow middle, then sloping outward slightly near the tip. This tapering is caused by the hairs breaking off. The upper portion of the dorsal surface is brown, the central portion white, the tip black. The underside is white except for the black tip. The hairs cannot be flared and the tail is not used for signaling, nor is it elevated when the deer runs, though the muley can flare its rump hairs when alarmed. The tail length from the rump to the bone tip is 7 to 8 inches (17.8 to 20.3 cm). From the rump to the hair tips it is 10½ to 11½ inches (26.7 to 29.2 cm).

This deer is a cross between a California mule deer and a Columbia blacktail. Such hybrids are characterized by a black stripe down the center of the tail.

Where the blacktail and muley interbreed, hybrids are most readily identified by the body size and coloration of the tail. A hybrid's tail is not much different in size from the tail of one parent or the other, but it has a black stripe, of variable width, running down the length of the dorsal side.

ANTLERS

Now let us look more closely at the opposite end of our subjects. A deer's tail may be of some interest anatomically and, in the case of the whitetail, behaviorally, but it lacks the romantic fascination of the antlers. The sweeping rack of a big buck has come to symbolize the majesty of the wilds for most people.

The words *horns* and *antlers* are often used interchangeably, as if they were one and the same thing, which they are not. Horns are found on sheep, goats, cattle, bison, and antelope. Horns are never shed. They grow continuously throughout the animal's life, but if they are broken off they do not grow back. Horns have a porous core and a hard, strong outer surface composed of keratin, the same substance as your fingernails. Horns are nourished by internal blood vessels and grow from the inside out. Usually, both the male and female of America's horned species have horns. The pronghorn is the only horned animal in the world that annually sheds a part of its horns. The outside layer, called the *casque,* is pushed off each fall by a new horn growing underneath. The pronghorn is also the only animal in the world with horns that branch, creating the prong for which the animal was named.

Antlers are found on the Cervidae family—deer, elk, caribou, and moose. Under normal conditions, antlers are shed each year. The caribou (and its European equivalent, the reindeer) is the only antlered species the females of which normally have antlers. The female roe deer of Europe, however, frequently have small antlers, and female whitetail, blacktail, and mule deer occasionally have antlers. Conversely, the Chinese water deer and the marsh deer of Asia are the only two deer species of which neither male nor female has antlers. These two Asian deer have tusks that are actually elongated canine teeth, projecting downward from the upper jaw for use in fighting and for protection.

Both horns and antlers are actually extended growths of the frontal skull plate. The core of the horn is a part of this frontal bone. The pedicel of an antler is also a part of this bone but provides only the base from which the antler grows. Antlers are true bone, composed primarily of calcium and phosphorus, with varying amounts of other minerals. But unlike most mammal bones, they are without marrow.

When a buck is born, he has two swirls of hair on his forehead, showing where the antlers will form as he gets older. At two or three months of age, the pedicels start to form as bony knobs on the buck's frontal skull plate. Without these bases, no antlers can grow. As has been known for many centuries, if a young buck is

A buck fawn shows the swirls of dark hair on his forehead between the ears and eyes where the pedicels will develop.
photo courtesy Irene Vandermolen

castrated before the pedicels have formed, he will never produce antlers. The great Greek philosopher Aristotle, who lived from 384–322 B.C., observed in his *Nicomachean Ethics* that "If stags be mutilated, when, by reason of their age, they have as yet no horns, they never grow horns at all; if they be mutilated when they have horns, the horns remain unchanged in size, and the animal does not lose them." Of course, like most hunters, he was not referring to horns but to antlers.

Everyone interested in antlers should read both Dr. Richard Goss's *Deer Antlers* and Dr. George Bubenik's chapter in Robert D. Brown's *Antler Development in Cervidae* (see Bibliography). Building on the pioneer work of Drs. Hartwig and Schrudde, done in Germany, Dr. Goss discovered that tissue, or *periosteum,* found beneath the skin covering the pedicels was antlerogenic. Both Hartwig and Goss have excised this tissue and placed it on various parts of the deer's body. Antlers have grown wherever the tissue was grafted. Dr. Goss grew antlers on a fallow deer's leg, hip, and ear. When supplied with blood, this transplanted periosteum caused the formation of a pedicel, and the subsequent growth of an antler. Although the antlers were rudimentary, they went through the complete stages of velvet, growth, hardening, peeling of velvet, and shedding of the antler in synchrony with those of normal deer exposed to the same growth-affecting periods of light and darkness, or *photoperiodicity.*

Abnormalities such as a third antler are uncommon but not really rare. Joseph Dixon recorded two muley bucks in Yosemite National Park, California, in 1927 and 1929, each of which had an extra pedicel base growing on the nasal bone in the middle of the head instead of on the skull. The extra antler on the first buck grew to be 2⅛ inches (53.98 mm) high. This was a real antler. While it grew, it was covered with velvet; it hardened, the velvet peeled off, and the antler was later dropped when the buck's normal antlers dropped off. This buck died in the winter of 1928. The following year another buck was seen with a third antler growing in the same spot, indicating that this characteristic was an inherited one.

In Colorado in 1964, Duwayne Statzer of California shot a four-point mule deer buck that had a 6¾-inch (171.45-mm) antler growing beneath its left eye. The normal antlers were polished, but the freak growth was still in velvet. There is also a record of a buck that had a small antler below his right eye. Prehistoric animals often had antlers growing from different parts of their

This is a three-antlered whitetail buck: the small third antler is in front of the right main beam.

heads. That this characteristic was not perpetuated through evolution proves that it was not advantageous.

In 1972, I photographed an eight-point whitetail buck in Hunterdon County, New Jersey, that had a third antler growing out of the frontal skull about an inch (25.4 mm) below the right main beam. The extra antler was only about ½-inch (12.7 mm) long but was polished. In 1973 the buck had a more massive body, and although he was still an eight-pointer, it was a large rack. The third antler was now over 3 inches (76.2 mm) long, ridged and polished, a perfect little spike. Unfortunately, the buck was shot that year so my investigation was ended. There have been occasional reports of three-antlered bucks from other parts of the country.

During the 1976 deer hunting season, Alan Burlett shot a nine-point buck near the town of Pomfret in Chautauqua County, New York. A nine-pointer is not that uncommon, except that Alan's buck was a typical eight-pointer with a third antler as its ninth point, growing right out of the top of the buck's head directly between its eyes.

Some of the third antlers found on bucks are really part of one of the main antlers, growing from a segment of the two normal pedicels that is joined beneath the skin. I believe this to be the case on the buck in Hunterdon County described two paragraphs back, and shown on an accompanying photo. The antler on the buck that Alan shot had its own distinct pedicel, and it was truly independent, hardened, and functional. It was 4⅘ inches (121.92 mm) long and ⁹⁄₁₀-inch (22.86 mm) in diameter at the base.

A seven-month old buck is usually only a button buck, but this one has tiny spikes showing.

The four-antlered buck: note all four antlers are growing out of two pedicels. photo courtesy Richard P. Smith

Philip Alquist of Kingsford, Michigan, shot a four-antlered spike buck in Menominee County during the 1980 gun season. The four spikes grew in a row across the top of the deer's head, the longest about 4 inches (101.6 mm) in length. Each pair of spikes came out of the same pedicel.

As mentioned above, under the swirls of hair on a buck fawn's forehead is the periosteum tissue. The fawn at this stage is far from being sexually mature. His testicles are not developed enough to produce testosterone. It has recently been discovered, however, that the endocrine system, stimulated by the pineal gland, causes the adrenal glands to produce a limited amount of testosterone. When the periosteum is activated by this hormone, it begins to lay down calcium deposits, forming the bony protrusions on the frontal skull plate above the eyes that become the antler pedicels. Without production of testosterone, as in the case of castration, these pedicels will not be formed, nor can antlers be grown.

By the time a buck fawn reaches 5 to 6 months of age, the pedicels are about ¾ inch (19.05 mm) long and have raised the skin up so that they are quite noticeable. A deer at this stage is known as a "button" buck because of the knobs. Usually, no further antler growth occurs until the following spring, when the buck will be about 10 months old.

The size and shape of the antlers depend on sex, food, age, and heredity. You are what you eat, and so is the deer. A diet containing high protein, the proper amounts of fats and carbohydrates, and adequate minerals such as calcium and phosphorus will produce large antlers. Many people, even today, think they can tell a deer's age by counting the number of points, or *tines,* on the antlers. It just isn't that simple. A buck does not reach his full physical growth until he is 4½ years old. From birth until he is 4½, the first demand on the food he consumes is for sustenance and body growth. After he has obtained his maximum body size, all nutrition not used for body sustenance can go to antler development.

The number of points on a buck's antlers may vary, depending on the food the deer eats that year. The diameter of the antlers, usually measured ¾ inch (19.05 mm) above the burr, or coronet, increases an average of ³⁄₁₆ inch (4.76 mm) each year but not at a constant rate. Furthermore, not all deer start out with antlers of the same diameter. *New Jersey Deer Report #3* gave the averages of the 1½-year-old bucks taken in the state in 1975. The antlers of bucks on the poorest soil in the Pines region averaged ⁷⁄₁₆-inch (11.11 mm) in diameter. The bucks from the rich farmlands had antlers averaging ⅞-inch (22.23 mm). A buck of the same age, taken in 1966 from the best of Pennsylvania's farm country, had 14-point antlers with a diameter of 1 inch (25.4 mm).

In a Michigan study of 260 whitetail skulls, the main beam diameter was taken ½-inch (12.7 mm) above the burr. The 1½-year-old bucks had antlers averaging below ¹³⁄₁₆-inch (20.64 mm). The 2½-year-old bucks averaged between ¹³⁄₁₆ and 1¹⁄₁₆ inches (20.64 to 26.99 mm). The 3½- to 4½-year-olds averaged between 1⅛ and 1⅜ inches (28.58 to 34.93 mm) and the 5½-year and older bucks averaged above 1⁷⁄₁₆ inches (36.51 mm).

The roughness, or pearlation, found on most antlers to 1 inch (25.4 mm) above the base makes accurate measurements difficult; so I tried to work out an aging system by measuring the diameter of the convex crown of the buck's shed antler. After measuring hundreds of known-age deer antlers, I found the system just wasn't accurate enough because the size of deer antlers—even those from the same subspecies—varies too much according to the nutrition available to the deer. I did discover that the diameter of the convex crown of most bucks' antlers measured between ¾- and ¹³⁄₁₆-inch (19.05 and 20.64 mm) for their first set. Thereafter, the antlers increase in diameter an average of ³⁄₁₆-inch (4.76 mm) per year and as much as ¼-inch (6.35 mm), depending on the amount and quality of food available. These antler bases are the only part of the antlers that increase in size every year. The rest of the antler mass and the number of points either increase or decrease according to the buck's diet in a given year.

Using my finding as a base figure, you can roughly calculate the age of the deer that shed any of the cast antlers you find. Gene Wensel, of Hamilton, Montana, has the cast left antler of a buck that lives or lived in Illinois. The antler is huge, having seven tines, including the brow tine. The main tine is 11 inches (279.4 mm) long. (Gene figured that, if the right antler was as good as this left one, the buck must have been the all-time number one typical whitetail buck.) I figure that the buck was 5 years old when it shed this antler.

Heredity is also important. The most recent research indicates that the shape and number of points may depend chiefly on heredity, while the size of the antlers is the result of diet. A 5-month-old buck usually develops only buttons. Most biologists claim that it is extremely unusual for a 5-month-old buck to develop anything more, but I disagree. It is not a common occurrence, but I think it is far more common than suspected. The antlers on a 5-month-old buck are nothing to brag about, but they can be antlers, not just buttons.

I do a lot of my deer photography in Hunterdon County, New Jersey. This is exceedingly rich farmland, and the soil has a high calcium content from limestone. Some of the largest deer in my state come from this region. I have seen a number of 5- to 6-month-old bucks sporting polished spikes ½-inch to 1 inch (12.7 to 25.4 mm) long.

Although most 6- to 7-month-old bucks are not capable of breeding, those that have peeled and hardened antlers at this age are. This was first discovered by Helenette Silver doing research with fawns for the New Hampshire Fish and Game Division. She had a number of fawns in one large pen. In the spring a number of the doe fawns gave birth to fawns of their own. No adult bucks had access to these fawns. But one of the little bucks had had peeled antlers when he was 7 months old, during the previous fall. Further testing by Silver and other biologists proved that once the nutritional level of a buck fawn is high enough to allow his first little antlers to harden and peel, he is also sexually advanced enough to have sperm and to be capable of breeding.

Antler Growth. In my home area, northwestern New Jersey, deer antlers begin to develop about April 1. The start of antler growth is a response of *photoperiodism,* regular cycles of light and darkness that affect growth.

Photoperiodism is tied to latitude, except near the equator, where some deer can breed at any time of the year. The antler cycle of these deer is also set randomly throughout the year, although it is still confined to a 12-month cycle.

From about the 32d parallel north to the 60th parallel, the breeding seasons and antler growth of most deer are synchronous. From the 31st parallel south, the breeding season is later and longer, although antler development may not lag as far behind. In the more southern or drier regions, both the breeding season and antler development is considerably delayed and is geared to the rainy season instead of the northern warm summer season. A cold, delayed spring in the north will adversely affect the antler development there, because it will hold back the growth of the vegetation needed for the antlers' growth.

With the winter solstice in December, the number of daylight hours gradually increases. The retina of the deer's eye functions rather like a photoelectric cell, transmitting messages about light received. As the hours of daylight increase, the eye is exposed to more light, and research has proved that the pineal gland, located in the brain and often called the "third eye," works in conjunction with the eye's retina. The pineal gland changes electrical impulses from the retina to a chemical command to the endocrine system. It is the pineal gland that determines when deer grow their antlers, when they cast them, and when they breed—as well as when birds migrate, when the woodchuck hibernates, and so on. The pineal gland more or less sets the biological clock that determines the timing and sequence of the activities of most creatures in nature. There are enough light hours per day in late March and early April to cause the pineal gland to become active.

At Penn State's deer research lab, the pineal gland was removed from a buck. In November, when all of the other bucks had hard, sharp antlers and swollen necks for the rutting season, and their gray winter coats of hair, the antlers of the buck without a pineal gland were still in velvet and still growing, and he was still in his red summer coat. The removal of the pineal gland did not prevent the changes characteristic of the rutting season from occurring but because of the absence of pineal gland's commands, they were delayed four to five months.

In a test of photoperiodism, control deer were subjected to the normal 10 to 12 hours of winter daylight, while the tested deer were kept in 16 hours of light by the use of electric bulbs in their pens. The deer exposed to more light were three to four weeks ahead of the control deer in all seasonal cycles. Their antlers started to grow earlier, they shed their winter coats earlier, their antlers were fully developed and polished earlier.

In another study, at the University of British Columbia, Ian McTaggart Cowan subjected mule deer to 12 hours of light and 12 hours of darkness daily year-round. These bucks were unable to shed their antlers and grow new ones. Other bucks, kept in continuous light, grew and lost three sets of antlers in two years.

Triggered by the growth hormone released by the endocrine system, bone salts are deposited on the antler pedicel by a network of blood vessels beneath the skin. The skin covering the growing antlers is called *velvet* because it looks and feels exactly like velvet cloth.

Velvet is actually a modification of the deer's regular skin that has a number of unique characteristics. The velvet consists of an outer layer of coarse collagen fibers and an inner layer of finer collagen fibers. Because of these fibers, which are composed of protein, it is almost impossible to strip velvet from an antler crosswise. The velvet must be stripped off longitudinally, as it was grown.

The skin at the end of each growing antler tine is known as the *apical epidermal cap*. Unlike the hairs on most mammals, which lie horizontal to the skin, the hairs on the deer's apical epi-

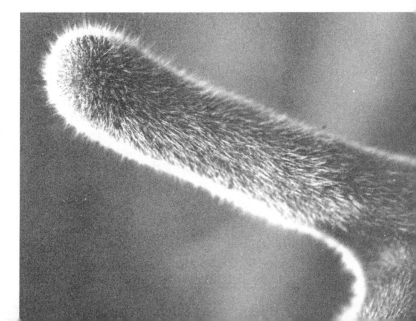

We call the fine hairs on the skin covering growing antlers "velvet" because it looks and feels like velvet material; velvet antlers are quite warm to the touch because of the rich supply of blood flowing through the velvet.

dermal caps stand on end. There are approximately 6,500 hairs to the square inch (6.5 sq cm). Every hair grows out of a sebaceous gland, which produces an oil that has some insect repellent properties, minimizing the amount of blood that would be lost to the hordes of bloodsucking flies and mosquitos that plague all wildlife. The oil gives growing antlers their shiny appearance.

We humans often remark about a scary situation making the hair on the back of our necks stand on end. They do—but at such a time, so do most of the other hairs on our bodies. Those hairs stand on end to detect danger. I have lost 94 percent of my hearing and constantly wear two hearing aids. The aids help, but even with them, I hear only a fraction of the sounds a normal person can hear. But when I am in a dangerous situation, my body hair stands on end, as if to compensate. This helps me to feel when a potential threat is near. The hairs act like radar detectors.

That is the same purpose for the erect hairs on the ends of deer's antlers. The entire velvet is a maze of nerve endings that originate from the trigeminal nerve, and the hairs on the ends of the growing antlers all stand on end as sensory receptors. To prevent damage to the soft antler tips, the hairs "feel" any obstruction and, through his nervous system, the buck is warned to avoid the obstacle. Even after the velvet has been shed, the buck may retain a memory of precisely where his antler tips are in relation to obstacles.

Like the deer's internal skeletal bones, especially the ribs, from which most of the minerals are withdrawn, the antlers are made up mostly of calcium and phosphorus, along with small amounts of magnesium and ash. These minerals will be replaced in the skeleton after the antlers are full-grown. Some of the minerals going into antler development are taken from the nutrients currently being ingested. This is the reason why bucks living in areas with a plentiful supply of these minerals in the soil, and hence in vegetation, will have the best chance of producing large antlers. If the soil in your area is lacking these minerals, fertilization of the soil will help. You can also put out specifically balanced commercial mineral supplements, such as Deer Lix. These minerals must be made available to the deer in the early spring when bucks are growing their antlers and does are carrying their young.

Matter cannot be destroyed. Any area that is fertilized, unless its vegetation is removed, as it would be in crop harvesting, will have most of the fertilizer returned to the land in the excrement of the wildlife living there.

Antlers begin to ossify, or turn to solid bone, from the base outward and up to the tips; this half-sectioned antler tip shows a ½ inch that is still soft and spongy.

The velvet is richly supplied with blood vessels to handle the tremendous deposit of minerals forming the antlers. It is the tremendous flow of blood carrying the minerals through the skin that causes the growing antlers to be "hot to the touch." No other part of a deer's body that I know of registers the deer's internal body temperature of 104°F (40°C) on its external surface.

Although it is probably incidental, there can be no denying that the growing antlers also actually regulate temperature. The temperature of the blood is lowered by the time it has coursed through an antler and returned to the body through the veins. The heat that reaches the external surface helps the antler cells grow at the fastest possible rate: Antler growth is one of the fastest known forms of tissue growth. A deer's antlers will often grow ¼-inch (6.35 mm) per day, an elk's antlers ½-inch (12.7 mm) per day, and a moose's antlers ¾-inch (19.05 mm) per day. Some biologists think it takes as much out of a buck's body to produce a large set of antlers as it does out of a doe's to produce a pair of fawns.

The arteries supplying this vast amount of blood to the deer's antlers have thicker walls than do the internal arteries. These thickened muscular walls allow the arteries to constrict and virtually shut off blood flowing from a cut in less than a minute, which prevents hemorrhaging. Approximately 12 arteries supply the blood for antler growth, and 12 veins return it to the body. The site of each of these vascular vessels is permanently imprinted on the growing antler, producing the longitudinal lines that can be seen on the hardened antler. Some blood is also supplied and returned through the center of the antler, but this supply is diminished and eventually eliminated as the basal portion of the antler begins to *ossify,* or harden.

Antlers grow like a twig on a tree, adding new growth from the base. As the epidermal cap is pushed upward, it lays down a mineralized cartilage that hardens into a *spongiosa,* or soft bone, as the antler elongates. Blood can flow through this spongy bone in both directions. Continued growth causes the basal end of the spongiosa to calcify completely. Eventually most of the antler base becomes solid bone, completely shutting off blood flow through the antler itself. Contrary to what might be expected, antlers become more brittle as they turn to complete bone just before they are cast. While the antlers have an outer cylinder of hardened bone but still have the spongy core, they can sustain a greater impact without breaking. More antlers are broken during the latter part of the rutting season even though more fighting is done in the first part of the rut. And, in the northernmost part of their range, bucks' antlers also become more brittle and will shatter in extreme cold.

The actual density of an antler varies throughout the antler itself, and also with the deer's age. Researchers have found that the hardest, that is, densest, part of the antler is the tip of the "eyeguard" or "brow" tine. Each succeeding tine is less dense, with the beam tip having the least density. One reason for this is probably that the buck's body supply of calcium is almost depleted by the time the antlers are full-grown. Also, the antlers have started to calcify at the base before mineralization of the beam tip has been completed.

The antlers of yearling deer are almost solid bone, with no spongiosa. Bucks usually grow larger antlers each succeeding year, and the spongy core becomes larger. Increased spongiosa means that the buck can grow larger antlers without a greater corresponding drain of minerals from its bones. As the deer matures, it becomes more dominant and does more fighting to prove this dominance. Having more spongiosa lessens the chances that the mature buck will break its antlers.

During the first couple of months of antler growth, while the bulk of the antlers is still soft, bucks become very reclusive and restrict their range to an area no larger than is needed to provide food, water, and shelter. During this time, when vegetation grows

The antlers on this New Jersey whitetail buck are just started to grow; by summer, a mature whitetail has a second tine forming on each beam.

By July second forks are developing on this blacktail buck's antlers; the fully-grown antlers on the mule deer at right are still velvet-covered in August.

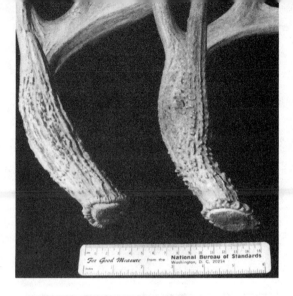

The base is the only part of the antler that always gets larger every succeeding year. The actual mass of the entire antler depends greatly on the buck's diet for that year; these two antlers are from the same buck when he was 3½ and 4½ years old. The amount of pearlation, the little knobs or ridges on the base, is genetic.

profusely and their needs can easily be met, the bucks virtually disappear. In July, as their antlers harden, the bucks will move about more freely.

Within three to four weeks after the antlers have started to grow, the tine will form. On a mule deer this tine seldom reaches 3 inches (76.2 mm). In another three weeks or so, a whitetail's second tine will grow, and the first fork will appear on a muley or blacktail. These tines grow longer as new ones are being formed. On a whitetail, the second tine formed is usually the longest. In four months, the antlers are fully developed and the burr starts to form.

Typical and Nontypical Antlers. In the generally accepted method of trophy measurement, a tine or point must be at least an inch (25.4 mm) long to be counted, and it must be longer than the base from which it grows. Antlers are usually symmetrical, bearing the same number of tines of roughly equal length on each side of the head. But asymmetrical antlers are also quite common. A fairly symmetrical head that has an extra point or two—or even three—on one side is considered "typical" for the purpose of trophy scoring, but differences between the left and right sides will lower its score. If there is more than 25 inches (62.5 cm) of abnormality, the antlers will be considered nontypical.

Truly abnormal racks—those having oddly shaped or located points—are sometimes fairly symmetrical and sometimes extremely asymmetrical. They are classified separately in the record book *North American Big Game*. Antlers in this nontypical category often have a great many more points than the typical tro-

Heads and Tails

phies, and such abnormalities would dominate the records if they were not classified separately.

A fantastic nontypical whitetail buck shot in Texas held the world record from 1892 until recently. It had 23 points on the right antler and 26 on the left. It was thought that its score of 286 would never be broken. In 1981 a huge whitetail buck was found dead in St. Louis County, Missouri, that had 19 points on the right antler and 25 points on the left. Its score was a staggering 333⅞. As if to challenge the "picked up" Missouri buck's claim, another amazing nontypical was discovered in 1983 in Portage County, Ohio. The buck had been found dead in the early 1940s and had been mounted and hung in a bar in Kent, Ohio, but had never been scored. The record book shows that the Ohio buck scored 328⅜ points as the number two nontypical. The top-ranking nontypical muley, killed in Alberta in 1926, was a 43-pointer and scored 355⅝ points. Nontypical blacktail records have not been kept, evidently

The drop tines and extra forks on the main beams qualify the mule deer at right as non-typical. Mule deer seldom have brow or eye tines as long as three inches. Unusual whitetails can have brow tines six inches or longer; the buck at left also shows a drop tine on his left antler.

because few nontypical blacktails have been sufficiently impressive. Cowes deer records are kept in both categories.

Freak antlers can result from injury during growth, but large nontypical racks bearing a great many tines are usually congenital oddities. I will examine the causes and characteristics of various abnormalities later, after discussing normal antler development. Although nontypical antlers are interesting and desirable trophies, symmetrical racks are considered to epitomize what a deer's antlers should be.

The world record non-typical whitetail was found dead on a fenceline in suburban St. Louis, Missouri. This amazing trophy shows almost every variation of non-typical antler formation, including 44 scorable Boone and Crockett points. photo courtesy Missouri Department of Conservation

When a hunter in the eastern United States speaks of taking a four-point buck, he means that the buck has a total of four tines—presumably two on each side. For many years, hunters in the West used one side only (the side with the larger number of tines if there was a difference) in giving the count. Western count does not include the brow tine. By western count, a four-point buck had four tines on one side, and he might or might not be an eight-point buck by eastern count. The old western system obviously lacks clarity. Fortunately, it is now losing favor.

A typical whitetail buck has main beams that project backward, then sweep out to the side, going upward, forward, and in some cases curving inward again. The first tines on either side are the usually short brow tines. The second points on the inside are usually the longest, and each succeeding tine is usually a little shorter. The tips of the main beam are also considered tines.

In a whitetail buck's second summer of life, if his diet is just barely adequate, he usually grows single, unbranched main beams known as "spikes." These spikes are usually 3 to 6 inches (76.2 to 152.4 mm) long and project straight up from the head. They are usually a sign of malnutrition. I don't say that the deer did not have enough to eat; it's just that his diet was deficient in protein.

My brother-in-law, Dave Markle, shot a spike buck that had long, gracefully curved spikes 14 inches (355.6 mm) long. James Hartman of Elverson, Pennsylvania, shot a buck that had 16-inch (406.4-mm) inwardly curving spikes. And I have photographed spikes varying tremendously in length and curvature. Most of these deer were of the same approximate age, the variations having resulted from heredity or diet or both.

When deer have spikes longer than 6 inches (152.4 mm) yet are in good condition, they are probably genetically inferior. This can only be determined if they are captive deer (so that their age can be verified), or if they are harvested. Any deer that is 2½ years old and still has spikes is genetically inferior, and it is unlikely ever to produce good antlers, even on a nutritious diet.

When each of a whitetail's main beams has just 1 tine branching from it, the deer is called a *forkhorn, four-pointer Y-buck, pronghorn,* or *crotch-buck.* A deer with more than four points is called a *rack buck.* Eight points is considered average for a whitetail, but there may be 10, 12, 14, or even 16 points. Or there may be odd numbers. Some racks are large but have few points; some are small but have more points. The combinations of sizes and shapes seem almost infinite.

A muley or blacktail buck at 6 months has just the skin-covered pedicels. I have never seen a mule deer or blacktail of that age with the little polished antlers I have seen on young whitetails, nor have I read of one in any of the literature.

When a mule-deer buck is 18 months old, it is usually a forkhorn. Each beam now has two tines, giving it almost a Y shape. A blacktail at that age usually has just spikes. At 30 months, a muley buck's main beams usually fork again. The rear fork now has the two points, while the forward section is still a single tine. A smaller brow tine may or may not be present. The blacktail at this time may be a forkhorn or it may have three tines on each beam. While it is usually easy to tell the antlers of a mule deer from a whitetail's because of the double forking of the main beams, this is not true of the blacktail. Often a blacktail, when it has three main points on each side, not counting the brow tine, has antlers indistinguishable from that of a whitetail. The main beam does not clearly divide and the three points appear to rise from the single main beam. To further confuse the matter of identification, western whitetail deer occasionally have antlers that appear to divide like those of the muley and blacktail. The mature mule deer's main beams generally fork and then fork again into roughly equal tines, looking like four good slingshots above the brow tines. This pattern is also considered ideal for a blacktail trophy but it is attained far less often.

With all of the deer, age, heredity, and food determine antler size. With a 16- to 18-percent protein diet and adequate calcium and phosphorus, bucks will develop from button bucks at six months to large rack bucks at eighteen months. This is a generalization, of course. Nutrition makes the difference with most deer but, as indicated in one Texas study, some bucks may fail to grow large antlers because they are genetically deficient. When genetically small-antlered bucks are put on the same diet as more impressive bucks, they still fail to grow comparably impressive racks.

When written up by the popular press, the Texas study created a tremendous furor. Everyone reading the articles wanted to eliminate all spike bucks, no matter where they were found. If spike bucks were the norm in habitat that was supporting fewer deer than its carrying capacity, so that sufficient, nutritious food was available, then I would agree. But if spike bucks were taken from habitat that was poor to start with, or had deteriorated, shooting the spikes would never guarantee the production of better deer. On lands where the deer population is far beyond the carry-

This young whitetail buck has unusually long spikes; in a fight, this "inferior" antler formation can be more deadly than fully-branched antlers. Since this buck appeared otherwise healthy, it's possible the spikes are in his genes, not a result of poor antler-growing nutrition.

ing capacity of the land, then the entire herd, both bucks and does, has to be reduced enough, and for long enough, to allow the habitat to recover. Then, and only then, will the quality of the deer improve.

Extensive testing done by Pennsylvania, New Jersey, and New York has proved time after time that if bucks have protein diets in excess of 18 percent, they will go from button bucks at 7 months to rack bucks at 19 months. Although these racks will have six or more points, they will not be massive. Mass comes with age.

I was so intrigued by the results of the Texas research that I read all the data I could find on it and then took a trip to the research station at Kerrville, Texas, where the tests were conducted. Their research proved that many of the deer they tested were definitely inferior deer and would never be anything but inferior deer, no matter what they were fed. What the report didn't emphasize, however, and nor did the popular literature, was that for many, many years Texans were forbidden by law to shoot a spike buck, no matter how long or short the spike. In effect Texas had created a breeding pool of genetically inferior deer. A 3½- or 4½-year-old spike buck would be a larger-bodied deer than would a 2½- or 3½-year-old rack buck. The tests at Kerrville proved that there was very little difference in the body weight of the spike and rack bucks of the same age. But because the spikes couldn't be harvested, many of them became much older, and probably much larger, than the rack bucks. Older and larger deer usually come into the rut before the younger, smaller deer and usually do most of the breeding. And in a fight, a buck with long spikes is more deadly than a rack buck with more numerous but shorter tines. The long spikes can slip right through a rack buck's antlers. So the law protecting spikes was turning those inferior bucks into breeding stock.

By contrast, 70 to 80 percent of all the bucks killed in my home state of New Jersey are taken when they are 17–18 months of age. About 27 percent of our bucks are spikes when they are harvested at 1½ years. Despite our liberal deer seasons, we still have many areas that contain more deer than the habitat can properly support. Even if some of those bucks are genetically inferior, and I am sure that some of them are, they have very little chance to pass on their inferior genes, because the larger bucks must prevent most of them from breeding.

The researchers at Texas agree that most spikes are nutritionally deprived and that, over time, this would create genetically inferior deer. The state of Texas is rectifying this stuation by allowing spike bucks to be harvested.

Al Brothers proved my point again by livetrapping a buck fawn in Kerr County, Texas, where most of the research on stunted

The entire pedicel and skull plate broke loose from the skull after the antlers had already formed, resulting in a freak up-and-down orientation of the antlers.

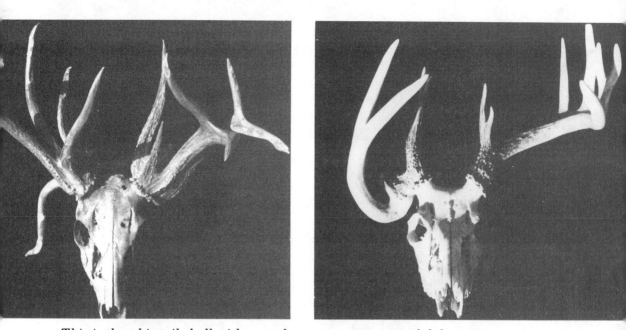

This is the whitetail skull with seven beams—a very unusual deformation in itself—and all the tines point in different directions. On the right, a deformed antler that apparently was bent after normal growth was completed, but before *the antler solidified. It hardened in this position.*

deer was done. The fawn was removed, before he was skeletally stunted, to Webb County, Texas, where he was released in fine deer habitat that provided excellent nutrition. The buck was 4½ when harvested, and he had developed both good body weight and good antlers.

Although antlers are usually symmetrical, abnormal shapes are quite commonly produced by mutations, injuries, metabolic or hormonal deficiencies, or heredity. I have found that injuries to the antlers are much more frequent among whitetails than mule deer. Perhaps this is because whitetails inhabit regions with more vegetation to bang into. Conversely, I have found more genetic deformities among mule deer, and I don't know of a theory to explain this.

When a deer's antlers are growing, they are so soft that they are easily bent or broken. I have seen whitetail bucks smash into fences and bend their antlers. As a rule, the antlers do not break off but are knocked out of shape. A buck shot in 1955 in Sullivan County, New York, had the right antler badly bent. The antler had continued to grow after the injury and it came to rest against the

The left antler on this Pennsylvania whitetail has split to form a third main beam. The deer at right damaged his right antler while in velvet, and the trauma will cause it to grow abnormally.

buck's mouth. The left antler had three points, polished and sharp, while the injured antler was still in velvet.

A six-point buck that was shot near Huntingdon, Pennsylvania, had well-developed, polished antlers. The right one was normal while the left one had been bent so that its tip pressed tight against the deer's upper jaw.

A buck mule deer killed near Laramie, Wyoming, had both antlers bent down toward its nose. The antlers then curved under the jaw, preventing the deer from opening its mouth more than ½-inch (13 mm). The antlers were still in velvet, and still growing. Although the deer was in good condition when shot, it probably would have starved to death as further antler development would have prevented its mouth from opening.

A forkhorn whitetail killed by an automobile in 1966 in Steuben County, New York, had a small but normal and polished left antler while the right one was bent down over the forehead and was in velvet. The Y of this antler fitted over the deer's nose and the right side of his face.

All of these deer were in good condition, indicating that the abnormality was caused by an injury to the antler, not to the body of the deer. I could enumerate many more such cases. If an antler is injured only slightly, it may harden and be polished as quickly as its normal mate. My friend Bob Elman showed me a typical example, the frontal skull plate of a spike buck shot in the Catskills in 1974. The left antler was normal, curving out, back, and up to a fairly sharp point. It measured 6½ inches (165.1 mm) long. The right antler curved in the opposite direction—forward and down to a blunt tip over the snout—and it measured 5½ inches (139.7 mm). In 1977 in Pennsylvania, an acquaintance of Bob's shot a buck with a fairly long, straight spike on one side and a forkhorn on the other. At the base of the fork was a large hole, clear through the bone. Again, both antlers were hard and polished.

The oddest antler formation in my personal collection was given to me by my friend Joe Taylor. The buck was killed in Warren County, New Jersey, by an automobile in early September. Most of the velvet had been peeled from the rack, although some shreds remained. The deer was in excellent condition. The pedicel on the left antler was 1¼ inches (31.75 mm) across, but the antler was 3¼ inches (82.55 mm) across. The 7 main beams have 13 tines

This unusual mule deer was shot by Frank Kacsinko along the Little Bow River in Alberta, Canada. There was nothing abnormal about the deer except the antlers, and the buck was fat and healthy despite the fact that the antlers' position would seem to make eating difficult, not to mention vision. The abnormal ridge along the skull suggests that the entire frontal skull plate was broken, tilting the antlers to this position.

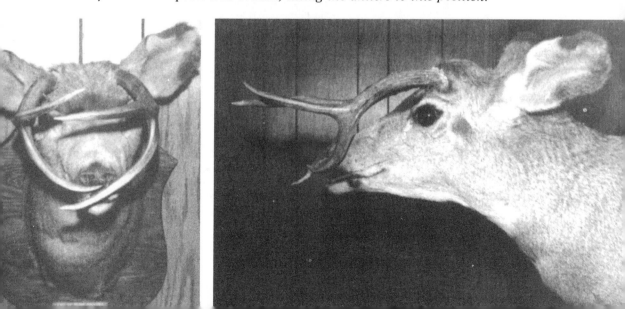

An injury sustained to its left rear leg caused the whitetail in the top photo to grow a deformed right antler the following year. The antler seems to respond to an injury to the leg that would normally bear its weight. The right antler of the mule deer, bottom, was broken off after it had started to grow, and the stunted antler remained in velvet.

Heads and Tails

that point in all directions. The accompanying photograph of this head shows the deformity better than I can describe it.

Such freak antlers are the results of trauma. The left pedicel was severely injured somehow just as the new antler growth started. The pedicel may have been split or even smashed. The effect on the deer's nervous system was so severe that the right antler, although uninjured, also had abnormal development. It was unfortunate that the buck was killed, because it would have been most interesting to see if the trauma was severe enough to have affected his antler development in subsequent years. A captive buck with the same kind of injury, raised by George A. Bubenik, grew similar freak antlers for the rest of its life.

I photographed another oddity in Pennsylvania. The deer's left beam split at the base to form two main beams. Usually this type of formation is genetic, occurring year after year. Unfortunately, I could not prove this with the two deer just described because neither lived beyond the year when I saw it. I did see a big bull elk in Yellowstone that had such a formation for three consecutive years. The elk's second brow tine formed another, unbranched main beam that was as large as the original main beam.

Most deer that are kept in captivity injure their antlers on fences. This usually results in split antler tips. Bucks that are captured or have to be transported to livetrap carrying cages almost always knock their antlers into grotesque shapes.

Occasionally deer antlers are seen that have formed a large knob with a hole in it. This is the result of a warble fly having laid its egg in the growing antler. In much the same way as an oak leaf or a stalk of goldenrod swells to grow around the larvae of a gall fly, the deer's antler swells around the warble-fly larva. When the warble emerges, a hole in the antler marks its exit.

There are many records of deer that have misshapen antlers due to bodily injuries. Usually, if a deer is injured on the right side of the body, particularly on the hindquarters, its left antler will show the damage, and vice versa. One theory is that the deer, in trying to lick the injury, is apt to bang its outside antler against vegetation. My own theory is that because the left part of the brain controls the right side of the body, and vice versa, damage done to one side is reflected by damage to the antler on the other side. One deer that I photographed in Hunterdon County, New Jersey, had a badly injured left hind leg that left it with a permanent limp. The following year this buck grew a normal antler on the left side but the right side had only a small, misshapen antler.

Louis Matthews and the cactus buck he shot in 1971. These unusual formations are related to incomplete testicle development which results in permanent masses of velvet-covered "antlers." photo courtesy of the Missourian Publishing Company

Some deformities are caused by unusual circumstances in which a buck's food supply is drastically reduced or no minerals are available. Sometimes injury to internal organs prevents a buck from properly utilizing or metabolizing its food, and this will be reflected in his antlers.

Like the bucks of Yosemite mentioned earlier in this chapter that grew freak antlers on their nasal bones, some deer have antler deformities as a result of heredity. For a number of years I made two or three annual trips to Yellowstone National Park at

The palmated brow tines on this whitetail are most unusual, although palmation occurs more frequently on the main beams.

different seasons for photography. I got to know many of the park's animals individually by their oustanding characteristics. One huge nontypical muley buck had the widest rack I have ever seen on a deer. Another not only had a widespread set of antlers, it always had a downturned tine on its left antler. Both of these deer had the same characteristics each year, indicating that the antler structure was inherited.

Palmation of antlers is another odd genetic characteristic. In some instances, the antlers are merely much wider than normal. Once this trait appears in a given area, many bucks in succeeding years will have antlers of this type. Occasionally the palmation is so extensive that the antlers look like those of a moose. Harold Throop of Odell, Oregon, shot such a mule deer in 1935. The antlers spread over 30 inches (76.2 cm) across and had 35 points, but the most interesting feature was the 14-inch wide (355.6-mm) palmations which gave the rack the appearance of moose antlers.

Leslie Robinette reports on 14 mule deer that had genetic antler deformities, all from the Oak Creek drainage of central Utah. Some of these bucks had no antlers or pedicels at all. Some had nothing at all on one side of the head and an exceedingly large antler on the other side.

In 1983 I photographed a good whitetail buck in central Pennsylvania that had normal antlers but long, palmated brow tines. One of the brow tines had four legal points growing in a row. Since I had never seen such an antler formation, I was most anxious to see if the buck would duplicate the unique growth the following year. He did. The same type of growth two years in a row, with no evidence of injury, proved that these palmated brow tines were of genetic origin.

A large mule deer buck killed by Ennes Aldredge of Ogden, Utah, was eight years old and weighed 204 pounds (92.5 kg) hog-dressed. The left antler was 25½ inches (64.7 cm) long, had a 1½-inch (38.1-mm) beam diameter and seven points. The antler on the right side was a mere bony protuberance ½-inch (12.7 mm) long.

Occasionally, deer have antlers that don't even look like antlers but resemble a mass of coral brought up from the ocean floor. Such deer are called "cactus bucks." Instead of having beams and points, they have a mass of knobs. Louis Matthews of New Haven, Missouri, shot such a mule deer in September, 1971, near Dixon, Wyoming. It had 69 knobs on its head. Such "cactus" formations are caused by atrophied testicles, and the antlers are usually retained for the rest of the animal's life.

If buck fawns are castrated while their antlers are growing and in velvet, the velvet never dries, nor are the antlers shed. However, these soft antlers are usually lost because the moisture in them freezes in the winter and the frozen pieces break off. Bucks that are castrated after the antlers are fully developed, solidified, and with the velvet peeled off, will shed these antlers within two or three weeks. Such bucks may grow another set of antlers the following year, but the antlers do not harden and are lost through freezing. If a castrated buck lives in an area that is not subject to freezing, the antlers will grow larger but not solidify, and the velvet will not be lost. Pieces of soft antler will continue to break off.

In Europe the small roe buck normally grows antlers during the cold winter months instead of during the summer. If a roe buck is castrated, its antlers will grow as a mass, not as elongated antlers, and will not harden. If roe deer grow their antlers in the summer, the cold weather in winter freezes off the antlers because of their moisture content. Castrated roe deer starting the growth of their antlers in the winter have all spring, summer, and fall for the antlers to grow before the cold of the following winter freezes them off. Roe deer, for some unknown reason, produce more antler growth than any other castrated deer. As the growth piles up on top of the deer's head, even cascading down the sides, the antlers may acquire a massed size much larger than the deer's head. The deer looks as if it is wearing one of those outlandishly high, bee-hive-style wigs worn by European women centuries ago. Such an antler formation is known as a *peruke,* which means "wig."

A great deal of research is currently being done on antlered does. At one time, Pennsylvania estimated that 1 out of every 18,000 antlered deer killed in that state was a doe. Recently it has been found that the percentage is much higher. As more states conduct this kind of research, the ratio is constantly being lowered. C. W. Severinghaus, of New York's deer-research work, found that in his state the ratio of antlered does to antlered bucks is 1 in 2,500 to 2,700. In 1959, J. Kenneth Doutt and John C. Donaldson of the Carnegie Museum in Pittsburgh, Pennsylvania, found that of 38,270 antlered deer killed, 17 were antlered does, producing a ratio of 1/2251, or close to the New York figures. Doutt and Donaldson continued their studies and in 1961 reported that out of 173,038 antlered deer killed in Pennsylvania in the previous four years, 43 were does—a ratio of 1/4,024. In checking through the records I find in the Pennsylvania study that no single area is producing a majority of these deer. The antlered does are randomly scattered throughout the state.

Antlered does fall into three categories. In the most common of the three, the antlers never harden, nor does the velvet ever come off. Does usually lose most or all of these antlers through freezing. Such a doe can breed and produce milk to feed her young. She is a true female, but her female hormones are usually not produced in sufficient quantity to suppress the somatotrophic growth hormones of the pituitary glands. The antlers are formed under the stimulation of the pituitary, but the doe, lacking testicles, does not produce the male hormone testosterone needed to harden these antlers and complete the cycle. Her neck does not swell as a buck's neck would, nor are the hock hairs stained as darkly.

A 4-year-old whitetail doe killed near Mercer, Pennsylvania, on April 19, 1967, was just starting to grow a new set of antlers. That this doe was definitely a female was proved by the triplets she was carrying.

The highest ratio of antlered does on record was reported by William Wishart from records of controlled hunts at Camp Wainwright in Alberta, Canada. The camp occupies 965 square miles (2499.24 sq km). The hunts took place between 1968 and 1982. In that time 665 adult bucks and 517 adult does were harvested. Of the 517 does, 8 had velvet-covered antlers ranging in length from 7/16-inch to 9½ inches (11.11 to 241.3 mm), a ratio of 1 out of every 65 does. None of these does ever shed their velvet. Since milk was obtained from all of them, they were breeding females. G. W.

This is an old muley buck as evidenced by the regressive characteristics of his rack: a wide spread but only a few short tines.

Mierau, working on mule deer in Colorado, reported on a velvet-antlered doe the female offspring of which also had velvet antlers, proving that this trait can be passed on genetically. Since the 8 does in the study reported from Camp Wainwright were taken over a 14-year period, and there is no record of two being taken in any one year, that high ratio is a false one. Most of the other studies were done in a single season and represent a true ratio. What the Camp Wainwright record shows is a genetic characteristic being passed on.

The second type of antlered "doe" is basically a male, but its sex organs are abnormal. Usually such a deer has both penis and vagina, but the scrotum is not visible because the testicles are up inside the body cavity. An animal of this sort never bears young. The antlers are like a typical male's, often well developed with tines. They harden and the velvet is peeled off. But the neck does not become swollen as a normal buck's neck would.

The third condition is very rare. It occurs when a tumor in the doe secretes male hormones. Both male and female reproductive organs may be present, and the antlers may or may not complete their development. Most often they remain in velvet.

Heads and Tails

A hunter who shoots any antlered doe should immediately contact his game department so that the deer and its reproductive organs can be saved and studied. It is hoped that in time, if enough specimens can be examined, our knowledge of deer with characteristics of both sexes, a condition called *hermaphroditism*, can be greatly increased.

When a buck gets old, its antlers may regress. It may retain a large, spreading rack while the number of points decreases. The rack will probably be smaller in the last years of life. A buck may then regress to being a forkhorn or spike, or he may not grow any antlers at all. This is caused by poor teeth, which prevent him from getting an adequate diet, and by dwindling hormones.

Calcification of the Antlers. June 21, the summer solstice, has more hours of daylight than any other day of the year at the Tropic of Cancer, as the sun reaches its zenith. From then on, the days get shorter. Photoperiodism started the antlers' growth, and photoperiodism stops their growth and causes them to harden.

Toward the end of July, the buck's antlers have just about attained their maximum size for that particular year. Antlers grow for just a little over 100 days. The new growth starts off slowly in late March, becomes very rapid from April through July, then tapers off until completed in mid-August. Though the antlers are not hardened when they reach their full size, there is some ossification at their bases.

As the days shorten, the pituitary gland, located at the base of the brain, produces a gonadatrophic luteinizing hormone that causes the testicles of the bucks to enlarge and descend from the body. The scrotum now becomes visible as it more than doubles in size, to 4 to 4½ inches (10.2 to 11.4 cm) in length, and about 6 inches (15.2 cm) in circumference. Whitetail bucks in the northern part of their range are not capable of breeding from about the end of April to the end of July.

As the testicles enlarge, starting around August 1 north of the 32d parallel, they begin to produce motile sperm. The production of testosterone starts to build in August, reaches a peak in late October and drops sharply in mid-December. The volume of semen and the sperm count reaches its height in mid-November and declines rapidly in mid-December.

Testosterone speeds up the calcification of the antlers, so that by mid-August they have completely hardened right up to the tips. The antler's internal blood flow is reduced and then completely eliminated as the antler solidifies. Then the antler burr grows outward, shutting off the vascular blood supply in the velvet. Although we say that the velvet has dried, there is still some residual blood left in velvet right up to the time it is shed. By the time the velvet has been shed, the bucks are capable of breeding.

From personal observation, I think that the drying of the velvet is irritating. Deer are anxious to get rid of it. I would liken it to our peeling off dead skin after a sunburn. Usually the velvet is peeled off within 24 hours. Yet I have seen one captive buck peel all velvet from his antlers in just 10 minutes. Nevertheless, I have also seen bucks that had dried velvet shreds on their antlers for several weeks. One buck I photographed had short strings of velvet that he could not get off the bases of his antlers. Sometimes deer will eat velvet as it comes off. It is not completely dry; it is usually quite flexible and the strands are bloody on the inside. Occasionally you can actually see where blood has run down an antler as the velvet was peeled off. It is this blood that colors the antlers brown. Sometimes there is some staining from the vegetation that the bucks rub against, but the brown color is chiefly from the blood. Once the velvet has peeled off, the antlers immediately start to bleach white from exposure to sun and rain.

Bucks peel their velvet off by rubbing their antlers against small, resilient saplings and bushes. Resilience is the key characteristic of the saplings and bushes they use. The deer want something with give, something that pushes back.

Many people have the impression that all buck rubs are made by bucks rubbing their velvet off. This is only the beginning of rub marks. Before a boxer fights, he exercises to tone up his muscles and build his strength. This is what deer are doing as they *fight,* not just rub their antlers against, vegetation all fall.

Most buck rubs are on saplings 1 or 2 inches (2.5 to 5.1 cm) thick, but many are on 3-inch (7.62 cm) trunks and occasionally on small trees 4 inches (10.2 cm) or larger. I have found a definite correlation between the size of the buck and the size of the sapling or tree he pushes against. The larger or heavier the buck, the larger the diameter of the trees he uses.

Bucks not only fight with the trunks of saplings, they also hook their antlers into branches. They even raise up on their hind feet to tangle with the higher branches. There is one record of a buck that actually entangled his antlers in overhead branches, became hung up and died. This action may have developed from food gathering. I have seen both deer and elk stand on their hind legs,

Rubbing behavior begins as the velvet dries out, but it does not stop when the velvet is completely shed.

slash their antlers into the overhead branches, and then feed on the pieces that they broke off.

During spring and summer, an average whitetail buck's neck, measuring 6 inches (15.2 cm) in diameter behind the ears, is about 16 to 17 inches (40.6 to 43.2 cm) in circumference. A blacktail buck's neck is about the same size; a mule deer's about 17 to 18 inches (43.2 to 45.7 cm) around. After the velvet has peeled, the constant battling with saplings and bushes strengthens and enlarges the neck muscles, making it easier for bucks to carry their antlers. The increased level of testosterone causes the blood vessels in the neck to enlarge so that they become engorged with blood, greatly swelling the neck. This engorgement acts as a shock absorber, cushioning the deer's neck and body from the tremendous impact of fighting. The 151-pound (68.5-kg) buck whose measurements I used earlier in regard to hoof surface and bearing weight, had a neck circumference of 21½ inches (54.6 cm). The largest buck's neck I have ever measured was a little over 28

During the rut, a buck's neck swells, as evidenced on this whitetail. Bucks often exaggerate this characteristic by flaring their neck hair.

inches (71.1 cm). I have seen larger necks, including one that looked as big around as my waist, 34 inches (86.4 cm). The two largest neck records I can find were of a mule deer and a whitetail, each with a 37-inch (94-cm) circumference.

Everyone knows that there is a specific rutting period for deer, but a lot of people don't know how to determine just when it occurs and when it is over. Bucks are in rut from the time their necks start to swell until the swelling goes down. But they are capable of breeding before their necks swell and for an even longer period afterward. For bucks in the northern states and in the provinces, the general rut is about 7 to 8 weeks long, but for deer in the southern states, the rut is longer. This is because the fawns of the northern deer have to be born from about May 15 to June 15 to take maximum advantage of the weather and the vegetation. Southern deer have the luxury of warmer climates and a more plentiful supply of food over longer periods, so the birthing time of their fawns is not as critical. The deer of the tropics can breed at any time of the year, and their fawns are born at any time of the year.

The rutting season for most bucks north of the 32d parallel begins about the third or fourth week of October and continues until the second or third week in December.

The majority of the does north of the 32d parallel will be bred between November 10 and November 25. The absolute peak of the rut usually occurs for a five-day period that may begin as early as November 5 or up to November 20. Although photoperiodism is the mechanism that triggers the rut, the precise timing is very variable and can be adversely affected by weather conditions and temperature. These latter factors affect not only the deer but their habitat, possibly limiting the food available, which in turn affects all of their life processes. I have found that exceptional cloud cover, prolonged fall rainy spells, and the dark phase of the moon all limit the amount of light reaching the deer's eye. These conditions will bring the peak of the rut a few days earlier than usual.

In the fall of 1985, we had the wettest November on record, and October, too, was very dark and stormy. The peak of the rut started about November 5 and was finished by November 10. Although there was still a lot of breeding activity afterward, all of the deer's activities had lessened by November 10.

And it was not just the deer that were affected. That year, more birds came down from the north earlier and flocked to my feeder than usual. Even the woodchucks retired to their hibernation dens earlier than usual.

LOSS OF THE ANTLERS

After the peak of the breeding, bucks' testosterone levels drop off markedly, due to the expulsion of testosterone in the semen and a drastic decline in the production of semen itself. Both are caused by the diminished activity of the pituitary gland in response to the decreased amount of daylight available each day. Combined, these factors cause the antlers to be cast. When the testosterone level drops, a layer of specialized cells at the base of the antlers, called *osteoclasts,* begin to reabsorb the calcium from the antlers. The solid mass of the antler becomes grainy until only chambered spicules remain. Eventually just the weight of the antler is sufficient to cause the remaining spicules to break and the antler drops off at the pedicel. The roughness at the base of the cast antler was caused by erosion of the spicules.

The largest, strongest bucks do most of the breeding, and these bucks drop their antlers the earliest. I have seen this happen year after year among both captive bucks and those in the wild. The faster breeding bucks use up their testosterone, the faster their antlers fall off. Large, strong bucks kept in captivity without a chance to breed will keep their antlers for a month or two longer than deer in the wild. The younger bucks, those with spikes or small racks, are kept from breeding by the mature bucks, and they, too, keep their antlers one or two months longer. This also holds true for elk.

The granulation process is fast. One day the buck's antlers are so solid they could not be broken off if you hit them with a pipe. The next day, after granulation sets in, one or both antlers may drop off by themselves. A friend of mine, Fred Space, had a captive buck that had not had a chance to breed, and therefore kept his antlers—and his aggression—on into March. One day Fred decided to knock the rack off to tame the buck down, and he hit the antlers with a piece of pipe. The blow knocked the buck off his feet, but the antlers remained intact. A couple of days later, the antlers dropped off of their own accord.

The loss of antlers causes the buck no pain: it is like the leaves falling from a tree. Both antlers may drop off simultaneously, or one antler may drop off one day and the other the next day, the next week, or even the next month.

A dropped antler has a convex base of about ⅛-inch (3.18 mm). This leaves a corresponding pit on the deer's head down to the top

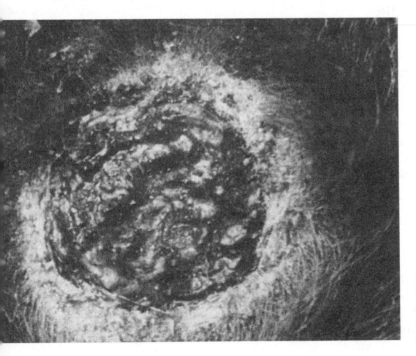

The pedicel shortly after the antler has been "cast": the scab-like skin will gradually grow over and then, in spring, erupt as a new pedicel growth.

Heads and Tails

This buck has just shed one antler: there is evidence that, in a stable environment, individual deer shed their antlers on a precise yearly schedule, although not both antlers at once.

of the pedicel. After the antler falls, the pedicel is slightly bloody and has a faint odor of decay. It looks as if the scab of a wound had been pulled away. This pit dries up in a few days as a scab is formed. The trauma of the antler casting causes the surrounding skin to grow over the pedicel, under the scab. When the skin has covered the entire pedicel surface, the scab is pushed off. The few hairs on this new skin, unlike the hair on the skin that prompted its growth, stand upright. This new skin will develop into the velvet that will start the growth of the new antler in the spring.

As the buck advances in age, the length of the pedicel becomes shorter as the skull grows up around it. In a young buck of 18 months, the pedicels may be ½-inch (12.7 mm) in length. In an old buck, the antlers appear to be growing right out of the skull.

Antlers, unlike horn, do not expand with the growth of the animal. After the shedding of the velvet, there can be no further growth, or replacement of broken antlers. But after the antlers are cast, the base of the succeeding antlers can grow larger each year. Antlers that may have been broken the preceding fall can be replaced, allowing each buck to go into the breeding season in the finest condition possible according to his genes, age, and habitat.

Antlers are not for protection from predators. This is proved by the fact that they usually drop off before the period of deep snow, when deer are most vulnerable to predation and to harassment by dogs. At such times, a deer protects itself by rearing up and slashing out with its front feet, and that is also how deer usually fight among themselves at all times of the year.

In August 1967, at the National Bison Range in Montana, I was photographing a young pronghorn antelope in a large enclosure when a muley buck approached. He presented the "ears-back-and-tucked-chin" sign of aggression, but I was not too concerned because he was in velvet. Bucks are seldom very aggressive while their antlers are soft and vulnerable to injury. I was caught off guard when the buck reared up and slashed at me with his front feet. I scrambled out of his way, alert now for trouble. Again the buck reared up, but I turned out of the way and pushed him off balance. Before he could rear up again, I was over the fence. That buck wasn't about to use his soft antlers, but he was far from being weaponless.

When a buck's antlers drop, he usually becomes less belligerent. The breeding season has tired him. He has to eat as much as possible in an attempt to gain back the weight he lost chasing and breeding the does.

Bucks are capable of breeding before the does are ready to receive them. In the wild a doe can run away from a buck, but in captivity she is often cornered. A lusting buck is not a gentle lover, particularly when he is constantly being thwarted. Under the influence of his surging hormones, he may fly into a murderous rage and vent his fury on the nearest living object. If there are no other bucks to fight, he frequently kills does he is penned with by goring them.

At Pennsylvania research stations, where many of the basic deer studies have been made, researchers carry shields of ¼-inch (6.35-mm) plywood when entering the pens of bucks during the rut. On several occasions bucks have attacked the researchers with enough fury and strength to pierce the plywood with their antlers.

Often, when a captive buck attacks or kills a doe or is exceedingly dangerous to his handlers, he is knocked out with tranquilizers, and his antlers are sawed off. Without his antlers, he usually becomes as docile as any of the does. No researcher has yet found the reason. The buck's testicles are still functioning; the male hormones are still in his blood stream, the pituitary is not affected to any degree; yet all belligerency usually dissolves. In fact, the buck

Heads and Tails

may not even attempt to breed. Losing his antlers seems to be an insurmountable psychological shock. But all rules have exceptions. Biologist Harry Laramie of New Hampshire removed the antlers of a 1½- and a 5½-year-old buck that had become very aggressive toward their handlers. Both bucks were later used for breeding and the does produced fawns.

Antlers have a social significance. In most cases the larger the buck's antlers are, the higher that buck is on the dominance scale. Though many writers portray the males as doing constant battle, antlered animals fight far less than is commonly supposed. There are fights, and some individuals fight a lot more than others; but bucks do a lot more posturing and bluffing than fighting. This is true of most horned species as well as antlered ones. Buck deer in their threat activity often shake or toss their heads at rivals. Big bull moose tilt their chins in, turning their antler palms upright. Then they slowly rock their heads in wide arcs, from side to side, so that rivals can see how big their antlers are. Elk, caribou, deer, and wild sheep all tuck their chins in as a sign of aggression and project their antlers or horns forward. The antlers or horns are now in position for fighting, but they also *show* to their greatest advantage. The larger the antlers or horns, the more dominant the animal is and the less chance that a fight will take place.

Deer throughout the United States and Canada all go through the stages of antler growth that I have described, but not all at the same time. The breeding seasons for all creatures in the world are keyed to a master plan. All wild mammals in the temperate zone have definite breeding seasons to insure that the young are born during a period that will allow for the greatest survival rate. In the United States, most animals breed so that the young are born in the spring to take advantage of warm weather, an abundance of food, and a lessening of predation. Some animals, such as wolverines and other members of the weasel family, breed whenever they chance to encounter a mate. Conception takes place, but implantation is delayed so that the fertilized egg does not become implanted into the wall of the uterus until the development will allow for the proper birthing time.

Breeding, birth, and antler development are inextricably tied together. As noted earlier, the timetables vary according to latitude. In the Northeast, bucks usually start to grow antlers during April. The fawns are born in the latter part of May and in June. Antler development is completed in the last part of August. By about September 5 most of the bucks will have the velvet peeled.

The peak of the breeding season occurs between November 10 and November 25. The antlers usually drop off from late December to the end of January.

On November 21, 1975, while at my blind in Hunterdon County, New Jersey, I saw a magnificent buck that had shed one antler. He shed the other two days later. This is the earliest dropping of antlers I have personally witnessed. The hunting season in New Jersey usually starts the second week in December. During our bucks-only season, a legally harvestable buck must have antlers at least 3 inches (75 mm) long. On several occasions I have seen big bucks that had shed their antlers prior to the season and could not be hunted. Pennsylvania's hunting season usually opens around the first of December. In a study conducted during the antlerless seasons there, it was found that on the average 0.8 percent of all bucks lost their racks before the opening of the bucks-only season. This 0.8 percent had to come out of the trophy bucks, because the big fellows shed their antlers first.

This pattern is almost universal for whitetails in most of their range north of Georgia and east of Iowa. According to Arnold Haugen's study of deer in Iowa, the whitetails there breed just a little later, and therefore shed their antlers a little later.

Records in Illinois show that whitetail bucks there carry their antlers a little longer than other deer in the northern part of the country. The antlers usually drop off from January 15 to February 15, with some bucks carrying their antlers until April 15. Illinois has extremely rich soil with a good limestone content. The deer of that state are farmland animals. They may inhabit the forested areas, but most of them feed in the fields. In Illinois you find no overbrowsing, no browse lines, no starvation. It has been proved that bucks fed an especially good, high-protein diet retain their breeding vigor for extended periods after the peak of the breeding season has passed. The bucks of Illinois have such a diet, and as long as the supply of testosterone stays high, the antlers stay on.

J. W. Farrar, a deer study leader in Louisiana, has written to me that the whitetail breeding season starts in late September and peaks in the second week in November over most of the state. That is an earlier start than in other sections of the country, but the most interesting fact is that the deer of the Louisiana lowlands and the delta peak in the middle of December, with the season extending into late January or February.

I was in Louisiana and south Texas photographing whitetails from January 7 to January 21, 1986. In both areas the rut was still in full swing, and the buck's necks were still swollen. In Louisiana I saw two does bred on the same day and much chasing of other does, indicating that this was the peak of the rut.

This relatively late season means that most fawns of the delta will be born in July and August. Delta bucks' antlers follow the same cycle, dropping off in February and March and starting to grow in May. The delta deer polish their antlers in October and November.

Deer that I've photographed in June in northern Florida had nicely developed racks but the antlers were not fully grown. These deer were about one month further advanced than the deer in New Jersey. They polish their antlers in early August, which would throw the breeding-season peak into October, with the fawns being born in the middle of April and the beginning of May.

These Florida whitetails were photographed in June. These warm-latitude bucks are about one month ahead of northern bucks in antler development because the antler cycle is triggered by lengthening daylight, or "periodicity" of light.

I have not had the opportunity to photograph whitetail bucks in the Everglades or most other portions of southern Florida. Charles Loveless, in his study of Everglades deer, found that the peak breeding season was in September, with the fawns being born in March. Antlers usually dropped off in November and new ones started to develop in February. By mid-July the antlers were completely developed, and the velvet was rubbed off in August.

The Key deer of Florida live under tropical conditions, in a continuously warm climate. Such animals have no regular breeding seasons, although they, too, have peak seasons. There is one record of a Key buck shedding one of his antlers in September, but this was considered abnormal. Most shed their antlers in March and April. The new ones start to develop in May, and the velvet is shed in late September. This would throw the breeding season back to the middle of December, with the fawns being dropped in July and August. From the records, it appears that most fawns are

Texas has more deer than any other state: over 3¼ million. These whitetail antlers were "cast" or shed on one ranch.

Heads and Tails

born in August, although spotted fawns have been seen in every month of the year. A beautiful six-point buck that I photographed on January 15, 1977, on Big Pine Key was still in the rut, for his neck was swollen and his antlers were still brown.

The Coues whitetails of Arizona's deserts drop their antlers in May and June. New ones begin to grow in July. This is a later date than for any other whitetail subspecies. The antlers are fully developed in November, and by December the velvet is cleaned off. The peak of breeding occurs in mid-January, and fawns are dropped in August and September. This later birthing coincides with the second rainy season of the year, between August 7 and September 4 in southern Arizona. The period of moisture produces the needed vegetation for the doe to be able to nurse her young.

In 1967 I made my first trip to the Southwest to photograph wildlife. For about a week I was a guest at the Gage Holland ranch at Marathon, just outside Big Bend National Park in Texas. That region is considered semidesert, averaging 12 to 13 inches (30.5 to 33 cm) of rain per year. There is an old saying about that section of Texas: "Everything stings, stinks, or bites." Cactus was all about. The plant life was the most inhospitable-looking I have ever seen—everything had thorns, stickers, spikes, or cutting edges. And I took care not to walk too close to the stalk-shrouded bases of the yucca, because the rattlesnakes hid there in the daytime.

I arrived during a terrific desert rain storm that broke a drought, made the temperature plummet, and brought the desert to life. Never had I seen so much wildlife in what appeared to be barren wasteland. There were flocks of quail, jackrabbits beyond counting, antelope, javelinas, and deer. I counted 80 mule deer.

I was amazed that such habitat could support any deer, let alone so many. It was the first time I ever saw small herds of bucks. Whitetail bucks often form little bachelor groups of 3 or occasionally 4. The muley bucks were far more sociable, foraging in groups of 6 to 12, with 8 being the average. This was in the last week of July, and I was amazed to see that the antlers on these deer were two months behind those of our northern deer in development. I learned, too, that the fawns were about to be dropped. This was my first experience with wildlife cycles geared to moisture rather than cold. The rainstorm that greeted me was the start of the region's vegetative growing season. The deer had adapted so that there would be enough vegetation and moisture to allow the does to nurse their fawns. This was the factor on which everything in the cycle pivoted. The bucks' antlers were polished in October, the breeding season peaked in January and February, the antlers were dropped in February and March, and the new growth began in May.

The mule deer and blacktails of California breed in December and January, with a peak from December 20 to January 15. The fawns are born in mid-July. The bucks' antlers are dropped in January and February, and new growth starts in April and May. The antlers are full-grown in September and are being polished during late September and the first part of October. The bucks start running the does in late November.

The mule deer of the northern Rocky Mountains and the Sitka blacktail deer of Alaska follow the same schedule as northern whitetails. Antlers start to grow in April, and fawns are born in May and June. The antlers are full-grown in August. Polishing begins in late August or early September. The breeding season peaks between November 15 and December 15. The antlers drop off in December and January.

There may be slight local differences in the schedules of the various deer, but the dates I've given encompass the time pattern for deer in all parts of the United States and Canada.

I am often asked why, if a deer's antlers always fall off, few are ever found. One reason is that there aren't that many bucks in any given area, and a great amount of ground per antler would have to be searched. Yet the chances of finding dropped antlers are greatly increased if you know the local wintering area of the deer and concentrate your search there. Most antlers are dropped during the winter concentration.

When a buck is carrying his antlers, they look impressive, but without the skull to provide the base and the arch, an antler lying on the ground doesn't stick up very high and is soon covered by leaves and other vegetation.

Of course, the main reason why antlers aren't found is that they are eaten. It is amazing how wildlife finds trace elements or minerals. Animals feed on vegetation grown on the richest soil whenever possible. They instinctively know what is best for them. It is almost impossible to find a bone or an antler out in the forest that has not been gnawed. And we don't often find such bones or antlers at all, because they are soon consumed. Mice, squirrels, hares, porcupines, chipmunks, foxes, and so on, all gnaw on bones for the calcium and phosphorus they contain. Even the deer themselves chew on dropped antlers. I have seen deer chew on a buck's rack while he still had it on his head. The other deer needed those minerals, and they were not about to wait until the antlers dropped off. As antlers or other bones are exposed to the elements, they become softer and are more easily consumed.

I brought an old moose skull down from Canada one summer and fastened it up on the garage roof. That garage became a magnet for all the squirrels in the area. They chomped through the antlers as if they were eating soft ice cream. In a little over two years, they had eaten all of the points and most of the palms from that very large set of antlers.

The antler cycle is a fascinating part of deer study, and there are many questions yet to be answered; others yet to be asked. For example, can the late casting of an antler affect a buck's health, or is the late casting a result of some other physical problem that manifests itself in the late casting? In 1985, Joe Taylor had two bucks on his deer preserve that fought a number of times for dominance. Finally one of the bucks became dominant, but it was challenged continuously. In its last fight as dominant buck, one of its antlers was broken off about 1½ inches (38.1 mm) above the base. The buck immediately became the underdog for the rest of the season. In late January he cast his good antler, but the broken stub stayed on. From that time, his health deteriorated steadily. He went off his feed and was reduced to skin and bones. Toward the end of February, he finally cast the stub; his health began to improve and eventually he was restored to fine condition.

5

External Glands

Three primary external characteristics—tail, antlers, and glands—identify each of three basic types of deer. All three have four major external sets of glands: preorbital (tear-duct), interdigital (between the toes), tarsal (on the inner sides of the hocks), and metatarsal (on the feet). Two other glandular areas are the forehead and the tail. As identifying characteristics, the metatarsal glands are very important and—unlike antlers—they are present on all three types of deer at all times, regardless of age or sex. To the deer themselves, of course, all the glands are important, and all are interesting to the student of nature.

THE PREORBITAL GLAND

The preorbital, or *lacrymal,* glands are the tear ducts located in front of the deer's eyes. These glands posed a difficult problem for old-time taxidermists because, after the head was mounted, the glands would dry, shrink, and leave big holes that were neither lifelike nor attractive. Modern taxidermy has overcome this problem. On the whitetail deer, each preorbital gland is about ⅞-inch (22.23 mm) long; on the blacktail, about 1¼ inches (31.75 mm); and on the mule deer, 1⁹⁄₁₆ inches (39.65 mm).

A preorbital gland is a trench-like slit of almost bare skin, dark blue to black. Frequently there is a residue in the bottom of the opening, formed by particles of hair and plant material held together by secretions from the eye. There are some sebaceous glands and sudoriferous glands located in the lip of the preorbital.

Sebaceous glands produce sebum, a fatty substance that lubricates the hair and skin to keep them from drying out and becoming brittle. Sudoriferous glands produce scent. These glands secrete a substance known as a pheromone through the hair follicles, causing a body odor. The word *pheromone* was coined from two Greek words—*pherin,* meaning "to carry," and *hormon,* meaning "to excite." Pheromones are essentially, although not exclusively, sexually stimulating scents. They are composed of steroids and fatty acids.

Biologists claim that the preorbital glands are not an important producer of scent. I wonder. On numerous occasions, I have seen whitetail bucks rub this gland against branches and other vegetation. They did not rub with the vigor that would be used if the spot were merely itchy. They rubbed carefully, as if they were marking the twig with scent. I have photographs of an African antelope species, the dik-dik, marking the edges of its territory with this gland. The rubbing perhaps explains how residue of plant material happens to be found inside the gland opening. Although the scent is not strong, this gland does emit a distinctive ammonia odor.

I think that the deer also uses its preorbital gland for self-marking. On numerous occasions I have seen deer carefully rub their preorbital glands on their forelegs. Bucks usually scratch an itch with their antlers or their feet, and I never saw an insect visible when the deer rubbed with its glands. By comparison, during the rutting season, elk are known to roll in wallows to get the odors of their own urine and semen all over their bodies.

The preorbital gland is a trench-like depression of nearly bare skin in front of the eye. In this photo, the waxy secretion is visible.

I have not seen whitetail deer raise the hair surrounding the preorbital glands, but when muleys and blacktails are startled or frightened, the hairs around these glands are raised and laid outward, making a large, very noticeable rosette. With the hairs erect, the preorbital trench opens and the gland becomes very visible. I do not know if scent is given off at this particular moment, but it would be logical to assume that the reason the gland opens is to release scent.

THE INTERDIGITAL GLANDS

The interdigital glands are located between the two center toes—the hoof lobes—of each of the deer's feet. The glands are easy to overlook because of the long hair that comes down over the hoofs. If the toes are spread apart and the hair separated, an opening will be found into which a match stick or twig could be inserted. Closer examination will show hairs inside the gland. These act as wicks for the yellow, waxy secretion that is constantly discharged from the many sudoriferous glands located under the skin. The hole, or *sac,* of the gland is much larger in the whitetail than in the muley or blacktail. Every time the deer puts its foot down, some of this scent rubs off on the ground or vegetation. The odor is strong (though I do not find it as offensive as some people claim it is).

There is more survival value for the deer in leaving a trail-marking scent than in not leaving it. In the United States, the cougar is probably the greatest predator of the blacktail and mule deer. Other members of the cat family—occasionally the bobcat and, in Canada, the lynx—also prey upon deer of all types. But since cats hunt primarily by sight, a trail-marking scent is no great help to them. Canine predators, on the other hand, usually hunt by scent until they locate their prey, but the glandular scent is only one component of the game odor they can detect. Once their prey is started, the canines either keep it in sight or trail it by body odor that hangs in the air and clings to vegetation. Predators that hunt by scent have such a keen sense of smell that they could detect or track prey even if there were no glandular odor, so the scent given off by this and other glands is no disadvantage to the deer. On the contrary, it is an aid to reproduction and to the protection and rearing of the young.

This gland plays an important role by helping deer to track one another. Each deer must have its own personal scent-gland odor, which allows it to be identified. Individuality is acquired through variations in the chemical components of the scent. It is the means by which the doe locates her fawn if the little one wanders away from the spot where it was bedded when the doe left it. A doe never tracks down another doe's fawn; she follows only her own. Each deer's scent is as individual as are our own fingerprints.

I have often watched a buck tracking a doe during the rutting season by the use of this scent. Bucks never track by "air-scenting," as dogs do. They always track with their noses held just a few inches above the ground. The bucks doing this are sexually stimulated, and usually hold their tails straight up in the air as they walk. Though interdigital scent might help predators, they do not need it. But deer do need it because they are primarily scent-oriented creatures.

THE TARSAL GLANDS

The tarsal glands are the most important scent gland. We are only beginning to understand how important these glands are to deer and the role they play in communication.

The tarsal gland is the large, tufted, discolored patch found on the insides of each of the hind legs on what is actually the deer's ankles. This patch of hair is about 3 to 4 inches (76.2 mm to 101.6 mm) across in the whitetail deer and 2 to 2 ¼ inches (50.8 to 57.15 mm) across in the mule and blacktail deer. Actually, the tarsal is not a gland in the true sense, as there is no opening or duct. Beneath the skin are sebaceous glands and sudoriferous glands connected to hair follicles which act as ducts to bring the secretions to the surface. The secretions are known as *lactones* and are individualized as to the age, sex, and physical condition of the deer from which they emanate.

Also beneath the skin are little muscles that control the long hairs of the tuft. Under normal conditions, the hairs are turned inward from the outer edge of the gland. When a deer is alarmed or frightened, the hairs are made to stand on end or are flared out. This action is very conspicuous, a visible signal to other deer.

The tarsal glands have pheromones but are primarily a depository for urine. All deer of all ages urinate on these hock hair tufts. The buck's penis sheath hangs loose instead of running along the belly as on domestic cattle. Under ordinary conditions, the unstimulated penis hangs down. To urinate on its hocks, a deer supports most of its body weight on the front feet and twists its hind legs

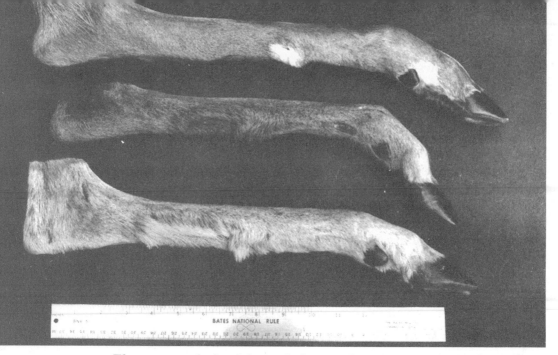

The metatarsal gland, located about midway between the hoof and the hock, is large on the whitetail leg on top, somewhat larger on the blacktail leg in the middle, and considerably larger and extending much farther up the leg on the mule deer, as seen at the bottom.

inward so that they appear to be "knock-kneed." A deer will often rub its tarsal glands against each other while urinating on them. Frequency of urination varies widely among individuals. The more dominant a deer is, the more important urinating on the hocks seems to be. It has also been proved that the more dominant deer have larger, more active tarsal and metatarsal glands than do subordinate deer.

Everyone has seen animals checking the identification of others of its own species. We see this most commonly with dogs. Most animals smell the anal, vaginal, or penile region. When deer meet, they check each other's hocks by smelling them, and they frequently lick each other's hocks. Deer show a great interest in their own hocks and often lick the tufts immediately after depositing urine on them.

The frequent urinating on the tarsal glands stains the hair tufts a deep, dark brown, and they reek with the odor of urine. Old-time hunters, and some modern ones, believe that the tarsal glands should be cut off as soon as a deer is shot to prevent tainting the meat. This is nonsense, because if the glands have not tainted the meat while the deer was alive, there is no chance of their doing

External Glands

so after the deer is dead. What can happen is that if the hunter, while dressing or skinning the deer, carelessly touches the glands and then handles the meat, the meat can be tainted. I always hang my deer by the head. When I am ready to skin it, I first saw off the hind legs several inches above the tarsal glands. Then there is no chance to contaminate the meat.

THE METATARSAL GLANDS

The metatarsal glands are an important means of species identification among the three types of deer. In the whitetail, each of these glands, including the hair tuft, is about an inch (25.4 mm) long and is found on the lower half of the deer's hind foot. The blacktail's metatarsal gland is about 2½ to 3 inches (63.5 to 76.2 mm) in length and is about midway down the foot. The muley's metatarsal is about 5 inches (127 mm) in length, starting well above and extending down to the middle of the foot. Interbreeding between mule deer and blacktail deer produces intermediate glands. Research has shown that although these glands are present in all whitetail deer in the United States and Canada, the metatarsals are smaller or sometimes lacking among whitetails south of the United States border.

Like the tarsals, the metatarsals are not true glands. The two main features of each metatarsal gland are the hard, cornified, or keratinized, hairless ridge in the center and the tufts of hair surrounding this ridge. Both sebaceous and sudoriferous glands are found in the ring of skin supporting the circlet of tufted hair. Some researchers claim that the metatarsal glands give off a garliclike odor. I think the metatarsals play a minimal role as scent glands, because I have not been able to detect any scent nor have I ever found any secretion from the hair follicles.

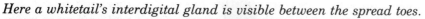

Here a whitetail's interdigital gland is visible between the spread toes.

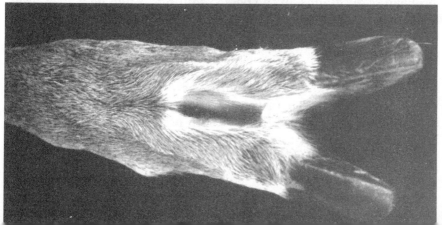

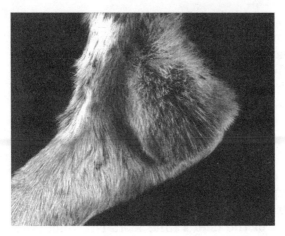

The tarsal gland is a tufted patch of hair inside each hind leg.

Some researchers have suggested that deer leave scent on the ground from these glands while lying down. (The glands are then in contact with the earth.) It has also been suggested that perhaps deer can pick up vibrations from the earth through these glands while lying down. I cannot verify or challenge either of these theories. No one really knows what use the metatarsals are to deer. The metatarsals may be glands that are atrophying, much like our human appendix, and which, being unused, will in time disappear.

FOREHEAD SCENT GLANDS

In 1972 I first photographed a whitetail buck rubbing its forehead on a stiff, dead weed stalk. It was not rubbing its antlers; it was rubbing the forehead skin between its antlers. The only explanation for this action was that the deer was depositing scent on the weed from forehead scent glands.

Later in the day, I photographed this same buck working out on several small saplings with his antlers. When he had rubbed the bark off with his rough antler bases, he proceeded to rub his forehead scent glands on the peeled saplings. I had no way of proving that the deer had forehead scent glands, but my many years of wildlife observation convinced me that this rubbing was done to deposit scent.

Since then I have taken many photographs of deer rubbing against sticks and saplings with their forehead scent glands, which are located both in front of and behind their antlers. The scent deposited is a stimulant to both other bucks and does. The bucks react to the scent by licking it and by rubbing the same spots

A mule deer urinates on its hocks; in this way, the tarsal glands become the primary depository for urine scent.

This mature whitetail is licking the scent from the tarsal glands of the button buck.

In 1972, I first photographed a whitetail rubbing his scent from his forehead onto a weed stalk. All deer, including the donor, are stimulated by the scent and lick it off, and sometimes repeat the cycle of rubbing and licking over and over.

with their own foreheads. On two occasions I have seen does, too, lick the rubs where the bucks had deposited scent. In my deer seminars, which I do all across the country, I now tell my audiences that they should be aware of this scent and that more time should be devoted to finding such rubs.

I have also noticed that the largest, oldest, most dominant bucks have darkened forehead areas during the rutting season, as if they were wearing skull caps. The dark W-shaped forehead pelage of mule deer bucks is well known, but that area remains dark all year long. The forehead region of the whitetail bucks probably darkened through the secretions of the glands, perhaps with some additional staining from the resins and saps of the saplings and trees that the forehead is rubbed on.

Back in 1936 S. Schumacher, working on the small European roe deer, discovered that these deer have active scent glands located in their forehead skin. Roe deer had been well known for their habit of rubbing their forehead on saplings, but Schumacher identified and quantified the apocrine glands that produced the deposited scent.

In 1971 Dietland Muller-Schwarze noted that both the blacktail and the mule deer, too, did considerable rubbing of saplings with their foreheads. But he did not find any appreciable increase

External Glands

in glandular activity in the forehead regions of either of these types of deer during the rutting season.

The landmark study of the forehead glands was done by Thomas Atkeson and Larry Marchinton in November 1982. They cut small strips of skin from the foreheads of captive deer and from deer that had been killed on the highway: both male and female, and both during the summer and again at the peak of the rutting season. They found that both sexes had sudoriferous glands in their forehead skin, with the male having many more. As expected, they found that these glands greatly increased in size and output during the rutting season. Although the forehead glands of both bucks and does increased in activity during the rutting season, those of the bucks increased proportionately much more. I cannot recall ever having seen a doe rub her forehead on a sapling, but I am on the lookout for it.

Although all of the bucks had increased glandular activity during the rutting season, the more dominant the buck, the more glandular secretions he produced. Fawn bucks had very little increase and yearlings just slight increases, not enough to stain their foreheads dark.

The researchers found that in their captive herd the 5½-year-old buck had the greatest glandular activity and the darkest forehead. Interestingly, they also found that a 2½-year-old buck taken from the wild had the same amount of glandular activity as did

The dominant buck will have the darkest forehead because he will be exuding more scent from the glands located there than the less dominant bucks.

Deer deposit scent from their saliva by licking the branches which over-hang primary scrapes. photo by Marilyn Maring

the 5½-year-old captive buck. The 2½-year-old buck had been taken from a heavily hunted area and was a dominant, breeding buck. But 2½-year-old captive bucks had been prevented from breeding by the dominant buck, and their glands had not developed as much as had those of their wild counterparts. These findings should help you to tell the social position of the bucks you see. A light-colored forehead indicates that the buck you are looking at is not a dominant buck; there are larger ones in the area.

While on the subject of deer glands, we should consider whether deer produce scent in their saliva. I can find nothing in the literature concerning this, but I think it is likely. The possibility is being studied, and scientific data will be available eventually. I base my conjecture about saliva scent on hundreds and hundreds of observations and on the many photographs I have taken of deer chewing and licking overhead branches, sticks, and saplings. They are not just licking forehead gland secretions, although they do this also. Very often deer lick and chew on overhead branches and twigs that are too high to have been rubbed with the forehead scent glands. They are not attempting to eat the overhead branches because they don't cut them off, although eventually many do break off. I think they chew on those branches to deposit their salivary scent on them, both as an attractant and a stimulant to other deer. Commercial scent makers will eventually flood the market with thousands of bottles of the stuff, and I am willing to bet that it will have a basic ammonia smell, to imitate the acid that is part of the deer's saliva.

External Glands

6

The Teeth and Eyes
Reveal the Age

T he age of a deer is of great interest to hunters and is crucial
to biologists if they are to manage deer herds properly. When
it was proved that antlers were no accurate indicator of age,
biologists took a cue from the old horse traders and began checking
on the teeth of deer. The saying "Don't look a gift horse in the
mouth" derives from the fact that the older a horse is, the more
worn down its teeth have become.

Using the idea of tooth wear and replacement, C. W. Severinghaus and Jack Tanck of the New York Game Department worked
up the system in 1949 that became the standard method of aging
deer. Burton L. Dahlberg and Ralph C. Guettinger of Wisconsin
refined the method in 1956. Additional modifications were made
by Jack Tanck in 1966.

An adult deer has 32 teeth. There are no upper teeth in the
front of the mouth, but there are six premolars and six molars,
three of each on each side. The lower jaw has 6 incisors, 2 modified
canines, six premolars, and six molars, a total of 20 teeth—10 on
each side. There are records of four Michigan deer having eight
molars each. I have the jawbone of a 2½-year-old deer that has
only two premolars instead of the usual three.

The canine teeth of deer are not the meat-piercing canines of
the dog family. A deer's canines so closely resemble the incisor
teeth that they are often classed and counted as such. In rare instances, deer have what are known as maxillary canine teeth in
the upper jaw. These are rudimentary and serve no purpose, since
there are no opposing teeth. Many times these teeth do not erupt
through the gums, and most people would be unaware of their
presence. Biologists locate unruptured canines by scraping the upper jaw.

C. W. Severinghaus found only 23 upper canine teeth in

18,000 whitetail deer he examined, or 0.1 percent. The farther south one goes, the more canine teeth are found in deer. Charles M. Loveless and Richard F. Harlow found four canine teeth in an examination of 95 deer in Florida, for 4.2 percent. At the Wilder Wildlife Refuge in Texas, 162 whitetail skulls disclosed 49 canine teeth in 29 of the animals. Some of the skulls had only 1 tooth, some had both. Twenty-six of the teeth were rudimentary and did not protrude through the gums. Among females, 18 percent had upper canines; among males, 17 percent had them. In Venezuela, E. Boelioni found four large canine teeth in 10 deer that he examined, a total of 40 percent. As noted in Chapter 4, the musk deer and the Chinese water deer have greatly elongated, functional, stabbing, maxillary canine teeth—tusks—and no antlers. The muntjacs, or "barking" deer, of southeastern Asia have both tusks and antlers. Biologists believe that the canine teeth diminished in prehistoric deer with the evolution of antlers. Some other members of the deer family, such as the caribou, also have canine teeth. Elk usually have well-developed maxillary canines, and thousands of elk were once killed so that these teeth could be worn as decorations by Indians and by members of the Fraternal Order of Elks.

Mule deer, blacktails, and whitetails are born with eight front teeth in the lower jaw. The six center teeth are incisors, while the two "corner" teeth are modified canines that look like incisors and are used and classified as such. The two central incisors are larger than the others and are referred to as *pincer* teeth. All of these eight teeth are milk, or "baby," teeth. Between the fifth and sixth month, the milk pincers are lost and replaced by permanent pincers, which are much larger and very noticeable. Between the tenth and the eleventh month, the other incisors are lost and replaced by permanent teeth. Incisors are accurate indicators of age up to one year and then are no longer used.

At birth, deer also have six molars, three to a side, top and bottom. These, too, are milk teeth and will be replaced. When the deer is 6 months of age, the first molars are fully erupted. By the ninth month, the second molars are usually fully erupted. At 12 months the third molars are usually partially erupted. When first erupted, these teeth are deeply ridged and have sharp points. The rear and highest ridge of the tooth is known as the *lingual crest*, the front and lower ridge is called the *buccal crest*. The first and second molars are bicuspid ("two-pointed") and the third is tricuspid.

At birth the six premolars also have sharp points, but these

Teeth and Eyes Reveal the Age

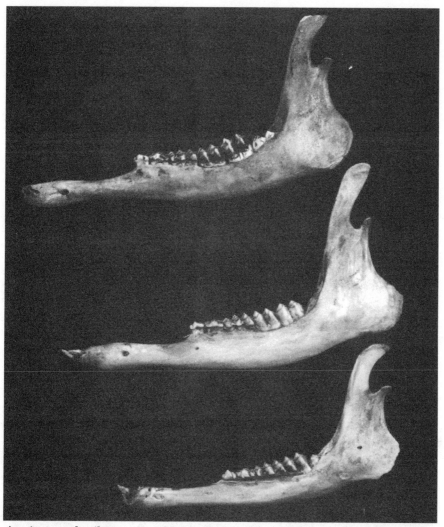

At six months (bottom), a fawn's first molar is fully erupted. At nine months (center), the fawn has a second molar. The jaw of a year-old deer (top) has a third molar partially erupted. For the first eighteen months— until adult teeth are fully developed—the molar teeth provide an easy, accurate way to determine age.

teeth are soft and wear rapidly. The first premolar has a single cusp, the second premolar has two, and the third has three.

At 17 months of age, the worn-down premolars are about to be replaced by permanent ones. While I was doing the research for my first deer book, a friend of mine arrived with a six-point buck he had taken in Maine. He told me his guide had stated that the

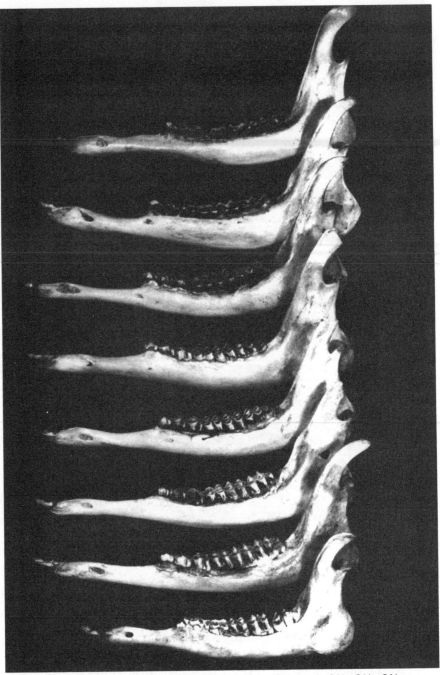

These jawbones were taken from deer 1½ (at bottom), 2½, 3½, 3½, 4½, 6½, 7½, 9½, and 11½ years old (at top). Note the progressive wearing down of teeth with age. Tooth wear above the gum line can be used by hunters and naturalists to judge age in the field.

128 Teeth and Eyes Reveal the Age

deer was 6½ years old. That guide may have been a good woods-man but he really didn't know much about deer; he was still count-ing antler points for years of age. A quick examination of the deer's teeth showed that the buck was exactly 17 months old. All of the premolar teeth were about to be shed and the third premolar still had the tricuspid cap in place.

At 18 months the deciduous premolars are lost, and the third premolar thereafter has only two cusps. Up to this point, tooth re-placement is a very accurate means of aging. From 18 months on, telling the age by the teeth becomes more difficult, because age must be determined by the wear on the teeth, and many variables enter into the picture.

I fully realize that the description below will be regarded as heresy by the experts. Nevertheless, I want to simplify the Sever-inghaus-Tanck aging system so that it is available to naturalists who want to use it without a lot of study. A caliper divided into millimeters is needed to take the basic measurements.

At 1½ years of age the deer's second molar is typically 10 mm from the gum line to the peak of the buccal crests. The deer's teeth wear down an average of 1 mm per year. Measure the remaining tooth on the molar between the peak of the buccal crest and the gum line, then look at the chart below.

Correlation Between Tooth Measurement And Age

mm	Approximate age	mm	Approximate age
10	1½	5	6½
9	2½	4	7½
8	3½	3	8½
7	4½	2	9½
6	5½	1	10½

Severinghaus and Tanck worked out their method using tagged, known-age deer living in the wild. Pen-raised deer could not be used because their diet would differ from that of wild deer. Diet is where the variables started to creep in. As a deer ages, the teeth are worn down by constant chewing of the cud. By the time a deer is 11 to 12 years old, its teeth are practically worn out—but then so is the deer's life expectancy. With age, the crests of the teeth are ground away, and more dentine becomes visible. To determine the animal's age, the teeth are measured above the gum line. The variables that lower the accuracy of this method are heredity, availability of food, access to minerals, and grit ingested with food.

Just as some humans have better teeth than others, so do some deer. And better teeth can often be traced to no other factor but heredity. Also very important is the availability of high-quality food. A deer with an adequate diet of nutritious food is apt to be a healthy deer with good teeth. Perhaps of the greatest importance is the availability of minerals. Deer on good limestone soil have larger antlers, stronger bones, and strong, long-wearing teeth. Another major factor is the amount of dust, sand, or other grit ingested with food. Hard substances cause greater tooth wear.

In a few instances, an oddity of jaw structure might also have some effect. I cannot substantiate this, but I can attest to the fact that deer vary in their jaw structure, just as humans do. Right after I wrote *The World of the White-Tailed Deer,* I was bombarded with fascinating discoveries and observations made by interested readers. I received word, for example, of a deer that looked as if it needed the services of an orthodontist. It was a buck killed by William Robbins, Jr., in Delaware County, New York. It was brought to my home so that I could examine and photograph it. The lower jaw was about 2 inches (50.8 mm) shorter than normal. The incisors, instead of nipping off twigs against the top mouth pad, had to function against the roof of the mouth. Apparently this was no handicap, as the buck was in good condition. There were no reports of similar deer from that area, but a short time later I received reports about a number of short-jawed deer taken during a period of several years in the area around Scranton, Pennsylvania. This anomaly has also been found in many parts of New York's herds. The short-jawed condition is probably a genetic characteristic, as it seemed to be increasing. Surely a severe case would alter the rate of tooth wear.

The great discrepancy in the wearing down of the teeth of the deer in south and northwestern New Jersey demonstrates the uncertainty of using teeth to measure age. South Jersey is very sandy, and along the coast there is a great deal of wind. At times the leaves of all vegetation are coated with dust particles—fine sand. Deer eating this vegetation ingest the sand, and the wearing down of their teeth is greatly accelerated. It has also been found that a buck's teeth wear out slightly faster than a doe's because the male is larger and consumes more food, producing more wear.

Teeth and Eyes Reveal the Age

Continued research has shown that the tooth-wear method is only about 43 percent accurate. Other, more accurate methods are currently being used, but the Severinghaus-Tanck method is still the only one that can be used in the field or by the average layman. Charts for the purpose, published by New York's Department of Environmental Conservation, are a must for anyone interested in aging by this method. They are available from the New York State Department of Environmental Conservation, 50 Wolf Road, Albany, New York 12233.

Another aging method developed was the weighing of the lens of the eye. Rexford Lord, working on cottontail rabbits, and other researchers found that eye lenses, including those of humans, thicken consistently with age. This method is definitely a laboratory technique: the lenses had to be dried in a special oven at a specified temperature for a specified time. Then, using a very fine scientific scale, the lens could be weighed and the age determined.

The most recent method again involves the teeth. The technique is to extract one of the incisors, slice it, and count the layers of cementum inside. This method was discovered by two Canadian biologists, E. Sergeant and D. H. Pimlott, in 1959, and was used

The most accurate method of aging deer is to cross section the tooth, view it through a microscope and count the number of dark lines indicating growth layers on the cementum, much as you would count the rings on a tree section. This tooth shows the deer to be 8½ years old.

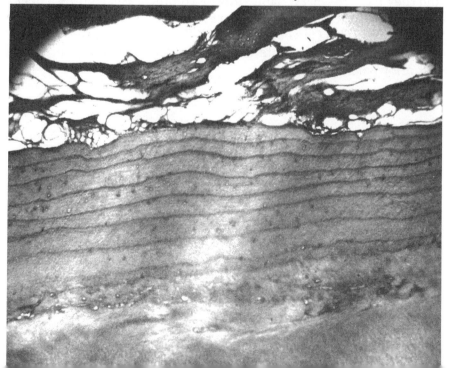

on other game species, such as moose. Only recently has it become the standard technique for aging deer.

The procedure is based on the fact that deer (and most other animals and plants) are subjected to a period of diminished growth during each winter. The diminished growth is what puts the dark annual ring on a fish scale, the tight series of rings on goat and sheep horns, the annual rings in a tree trunk. To age any of these, you merely count the dark annual rings. To age deer, you count the dark rings between the layers of cementum in the root portion of the primary incisors. To make the rings more easily visible, the tooth is decalcified and then sliced or ground latitudinally. The sections of the teeth are then stained with various solutions to make the rings more conspicuous. The sections are mounted on glass slides and viewed through a microscope.

There are many advantages besides accuracy to using the incisor tooth sections. The main one is cost. Kansas found that when all hunters were required to bring their deer to checking stations, it cost the state $12 per deer to check the teeth by the older method, for a total cost in 1969 of $20,000. In 1971, the state asked all hunters to send in the two center or pincer incisors. The total cost that year was only $1,115. The work on aging the deer was accomplished in about 400 hours, compared to the 3,200 to 5,700 hours it had taken to operate checking stations.

The hunters were happy to cooperate because it meant they no longer had to haul their deer long distances for checking, nor did they have to worry about ruining trophies. To examine a deer's teeth for wear, the usual procedure is to slit the cheeks back to the massiter muscles (jawbone muscles). This is fine for the biologist, but any hunter who wants to have his deer's head mounted will not allow such mutilation.

7

Life Span

F ew deer ever live out their potential life span of 11 or 12 years. Very few bucks even reach their prime of 4½ years.

Many states have a "bucks-only" season with a special doe day or season in addition. Only a few states allow the hunting of either sex on a regular basis. In the first instance, the preponderance of does over bucks will be great. Even where either sex can be shot, more bucks are taken because they are desirable trophies and because some hunters still cannot bring themselves to shoot a doe. Consequently, in almost every state the females make up the bulk of the population, and hence they have the greater chance of living out their life span.

In New Jersey, very few bucks live beyond 2½ years. Most are killed the first year that they bear legal antlers—at age 1½. In 1975 in one of New Jersey's management zones, 91.8 percent of the bucks killed were 1½ years old. C. W. Severinghaus figured that the average life expectancy for bucks in New York's western counties to be 1⅕ years; in the Catskills, about 1⅗ years; in the Adirondacks, about 2 years. The records of Pennsylvania's Game Research Division show that 67 percent of the bucks killed are 1½ years old, 20 percent are 2½ years old, 9 percent are 3½ years old, 2 percent are 4½ years old, and 2 percent are 5½ years old.

Does reach a greater age in the wild and also fare much better in captivity. In 1931, extensive studies on the life span of deer indicated that they have specific longevities of 8 to 12 years and potential longevities up to 23 years, almost twice the age cited by most authorities. The larger species of deer have the longer life spans. Averaging the life span of 200 deer, researchers came up with 10 years, 4 months, 9 days for bucks and 10 years, 10 months, 27 days for the does. According to the records of friends of mine who have deer preserves, does live quite a bit longer than bucks. Females of almost all species of mammals, including humans, tend to live longer. The males of most species are usually 20 percent larger than the females. Perhaps males are worn out sooner by this extra weight and the extra food that has to be eaten and processed to achieve and maintain this weight.

The greatest longevity recorded for a mule deer is that of a blacktail doe on Gambier Island, British Columbia. In the fall of 1918, this doe and single fawn became tame and would come to feed at a farm. The blacktails of that area do not usually breed as fawns and usually have a single fawn at their first birthing, so it was concluded that this doe was a little over two years old when she first appeared. She stayed on the farm until she died in the winter of 1938–1939 at over 22 years of age.

One of the oldest captive whitetail deer that I can find a record of was a doe called Lady, owned by R. B. Howard of Putney, Vermont. She was born in captivity on June 8, 1932, and died on December 7, 1951, when she broke her neck jumping out of her pen. She had stayed in good health up through her 19th year, with no sign of feebleness. Her weight was normal, and her appetite was good. Her teeth were worn out, however. They lasted as long as they did because for most of her life she had been fed soft, easily chewed commercial feed. Her health started to fail during the last 4 months of her life, her coat becoming rough and patchy, and she was beginning to stiffen up. There is another claim of a doe in Wisconsin that lived to be 19½ years old, but no data is available. C. W. Severinghaus and Jack Tanck had a set of teeth in their collection that came from a doe that lived to be 16½ years old, but, again, I have no further information.

R. S. Palmer reported a semi-tame whitetail doe called Diana at Tomhegan Camps in Maine. She was killed by a rutting buck on November 13, 1952, when she was 18½ years old. (I will talk more about this remarkable doe in the chapter on birth and fawns.)

Several semi-tame blacktails that lived in the wild but came to farms to feed on Hardy Island, in British Columbia, reached ages of 15 or 16 years. A captive blacktail buck owned by R. O. Ramport of Ukiah, California, was 16 years old when it died. Note that this is an impressive longevity record for a buck.

Richard Smith, writing in *Deer & Deer Hunting Magazine*, tells of a whitetail buck that lived most of its life on the Hercules Powder Plant grounds in Ishpeming, Michigan. The buck's mother was accidentally made captive when a fence was built around the plant as a security measure since it was producing munitions during World War II. In the spring of 1944, she gave birth to a pair of buck fawns, one of which was nicknamed Pete.

Although no one collected the bucks' antlers, a plant employee, Edwin Rosewell, said that the bucks had their biggest antlers between three and eight years of age, although all of them were big. Pete's brother was killed by dogs when he was between 12 and 14 years old. No one kept exact records on the deer; they were just there. The birth date of the two bucks was always easy to remember because it occurred the first spring after the fence had been constructed.

The Hercules plant was closed down in 1961. Researchers from the Cusino Wildlife Research Station livetrapped Pete and took him to their facility. Louis Verme recalls that the 17-year-old buck had spike antlers that were still in velvet late in the fall, when he was captured. The buck died that winter at Cusino when he was 17 years, 8 months old.

Benjamin Tullar, Jr., writing in the *New York Fish and Game Journal* in January 1983, tells of an N. A. Gordon who, in 1975, estimated the age of a whitetail doe on an estate in Orange County, New York, at between 19 and 23 years. When she was shot because of severe hair loss on her head, neck, and ears, she was carrying an 80-day-old fetus. Although her teeth were sectioned and stained, the researchers could not accurately age her.

My good friend Joe Taylor, who runs a camground, Camp Taylor, at Columbia, New Jersey, has kept a captive deer herd since 1959 for the enjoyment and education of the campers. On May 20, 1960, one of his does gave birth to a doe fawn that became beloved of all of the campers because of her tameness, and then famous because of her longevity. She was called "Mommy." The name Mommy was especially apropos, because she behaved like a mother to all of the fawns in the pen, allowing them all to nurse. Later in life, Mommy even adopted a fawn in the wild.

In 1964 some dogs running around the outside of the pen caused her to panic, and she broke her right front metacarpal bone. Joe did not splint the break, which mended itself in three months, although it knitted with a lump 1½ inches (38.1 mm) in diameter. In 1965 dogs chasing around the outside of the pen again caused her to rebreak the same bone in the same exact spot. Again it took three months to heal. But this time the bone knitted with no discernable lump at all, although she did favor the leg slightly when she walked.

Mommy's condition went downhill fast when she was 16; her coat was very poor, she became thin, and she limped badly. Be-

L. L. Rue with Joe Taylor's doe, "Mommy," during the last year of her life when she was over twenty years old. photo courtesy Irene Vandermolen

cause of her tameness, Joe decided just to turn her loose on his property. The access to all of the good natural browse and nongrass plants rejuvenated her, and her winter coat came in sleek. She fattened up, and most of the stiffness went out of her walk. From that time on, Joe turned her out each spring and left her out until hunting season in December, when he returned her to the campground.

Mommy gave birth to a single fawn when she was a year old and to twins every year thereafter until she was 17 years old, when she reverted to bearing a single fawn. She lost that fawn due to a prolonged cold, rainy spell. At 18 she gave birth to a single fawn and then later adopted a wild fawn. She nursed and raised both fawns that year. In her 19th year, she gave birth to a single fawn, which she raised. As tame as Mommy was, we could never find her fawn during the first two or three weeks of life, when she kept it hidden. Both Joe and I would watch her for long periods of time, but she would never betray its location. We finally saw the fawn only when she brought it out of hiding at about 3 weeks of age. She evidently only went to her fawn at night.

When Mommy was 20, she was barren for the first time, and her advanced age showed in her gauntness. She died in December 1980, having lived to be 20 years and 7 months old.

Life Span

As Mommy grew older, I implored Joe to save her lower left jawbone for my collection when she died. But when the old doe died, Joe was unable to remove her jawbone: it crumbled to pieces in his hand because of her severe osteoporosis. All of her teeth were worn down below the gum line. Mommy would not have been able to eat at all if she had not been fed a ground-grain diet. She not only had outlived most other deer, she had long outlived her own teeth.

The oldest captive whitetail on record was written up by my good friend Charlie Alsheimer in *Deer and Deer Hunting Magazine*. Charlie told of a whitetail buck named Henry II, raised by Ben Lingle at Linglewood Lodge in Clearfield County, Pennsylvania, that lived to be 20 years, 10 months old. Henry II was born in May 1962 and lived in a large wooded compound. The big buck died in March 1983.

This buck, it must be remembered, lived a pampered, sheltered life and had access to a constant, soft, highly nutritious diet. With these advantages, he set two records. Henry II lived longer than any whitetail buck anywhere, anytime, that I know of. Even more unusual, his finest set of antlers was grown when he was 15½ years old—long after he should have reached his prime.

The accompanying photograph shows the buck mounted with his largest growth of antlers. Unfortunately, Mr. Lingle did not

Ben Lingle with 17 of the 20 sets of antlers produced by his buck, "Henry II." The buck lived to be 20 years and 10 months old, making it the longest-lived buck on record. photo courtesy Charles J. Alsheimer

keep the buck's first three years' antlers because he raised many deer and there was nothing really outstanding about Henry II's rack until his fifth year. Please note that all of Henry II's antlers are large, and the best sets were grown from 1973 to 1981. Notice, too, that the shape of the antlers is constant throughout the buck's life, demonstrating that normal antler conformation is genetic.

In 1969 John Ozoga of Michigan reported some very interesting longevity records of wild whitetail does living under natural conditions. These deer had been tagged as fawns and retrapped later. One doe was 14 years, 9 months old and a second was 14 years, 7 months when trapped and released. Both were then in good health. It is unfortunate that their tags were not recovered to complete the record.

Undoubtedly, where deer are subjected to little or no predation or hunting pressure and the food supply is good, quite a few of them live out their potential life span. With the greatly improved accuracy achieved by counting the cementum rings to determine the age of deer, more longevity records probably will be set.

Two outstanding longevity records were reported in the *Alabama Conservation Magazine* in 1984. A doe that had been tagged and stocked as a juvenile in 1960 was shot in 1981, making her 21 years old. Even more remarkable, another adult doe that was tagged and stocked in 1961 was shot in 1983. This doe had to be at least 23 years old. Neither of these does ever strayed beyond the area where they were stocked. It makes one wonder just how much longer they might have lived if they had not been harvested, and marvel that they had managed to survive so many hunting seasons. Would these does have been able to survive so long in the North, where they would have had to contend with extreme cold and short rations? We will never know, but we do know that they lived longer than any other whitetails on record, and they did it on their own in the wild.

8

Calculating Live and Dressed Weights —and Conserving Venison

E ither a deer's size is mighty deceptive or else most hunters (honest though they may be in other respects) are fishermen at heart. A hunter telling about a big one that got away usu ally holds a hand out, shoulder-high, to show how big the deer real-ly was. A hunter who holds a hand that high ought to be describing an elk, not a whitetail or mule deer. A hand held out about belt-high would be just about on target to indicate the shoulder height or to pat a typical adult whitetail on the back. It would be held only a trifle higher even for a husky Rocky Mountain mule deer. The next chapter will deal more specifically and in some detail with weights and measures for various types of deer. Meanwhile, I will make a valid generalization regarding size. A good whitetail buck in most regions of the country stands only a little over 3 feet (91.44 cm) high at the shoulder, and a good muley stands only a couple of inches higher.

Whether or not a trophy is the goal, a hunter is likely to be interested in how handsome a pair of antlers a buck has. Next, the hunter is most interested in how big the deer is. For that matter, anyone who is curious about wildlife would surely like to know the weight of a typical deer or a specific deer that has been harvested. Weights, like shoulder heights, tend to be grossly exaggerated. This is understandable, because few people have sufficient biologi-cal knowledge to judge an animal's weight simply by looking at it. And equally few know how to weigh an animal or estimate its weight.

The job of weighing a deer is complicated. Deer are seldom shot where they can be conveniently and accurately weighed. If the deer is really a huge one, the hunter usually has no way of getting it to a scale so that its live weight can be recorded. If the deer is average, the hunter may be curious about the weight but isn't going to drag the deer out whole. So most deer are eviscerated and then taken out.

Few hunters have the opportunity to handle and weigh enough deer to become proficient at estimating live weight. It has been found that they usually overestimate the weight of their deer by 20 to 25 percent. I have found that deer seem to gain weight in direct proportion to the time, distance, and difficulty involved in getting the carcass out of the woods.

There is a substantial difference between a deer's live weight and its weight after field-dressing or hog-dressing. Therefore, before delving into the problem of weighing a deer—or estimating its weight—I had better explain the procedures involved in dressing and caring for the venison once the deer has been harvested. In reading the weight records, you will often notice the terms *hog-dressed* and *field-dressed*. Frequently, the terms are used interchangeably, but they mean two different things.

When a deer is field-dressed, the skin on its belly is opened and its paunch, intestines, and reproductive organs are taken out. The liver, heart, and lungs are left in until the hunter reaches camp or home, and the diaphragm is left intact to help hold the organs in place and keep the chest area clean.

When a deer is hog-dressed, the carcass is usually split open from the throat or upper sternum to the pelvic arch. The windpipe is cut off as high as possible, and the heart, lungs, diaphragm walls, liver, paunch, intestines, and reproductive organs are all taken out. Probably 95 percent of the deer taken are dressed in this fashion. In both cases, the feet are left on, and in both cases the blood drains out.

I always hog-dress my deer and can do the job in about seven minutes. I cut completely around the anus first, then roll the deer on its back and straddle the chest, facing to the rear. Pulling up on the belly skin, I insert my knife below the breastbone. By pulling up on the skin, I avoid cutting into the paunch or intestines. I then cut the belly skin down to the pelvic arch. I always carry a 6-inch (15.2 cm) Randall knife that is heavy enough to allow me to slice through the ribs where they join the sternum up to the neck. After severing the windpipe, I cut out the diaphragm walls, pull out the entrails and pull the anus, which I have already freed, in through the pelvic arch. I then roll the deer up on its belly with its legs spread wide to drain out all the blood.

Then, if I have to carry or drag the deer, I do something that few hunters bother about. I sew up the body opening with a burlap-bag sewing needle and heavy cord to keep it clean. As soon as I get the deer to my destination, I hang it up by the head and prop the body cavity open with a stick in the rib cage.

Many hunters leave their deer hanging to "age" the meat. Meat will age properly only at temperatures between 36° F and 38° F (2.2° C and 3.3° C). If the temperature gets above 40° F (4.4° C), the meat will start to spoil. Most deer carcasses are hung out in cold weather and freeze solid. When meat freezes, it is not aging, and the hunter is merely wasting time. If the deer cannot be butchered promptly, freezing keeps it from spoiling, but no one should believe the meat is aging.

In warm weather the carcass should be covered with a cheesecloth deer bag to prevent flies from laying eggs on the meat. If no such bag is available, a liberal application of black pepper will help to keep the flies off the meat. But when I can't age my venison properly, I butcher the carcass as soon as the body heat is out of it. A deer's body temperature is about 104° F (40° C).

Though hunters are always interested in what their deer weighed, biologists are even more interested because the body weights are the key to the condition of the deer herd and its habitat.

The easiest method of "weighing" a deer in the woods is to use a tape measure. I carry a 10-foot (3-m), ¼-inch (6.35-mm) tape measure in my pocket. When I go afield, I am always finding something of interest that I just have to measure. That tape has helped with a lot of the data in this book. Take your tape and measure the deer's "heart girth"—its girth just behind the front legs. That figure can be used to give you both the approximate live and dressed weights of your deer.

For many years it has been known that there is a definite correlation between heart girth and body weight. The agricultural colleges worked out a formula that has long been standard in giving approximate weights of domestic cattle. I have a tape, made by my good friend Bob Hults, marked with the equivalents giving the live weight of the deer, its dressed weight, and the amount of edible meat. I carry this tape with me at all times. Charles Smart, Robert Giles, Jr., and David Gwynn, all of the Division of Forestry and Wildlife Resources at Virginia Polytechnic Institute and State University at Blacksburg, developed a chart that they have permitted me to include here. All you have to do is measure the deer's

Weights and Heart Girth

Heart girth		HOG-DRESSED WEIGHT				LIVE WEIGHT			
		Fawns		Adults		Fawns		Adults	
in.	(cm)	lbs.	(kg)	lbs.	(kg)	lbs.	(kg)	lbs.	(kg)
20	(50.8)	27	(12.2)			36	(16.3)		
21	(53.3)	30	(13.6)			40	(18.1)		
22	(55.9)	33	(15.0)			44	(20.0)		
23	(58.4)	37	(16.8)			48	(21.8)		
24	(61.0)	40	(18.1)			52	(23.6)		
25	(63.5)	43	(19.5)			57	(25.8)		
26	(66.0)	46	(20.9)	46	(20.9)	61	(27.7)	60	(27.2)
27	(68.6)	50	(22.7)	52	(23.6)	65	(29.5)	68	(30.8)
28	(71.1)	53	(24.0)	58	(26.3)	69	(31.3)	75	(34.0)
29	(73.7)	56	(25.4)	64	(29.0)	73	(33.1)	83	(37.6)
30	(76.2)	59	(26.8)	70	(31.8)	77	(34.9)	90	(40.8)
31	(78.7)	63	(28.6)	76	(34.5)	81	(36.7)	98	(44.5)
32	(81.3)	66	(29.9)	82	(37.2)	85	(38.6)	106	(48.1)
33	(83.8)	69	(31.3)	88	(39.9)	89	(40.4)	113	(51.3)
34	(86.4)	73	(33.1)	94	(42.6)	93	(42.2)	121	(54.9)
35	(88.9)			101	(45.8)			128	(58.1)
36	(91.4)			107	(48.5)			136	(61.7)
37	(94.0)			113	(51.3)			144	(65.3)
38	(96.5)			119	(54.0)			151	(68.5)
39	(99.1)			125	(56.7)			159	(72.1)
40	(101.6)			131	(59.2)			166	(75.3)
41	(104.1)			137	(62.1)			174	(78.9)
42	(106.7)			143	(64.9)			182	(82.6)
43	(109.2)			149	(67.6)			190	(86.2)
44	(111.8)			155	(70.3)			197	(89.4)
45	(114.3)			161	(73.0)			205	(93.0)

heart girth and read the hog-dressed and live weights from the table.

I have found that this tape and table also work out pretty closely on men. I have a 42-inch (106.7-cm) chest, and, according to the chart, I should weigh 182 pounds (82.6 kg). Actually I weigh about 186 pounds (84.4 kg). But some of my chest seems to have slipped down toward my middle. So I checked some friends, and the tape and table came very close to their scale weights.

Of course, nothing really takes the place of scales for accurate weights. If you finally get your deer to a scale, you will have either the field- or hog-dressed weight, depending on how you dressed your deer. In the next chapter, which deals with average weights and record weights, I will include some estimates of live weights. You will have to allow for slight inconsistencies in some instances because there was no way to ascertain in every case how a deer was dressed out before it was weighed. Most reports specify *hog-dressed,* but some use the term *field-dressed* or—confusingly—just *dressed.*

To calculate the live weight of a deer when you have the hog-dressed weight, you can use a dependable formula. Divide the hog-dressed weight by 4, then add that quarter to the hog-dressed weight to come up with the live weight. If you know the live weight of your deer and want to know the hog-dressed weight, divide the scale weight by 5 and subtract that fifth.

If you want to use a simple mathematical formula instead of dividing and adding, just multiply the hog-dressed weight of your deer by 1.25. Some state game departments and some biologists multiply the dressed weights by 1.275 or 1.30. If you check some of the weight records in the next chapter, you will see that these higher figures were used—why, I'm not sure. Using any of these figures is only an educated guess. Personally, I think that a figure slightly lower than 1.25 should be used on big deer, because a small deer has more "innards" in relationship to its total body weight than the great big fellows do.

At any rate, a deer's blood and entrails account for roughly a fifth of its total weight. It has been calculated by Dr. Monica Reynolds, of the University of Pennsylvania's School of Veterinary Medicine, that a deer has approximately an ounce of blood for each pound of its total weight. Thus, a 125-pound (56.7-kg) deer would have 125 fluid ounces (3.7 liters) of blood, which is almost 8 fluid pints. Bucks have approximately 8 percent more blood than do does of the same body weight.

According to the time of the year, fluid, including the deer's blood, makes up anywhere from 63 to 73 percent of the deer's total body weight. The fluid content varies at different times of the year according to the amount of fat stored in the body, with the greatest fat accumulation for bucks being in mid-October, just before the onset of the rut. Does reach their greatest body weight in late November or early December, just before the onset of snow. Fat accumulations may comprise as much as 17 to 20 percent of the deer's total body weight. The bones of a deer are about $\frac{1}{10}$ of the total body weight. I have weighed the organs of an adult buck deer in the 140-pound (63.5-kg) class and found that the buck had a 2-pound (0.9-kg) heart. This is twice the weight of the heart of a typical human of the same body weight. The lungs weighed 3 pounds (1.36 kg). The liver, which is the heaviest organ in the body, weighed 5 pounds (2.27 kg).

The next question most hunters ask is, how much edible meat will they get from a deer? A lot depends on where the deer was hit, what damage the bullet did, and how good a butcher the hunter is.

Some years ago I did a photographic step-by-step series on butchering deer for *Outdoor Life* magazine. Evidently it was useful; it was reprinted three times. I called the article "A Deer in a Dishpan." When I butcher a deer, I remove all fat, tissue, and bone, leaving nothing but pure meat. This reduces an average deer—125 pounds (56.7 kg) live weight—to about 46 or 48 pounds (20.9 to 21.8 kg) of pure meat. At that rate, you can carry your deer meat in a large dishpan.

A rough formula often given for finding how much meat you will get from a deer is, again, to divide your deer's hog-dressed weight by 4 and subtract that quarter. You may also multiply the hog-dressed weight by 0.75. According to this formula, if your deer weighed 120 pounds (54.4 kg) hog-dressed, you should end up with 90 pounds (40.8 kg) of meat. But these figures include the bone in a number of cuts—just as bone is included in such store-bought beef cuts as spare ribs and shoulder roast with the bone in. On a 125-pound (56.7-kg) live-weight deer, the skin weighs about 10 to 12 pounds (4.5 to 5.4 kg). The feet weigh 6 to 8 pounds (2.7 to 3.6 kg), depending on where you cut them off. The head weighs another 6 to 8 pounds (2.7 to 3.6 kg), and the bones weigh 16 to 20 pounds (7.3 to 9.1 kg). I usually figure that just about 50 percent of the deer's hog-dressed weight—or roughly 40 percent of the deer's live weight—will be pure meat. But as I said, a lot depends on the butcher. I have a friend who uses the entire neck not for stew but for a roast, with all those bones left in it. He happens to be a fine chef, and his dinner guests describe his neck roast as a gourmet's feast. Or so he claims.

9

Lightweights,
Middleweights,
Heavyweights

A s I noted in the last chapter, the shoulder of adult deer (the highest part along the back) would be only about belt-high on an average man. To some people, that seems incredibly small. To support that claim, here are some figures based on a consensus of sources ranging from the studies by Seton to game-department reports and my own measurements and weighings. As you consider these measurements, bear in mind that does usually are substantially smaller and lighter than bucks.

Adult Key deer seldom stand more than about 28 inches (71.1 cm) high or weigh more than about 80 pounds (36.3 kg). Often they are much smaller. Among the little Coues deer, or Arizona whitetails, a buck is apt to stand 31 inches (78.7 cm) at the shoulder and weigh perhaps 98 pounds (44.5 kg). A northeastern whitetail buck generally stands less than 40 inches (101.6 cm) high and weighs less than 160 pounds (72.6 kg), although some bucks and even some does become much heavier under favorable conditions. The Columbian blacktail is slightly smaller than the whitetail. After studying the blacktails in Washington's Olympic Mountains, Seton concluded that the bucks seldom weigh over 150 pounds (68 kg). According to the latest reports, the average is closer to 100 pounds (45.4 kg). My own measurements of an admittedly rather small Columbian doe showed her to stand only 29 inches (73.7 cm) high, and she weighed only 83 pounds (37.6 kg). Don McKnight, a research chief with Alaska's Department of Fish and Game, has described Sitka blacktails as a trifle smaller and shorter-legged than Columbian blacktails. Their weight is about the same. Even a mature muley buck of the Rocky Mountain variety—the largest subspecies—is apt to stand no higher at the shoulder than about

42 inches (106.7 cm) and weigh no more than about 200 pounds (90.7 kg). Of course, some record muleys—and whitetails, for that matter—have weighed more than twice as much, but a very large mule deer is likely to have a shoulder height of no more than 44 inches (111.8 cm).

The averages just given do not take regional differences into account, and even within a single subspecies, or a single herd, some deer are far more massive than others. How big can a deer grow? That depends on species, subspecies, geographic location, age, sex, and the amount and quality of food available in the area. Individuals vary greatly in body and limb size and general conformation. One hears about long-legged "ridge deer" and short-legged "swamp deer." Some people must think that running up mountains stretches a deer's legs. I'm always tempted to say that I would think it would wear the animal down and make it short-legged. Some deer may be short-legged genetically, or stunted from malnutrition. Some deer are long and lean, others blocky; some have short muzzles, some have long muzzles, some have rounded "Roman noses." Apart from the record sizes and weights, compilations of statistics are merely averages. I will list the significant findings of specific studies to provide a clearer and more detailed concept of deer sizes, and I will also give the record sizes and weights—some of which are downright astonishing.

Before examining some figures from different parts of the country, let's clarify the picture by discussing body growth and the factors that affect it. Those factors are in operation even before a deer is born—that is, during gestation. The gestation period for North American deer is usually 200 to 205 days (about 6⅔ months). The length of the gestation period is determined primarily by the food available to the mother. Records for gestation run from 187 to 212 days. Does that drop their fawns in the shorter range of the time are those that have received adequate food. Does on an inadequate diet carry their fawns for a longer period because this is nature's way of trying to get all fawns off to the best start possible. A longer gestation period allows the fawn more time for total development before birth. Most interestingly, except in cases of severe malnutrition, most fawns are born weighing in at about the average for their subspecies, regardless of the physical condition of the doe.

If food is plentiful, the doe gets sustenance plus the nutrients needed to feed the developing fetus. Under normal conditions an adult doe usually has twins. If there is a severe shortage of food, she will produce only one fawn, or if two fawns have been conceived, one will die during gestation and be reabsorbed. An extended period of starvation will cause the second fawn to die. The fetus may simply be aborted or it may be reabsorbed back into the doe's body so that she can utilize the nutrients. At this stage, nature is more concerned with saving the life of the mother than the life of the young.

As pregnancy advances, this situation is reversed. After the fourth or the fifth month, the needs of the fetus are met first. If food is scarce in the early spring, the needs of the fetus are filled and whatever is left over goes to sustenance of the doe. If the doe has no body reserves left, her own body is catabolized, with the nutrients being drawn from her own tissues. The doe may be emaciated but her fawn or fawns will be about normal in weight and size for her subspecies in her particular area or only slightly less.

I have several deer at my home that I use for research and observation. I was given an 8-month-old doe fawn that was bred. This doe was very tame with humans but very timid with other deer. Consequently she was at the very bottom of the hierarchy. She had been part of a large herd of captive deer but was very thin because the other deer drove her away from the feed. When I got her, I fed her an unlimited diet of 16–18 percent protein feed and choice second-cut alfalfa, and there was a substantial amount of browse available in my pen.

I had to leave for Chicago on a Friday night to do a deer seminar. When I fed the doe that Friday morning, I noticed that her milk bag had swollen. From past experience I knew that within 24 to 36 hours after a doe's udder fills up with milk, she will give birth to her young.

When I got back on Monday I was told that a single fawn had been born on Saturday. I hunted through the compound for several hours before I finally found where the doe had hidden the fawn. The fawn was a little buck, strong and lively, but I was amazed to find that it weighed 10¾ pounds (4.9 kg) just two days after birth. That meant that it must have weighed at least 10 pounds (4.5 kg) when born. I fully realize that does giving birth for the first time usually have a single fawn and that single fawns are usually quite a bit larger than either fawn of a set of twins. Nevertheless, because of the female's poor condition, I would have expected the fawn to be a small one. Most of what the mother ate must have gone directly to the nourishment of the fetus. The doe remained thin during her lactation period but then became sleek and fat.

The fawn kept his good head start and I really expected him to produce little antlers his first year since there were protrusions above the pedicels of about ½-inch (12.7 mm) that solidified as antlers. But, they did not become pointed, just raised, peeled bumps.

Male fawns weigh about 20 percent more than females at birth, and they maintain this ratio (or increase it) for the rest of their lives. Whitetail fawns in my area of New Jersey usually weigh between 5 and 7 pounds (2.3 to 3.2 kg) at birth. Last spring I had a chance to weigh two full-term buck fawns taken from a doe that had been killed on the highway. These little bucks were identical in size and weight. At 7¾ pounds (3.5 kg) each, they were the heaviest twin fawns I have ever personally weighed. (The deer in my area are not as large as they should be because our herd population exceeds its food supply.) The bucks measured 29 inches (73.6 cm) in overall length from nose tip to tail tip, the tail being 4 inches (10.2 cm) long. They stood 18½ inches (47 cm) high at the shoulder, had an ear length of 3½ inches (8.9 cm), a crown-to-nose length of 6 inches (15.2 cm) and a heart girth of 12 inches (30.5 cm).

In 1950 Arnold Haugen and L. Davenport published the birth weights and measurements of 29 fawns from upper Michigan. The males averaged 7 pounds, 7 ounces (3.4 kg) and the females averaged 5 pounds, 11½ ounces (2.6 kg). The minimum male weight was 4½ pounds (2 kg) and the maximum was 14½ pounds (6.6 kg). This last is the heaviest whitetail fawn that I can find on record. The minimum female weight was 3¼ pounds (1.5 kg) and the maximum was 8¼ pounds (3.7 kg). C. W. Severinghaus reported a female whitetail fawn in New York that weighed 10 pounds (4.5 kg), which is the heaviest record I can find for a female.

Raymond Hall gave the measurements of a California muley fawn: a total length of 23¼ inches (59.1 cm), a tail length of 4⅓ inches (10.9 cm), a hind-foot length of 11¼ inches (28.6 cm), and an ear length of 4¾ inches (12.1 cm). Paul Hudson and Ludwig G. Browman gave the weight and measurements of a mule deer thought to be about 7 days old. The overall length was 30½ inches (7.5 cm), hind-foot length 11 inches (27.5 cm), ear length 4½ inches (11.4 cm) and weight 15¼ pounds (6.9 kg).

Fawns grow extremely fast. They double their birth weight in 15 days and double that total again in another 15 days or so. In 6 months their weight increases tenfold. Skeletal growth is very rapid up to the age of 7 months, when it stops for the winter. Suspended growth is nature's way of allowing food intake to be converted to the fat reserves needed for winter survival. At 6 to 7 months of age the fawns are about two-thirds the size of their mothers, whom they accompany.

Joseph Dixon gave the measurements of a 5-month-old muley fawn: total overall length was 39 inches (99.1 cm), tail 5 inches (12.7 cm), hind-foot length 14 inches (35.6 cm), ear length 6½ inches (16.5 cm), and height at the shoulder 25 inches (63.5 cm). Unfortunately, he did not give the weight, but it would be in the vicinity of 60 pounds (27.2 kg).

A 5- to 6-month-old whitetail buck that I measured in New Jersey yielded the following data: overall length 43 inches (109.2 cm), height at shoulder 27 inches (68 cm), ear length 5½ inches (14 cm), hind-foot length 15 inches (38.1 cm), and a heart girth of 26 inches (66 cm), which would figure out to about 60 pounds (27.2 kg) live weight.

The New Jersey Division of Fish and Game has done considerable deer research and it is from their *Deer Report #3* that I have taken the following material. In New Jersey in 1977, we had an estimated deer population of 75,000. Our habitat for deer runs from very poor in some of the Pine Barrens area, to good in the northwestern section, to excellent in the southwestern and west-central sections. The habitat determines the deer's weight. In our poorest Pines section, #21, the average live weight for a 5- to 6-month-old whitetail buck is 45 pounds (20.4 kg), and the average for does is the same. The rich farmland of Salem County has just emerged as our top deer-producing area, edging out Hunterdon and Mercer counties. In Salem County the average live weight of a 5- to 6-month-old whitetail buck is 93 pounds (42.2 kg), and the does average 64½ pounds (29.3 kg).

It cannot be stressed too often that deer are what they eat. Illinois has no overbrowsing. There is a superabundance of food from some of the richest soil in our nation, and the results show. In their book *Prairie Whitetails,* John Calhoun and Forrest Loomis tell of four records of giant Illinois whitetail buck fawns. The largest, 5 to 6 months old, weighed 176 pounds (79.8 kg) and was shot in Mason County in 1966. The second-largest weighed 170 pounds (77.1 kg), the third weighed 166 pounds (75.3 kg), and the fourth weighed 164 pounds (74.4 kg). (All are live weights.) These fawns were exceptional even in Illinois, where the average buck fawn of comparable age weighs 90 to 100 pounds (40.8 to 45.4 kg).

The importance of food is also shown in the statistics compiled by the state of Nebraska, which has populations of both whitetail and mule deer. As a rule, mule deer average quite a bit larger than whitetails, but not in Nebraska. There, the record muley outweighed the record whitetail, but an average Nebraska whitetail of a given age actually outweighs an average Nebraska mule deer of the same age. The heaviest mule deer ever recorded in Nebraska was a buck shot in Garden County in 1957. It had a scale-weight of 310 pounds (141.6 kg) hog-dressed, and its live weight was estimated to be about 380 pounds (about 172.4 kg). This record muley weight is only a little greater than that of Nebraska's record whitetail buck, killed in Cherry County in 1957. It weighed 287 pounds (130.2 kg) hog-dressed, which means it had a live weight of over 350 pounds (over 158.8 kg). Nebraska's statistics are very interesting because they include the average weights for the whitetail and the mule deer by age class, as shown in the following.

Average Weights of Nebraska Whitetails

AGE CLASS	MALE		FEMALE	
	lbs.	kg	lbs.	kg
6 months	87	39.5	81	36.7
1½ years	156	70.8	128	58.1
2½ years	192	87.1	137	62.1
3½ years	217	98.4	144	65.3
4½ years	238	108.0	151	68.5

Average Weights of Nebraska Mule Deer

AGE CLASS	MALE		FEMALE	
	lbs.	kg	lbs.	kg
6 months	75	34.0	70	31.8
1½ years	135	61.2	113	51.3
2½ years	173	78.5	122	55.3
3½ years	201	91.2	128	58.1
4½ years	213	96.6	126	57.2

The surprising discrepancies between the whitetails and mule deer in the Nebraska statistics are due solely to food. The whitetails inhabit the rich bottomlands, the farmlands, while the mule deer are found in the rougher, poor-soil areas of the state.

Pennsylvania, too, has discrepancies between the deer harvested on the rich farmlands and those taken in the poor woods areas of the north-central portion of the state. The statewide averages are considerably smaller than those of Nebraska's whitetails. At 6 months, both male and female fawns in Pennsylvania average 54 pounds (24.5 kg). At 1½ years the bucks average 102 pounds (46.3 kg), the does 92 pounds (41.7 kg). At 2½ years the bucks average 117 pounds (53.1 kg), the does 100 pounds (45.4 kg). And at 3½ years the bucks average 127 pounds (57.6 kg), while the does still average 100 pounds (45.4 kg).

For years a really big buck in Pennsylvania was called a "Michigan buck" because in the early 1900s Pennsylvania imported deer from that state. The imported deer were slightly larger than those in Pennsylvania at the time. Pennsylvania habitat was then a deer's heaven. The deer population was low, and thousands upon thousands of acres of forest were growing back into sprouts and other nutritious deer food. The whitetails responded by increasing in numbers, weight, and body size. Today, the really big-racked, heavy bucks from Pennsylvania's rich farmland are Pennsylvania deer, and so are those in Pennsylvania's northern wooded areas that are getting progressively smaller and developing ever smaller antlers.

As for Michigan deer, many years of poor forestry management in parts of that state wrought havoc on the whitetails, but management practices have changed, and it is possible that Michigan will again produce sizable numbers of sizable deer. Several other states, such as Missouri and Illinois, have fine habitat and big whitetails. Still other states, such as Missouri and Illinois, have fine habitat and big whitetails. Still other states, from California to the crests of the Rockies, produce big muleys, though the herds have declined in recent years. Habitat largely determines average size.

In New Jersey, 1½-year-old bucks on the best land average 115.3 pounds (52.2 kg) while those on the poorest land average 70 pounds (31.8 kg). Females of the same age on the best land average 116 pounds (52.6 kg), while those on the poorest land average 60.5 pounds (27.4 kg). Two figures really stand out here. The most important is that the does living on good soil weigh almost *twice* the weight of does inhabiting the poor regions. The underweight herd in those poor areas will continue to drop in population as well as size until corrective measures are taken. Does on the poor areas will not breed as fawns and, if they do breed at 1½ years of age, they will drop one fawn instead of two—for a net loss of two potential deer from each doe in two years. We know the problems and we know the answers, but much of the time the public living in the poor deer areas will not allow our game managers to manage the game and correct the existing problems. The New Jersey "game doctors" recognize the sickness and have written out the prescription, but the public won't take the medicine. Deer herds must be managed to keep the population in balance with the food available, either by a heavier harvest of the deer or by sufficient expenditures on habitat improvement to provide more food. I will discuss these problems more fully in the chapters on wildlife management.

The second statistic that stands out is that on our best land, the 1½-year-old does actually outweigh 1½-year-old bucks. This is most unusual and can only be attributed to food. Of course, the bucks eventually outgrow the does. In fact, they pull away from the does in both size and weight with each passing year.

New Hampshire has done considerable work on the weights of the whitetail deer of different age classes. The state's biologists also take cognizance of the weight loss sustained by deer during the hunting and breeding seasons. They have prepared charts showing the average weights for bucks and does in the five major

Average Weight and Weight Loss of New Hampshire Whitetails During Breeding and Hunting Season

AGE CLASS	WEIGHT AT BEGINNING OF NOVEMBER		WEIGHT AT END OF DECEMBER	
Male	lbs.	kg	lbs.	kg
6 months	80	36.3	75	34.0
1½ years	141	64.0	120	54.4
2½ years	195	88.5	166	75.3
3½ years	235	106.6	177	80.3
4½ years and older	245	111.1	186	84.4
Female				
6 months	70	31.8	65	29.5
1½ years	125	56.7	100	45.4
2½ years	138	62.6	125	56.7
3½ years	150	68.0	137	62.1
4½ years and older	145	65.8	131	59.4

age classes during each week of November and December. I will summarize the data here because no other state, to my knowledge, has attempted to prove this significant weight loss.

In giving average age-class weights, the New Hampshire study shows the yearly increase in weight of the bucks. Of even greater importance is the weight loss during the rut. It was generally thought that whitetail bucks lost about 20 percent of their body weight because of the breeding season. This study proves that the weight loss is at least 25 percent for bucks 3½ years old or more, the bucks that do most of the breeding. The actual weight loss is even greater than indicated here because bucks taper off their food consumption for about a month prior to the main breeding season. The bucks undoubtedly weighed more at the beginning of October than they did on the first of November, but before the hunting season opened there were not enough bucks available for weighing to make a sampling.

Data from the Arizona State Game Department includes average weights for the Coues whitetail deer taken in the 1974 buck season. The 1½-year-old bucks average 68 pounds (30.8 kg); 2½-year-olds averaged 79 pounds (35.7 kg); 3½-year-olds averaged 100 pounds (45.4 kg); 4½-year-olds averaged 107 pounds (48.5 kg), as did the even older bucks.

The Coues is one of our smallest deer. Jerome Pratt kept some very interesting records while he was wildlife manager at Fort Huachuca, Arizona. The 881 Coues bucks harvested at the fort between 1958 and 1963 averaged 98 pounds (44.5 kg), and the does averaged 72 pounds (32.7 kg). Coues fawns generally weigh 3½ pounds (1.6 kg) at birth and 31 pounds (14 kg) at 3 months. At that age they stand 23 inches (58.4 cm) high at the shoulder. Average bucks stand 31 inches (78.7 cm) at the shoulder.

In 1963 Charles Walker killed a Coues buck that weighed 132 pounds (59.9 kg) at Fort Huachuca. The following year, William Culbertson killed one that weighed 137 pounds (62.1 kg), and three weeks later Mrs. Diane Walton broke all records with a buck that weighed 145 pounds (65.8 kg) field-dressed—for an estimated live weight of 181 pounds (82.1 kg). It is thought, however, that her heavyweight trophy was a descendant of a cross between the Coues deer and a mule deer. Two known crosses had been previously released in the area where the Walton buck was taken.

Key deer are even smaller than the Coues. John Dickson, in his study of Key deer, states that adults measure 26 to 28 inches (66 to 71.1 cm) high at the shoulder and weigh 35 to 80 pounds

(15.9 to 36.3 kg). The average weight for adult bucks is about 45 pounds (20.4 kg).

Ellsworth Brown of Washington state gives some age-class weights for Columbian blacktails, and Don Robinson of British Columbia's Game Department has recorded some for the Columbian blacktails on Vancouver Island. The accompanying table shows the averages compiled for those two areas. The samplings from Washington include older deer than those sampled on Vancouver Island. It is interesting to note that the Washington does lost weight slightly but consistently after reaching 5½ years of age. The Washington bucks lost weight at 6½ years but then regained it and continued to gain. Their 6½-year dip may have reflected a very poor forage year when the samples were taken. It is also interesting to note that the Vancouver blacktails are much lighter in weight than those in Washington (and, I should add, those in Oregon). This, again, is a matter of forage. Blacktails of the California chaparral are even lighter than Vancouver's deer.

Average Weights of Columbian Blacktails in Washington

AGE CLASS	MALE		FEMALE	
	lbs.	*kg*	*lbs.*	*kg*
6 months	57	25.9	55	24.9
1½ years	107	48.5	95	43.1
2½ years	137	62.1	110	49.9
3½ years	161	73.0	117	53.1
4½ years	165	74.8	118	53.5
5½ years	207	93.9	129	58.5
6½ years	192	87.1	127	57.6
7½ years	211	95.7	124	56.2
8½ years	232	105.2	124	56.2

Richard D. Taber and Raymond F. Dasmann, in a study of Columbian blacktails in California, were unable to compile weight averages for 5½- and 6½-year-old bucks but, with those exceptions, they recorded the weights of both sexes from 1½ to 7½ years. California's blacktails are very light, and the does, like those in Washington, lose weight after their prime years. Another weight table summarizes the findings.

The weights and sizes of Columbian blacktails in the California chaparral closely parallel the weights and sizes of Sitka black-

Lightweights, Middleweights, Heavyweights

Average Weights of Columbian Blacktails on Vancouver Island

AGE CLASS	MALE		FEMALE	
	lbs.	kg	lbs.	kg
1½ years	78	35.4	75	34.0
2½ years	93	42.2	84	38.1
3½ years	111	50.3	86	39.0
4½ years	135	61.2	92	41.7
5½ years	161	73.0	99	44.9

Average Weights of Columbian Blacktails in California Chaparral

AGE CLASS	MALE		FEMALE	
	lbs.	kg	lbs.	kg
6 months	33	15.0	26	11.8
1½ years	62	28.1	57	25.9
2½ years	95	43.1	80	36.3
3½ years	101	45.8	79	35.8
4½ years	127	57.6	88	39.9
5½ years	—	—	79	35.8
6½ years	—	—	79	35.8
7½ years	144	65.3	78	35.4

tails in most areas. The *Alaska News* states that, in October, Sitka blacktail bucks average about 150 pounds (68.3 kg), the does 100 pounds (45.4 kg), but other studies indicate weights as low as those for the California deer. Of course, much depends on such factors as the forage in a study area and whether the deer are studied before, during, or after the rut.

A. Starker Leopold, Thane Riney, Randal McCain, and Lloyd Tevis Jr., in a study of the California mule deer in the Jawbone herd, summarized some of the weights. At 6 months, buck fawns average 49 pounds (22.2 kg), does 47 pounds (21.3 kg); 1½-year-old bucks average 85 pounds (38.6 kg), does 77 pounds (35 kg); 2½- to 5½-year-old bucks average 135 pounds (61.2 kg), does 102 pounds (46.3 kg); bucks 6½ years and older average 140 pounds (63.5 kg), does 101 pounds (45.8 kg). Again the records prove that bucks continue to gain weight until they reach extreme old age while the does lose weight after 4½ or 5½ years. This decline was also borne out by C. W. Severinghaus in a study of whitetails in New York's Adirondacks.

Some does become sexually mature at 7 months of age if they have a nutritious diet. Far more whitetails than mule deer breed as fawns. Most 1½-year-old does are sexually mature but, again, whitetails at this age produce more fawns per doe than do the mule deer. Some blacktail does do not become sexually mature until they are 2½ years old. Most does attain their maximum growth at 2½ years, although they gain a little more weight with age.

Among all three types of deer, most bucks reach sexual maturity at 1½ years, though most of them are prevented from breeding at that age by the older bucks. They do not reach their full skeletal size until they are 4½ to 5½ years old. They continue to gain weight after that until they are about 8½ years old, when their weight and antler development usually begins to decline.

I have measured and weighed many deer. A whitetail (*O. v. borealis*) buck that I recently worked on was about average for the northeastern states. The buck weighed 151 pounds (68.5 kg), live weight. It measured 68 inches (172.7 cm) from nose tip to tail tip and stood 39 inches (99 cm) high at the shoulder. Most whitetail does of this area are 34 to 36 inches (86.4 to 91.4 cm) high at the shoulder, most bucks 38 to 40 inches (96.5 to 101.6 cm).

Ernest Thompson Seton recorded the measurements and weights of quite a few outstanding whitetail bucks, and some of his figures merit inclusion here. For example, he reported that a buck shot by John Denny in 1877 in New York's Adirondacks weighed 286 pounds (129.7 kg) hog-dressed, which would give it a live weight of 357 pounds (161.9 kg). John W. Titcomb recorded a buck killed in Vermont in 1898 that weighed 370 pounds (167.8 kg) live weight. In Warren County, New Jersey, in 1895, Warren S. Potter killed one that weighed 318 pounds (144.2 kg) hog-dressed, or 400 pounds (181.4 kg) live weight.

Seton's statistics get better as they go on. Henry Ordway killed a *giant* buck near Mud Lake in New York's Adirondack Mountains in 1890. It was scale-weighed before dressing at 388 pounds (175.9 kg). The deer had been bled and had probably lost at least 12 pounds (5.4 kg), so its live weight was calculated to be 400 pounds (181.4 kg). Even more remarkable were the deer's measurements: overall length from nose tip to tail tip, 115 inches (292.1 cm); height at shoulder, 51 inches (129.5 cm); neck circumference behind the ears, 37 inches (94 cm). There are no measurements anywhere for other whitetail deer that come close to this record. The buck had 9 tines on one beam, 10 on the other side. The longest tine was 13 inches (33 cm).

A Georgia buck's shoulder height takes the number two record, so far as I can ascertain. In 1972, Boyd Jones, from Florida, bagged a monster buck in Worth County, Georgia, whose dressed scale-weight was 355 lbs. (161 kg). A biologist, Jimmy McDaniel, from Florida Game & Fresh Water Fish Commission, witnessed the weighing and measured the buck's shoulder height at 45 inches. The buck would have had an approximate live weight of 443 to 452 pounds. (201 to 205 kg), according to which multiplier was used. This buck was an offspring of the deer that the Worth County Wildlife Club had imported from Wisconsin a decade before.

The world-record weight for the whitetail was held for years by a buck killed by Albert Tippett in 1919 near Trout Creek in Michigan's Upper Peninsula. Railroad scales showed it to weigh 354 pounds (160.6 kg) hog-dressed. The estimated live weight was 425 pounds (192.8 kg). The weighing was witnessed by a number of people.

Otis Bersing, in his book *A Century of Wisconsin Deer*, described two whitetails that top the Tippett buck, but neither is considered an official record because the weighings were not officially witnessed. In 1924 Robert Hogue shot a buck in Sawyer County, Wisconsin, that had a dressed scale weight of 386 pounds (175.1 kg) and an estimated live weight of 491 pounds (222.7 kg). In 1941 in Iron County, Wisconsin, Arnold Peter shot one that had a dressed scale-weight of 378 pounds (171.5 kg) and an estimated live weight of 481 pounds (218.2 kg).

Lest anyone get the impression that all the big whitetail bucks have disappeared since the "good old days," let me cite some more recent reports. In 1962, Dean Coffman and his friend John Ryan were hunting near Blencoe, Iowa. Using a pump shotgun (because only shotguns were legal in Monona County that year), Coffman dropped a very big buck with a rifled slug. The deer was scale-weighed before witnesses. Bill Welker, a biologist with the Iowa State Conservation Commission, verified the live weight at 440 pounds (199.6 kg). The buck was 4½ years old and had six points on each main beam. It was the largest ever taken in Iowa, where whitetails weighing 250 pounds (113.4 kg) are common.

On a rainy day in November 1955, Horace Hinkley and his wife, Olive, were hunting on the Kennebec River near Bingham, Maine. It was a good hunting day because a hunter could move through the woods silently and there was only a light breeze to move scent around. Mr. and Mrs. Hinkley took stands on opposite

At left, Dean Coffman with the 440-pound (199.3 kg.) Iowa buck he shot in 1962. In November 1955, Horace Hinkley shot the heaviest buck ever officially recorded in Maine. The monster was estimated to have weighed 450 pounds (204.1 kg.) live weight. photo courtesy Maine Fish and Game

sides of a ridge. At about 9:20 that morning, Hinkley fired at a buck but missed. A few minutes later, Olive Hinkley's rifle cracked, and after a few moments she shouted that she had downed a big buck. Hinkley, certain that there were more deer in a thicket of beech where he had missed the first one, remained where he was and did not respond. Suddenly a huge buck came crashing out of the brush toward him. Hinkley dropped the animal with one shot. It was so heavy that the Hinkleys had to get help to haul it out.

The buck was not officially weighed until three days later. The weighing was performed by Forrest Brown, an official state sealer of weights and measures, and there were two witnesses. Hog-dressed, and after three days of drying out, it still scaled 355 pounds (161 kg). Bob Elliot, of Maine's Department of Game, calculated the live weight to be at least 450 pounds (204.1 kg) and more probably 480 pounds (217.7 kg). The buck's rack was excellent, but not in keeping with its body size. There were eight points on each side, with a spread of 21 inches (53.3 cm) and a beam length of 24½ inches (62.2 cm).

Minnesota, however, has had a claim to the all-time record weight since a bitterly cold November day in 1926, when Carl J. Lenander, Jr., went hunting with his father near Tofte. Lenander had been on his stand only a short time when he saw a monstrous buck walking into range. He dropped it with a single shot. Field-dressed, it scaled 402 pounds (182.3 kg). The Conservation Department calculated its live weight to be 511 pounds (231.8 kg). In 1961 Minnesota officially recognized the Lenander buck as the largest whitetail ever killed in the state. For that matter, no heavier has ever been officially recorded in North America although the record was matched in 1981 by another Minnesota buck.

The new record buck was taken by George Himango, an Ojibwa Indian from Duluth. Himango took the deer in November 1981 on the Fond du Lac Indian Reservation.

Although George's brother had taken a nice buck in the morning, George and his cousin had hunted most of the day with no luck. About 3:00 in the afternoon, George heard a deer coming through the woods behind him. The brush was so heavy that George could not see the animal although he could easily hear the footsteps. Then there were no footsteps; the deer stopped, because it was either suspicious or curious. George had helped his brother dress out his deer in the morning. He had carefully removed the bladder full of urine and had spread it, and some deer scent, all around the area of his stand.

When the buck did move, it was moving toward the only opening in the brush. George prepared to fire when the buck moved into the clear. Upon reaching the opening, the buck turned and started right toward George. Scenting the buck shot by George's brother, it then snorted several times and began to paw the ground as a challenge. George fired while the buck had his head lowered; the bullet, glancing off the brow tine, went through the neck and into the animal's chest, dropping it in its tracks.

George had known that the buck was a big one when it stepped into the clear but realized he had shot a monster only when he got over to where the buck lay. Even with the help of both his cousin and brother, it took 3½ hours to drag the field-dressed carcass ⅛ of a mile to the edge of the lake where their canoe was tied.

Pulling the canoe up to the edge of the lake, they rolled the deer into the canoe and then pushed the canoe out into the water.

George's uncle, who is 70 years old, has hunted the reservation all of his life, and had never seen such a big deer. When the deer was taken to a butcher shop, it weighed in at 402 pounds (182.3 kg), a record verified by six additional witnesses. With a hog-dressed scale-weight of 402 pounds times .1271 the live weight of the deer by formula was 511 pounds (231.8 kg).

Although the deer was huge and the rack was a good one, it was not big enough to get into the Boone and Crockett records. The head scored 154 points. Carl Lenander's record had not been broken, but it had been equaled.

Maine has held an annual Big Bucks Contest for the past couple of decades. One of the largest whitetails killed there was taken in 1964 by Lovell Barnes of Hiram. Field-dressed, it had a scale-weight of 312 pounds (141.5 kg), and the live weight was estimated at 406 pounds (184.2 kg). Two other bucks weighing over 400 pounds (181.4 kg) have been taken in the state in the last two decades, but the reports lack details.

In Illinois, another state that has a healthy population of good-sized whitetails, an archer took the largest recorded buck. Killed in 1970 in Carroll County, it had a live weight of 370 pounds (167.8 kg).

Although the average size of whitetails tends to be largest in more northerly latitudes, Missouri is another state known for big bucks. Missouri's official weight record is held by a buck killed by Clifford Davis in 1954 in Livingston County. Its live weight, confirmed by scale, was 369 pounds (167.4 kg). A bigger one was killed in 1970 by Dwain Perrige of Edina, Missouri, but no game-department officer was on hand to supervise the weighing and make it official (though the scales were carefully checked and witnesses were present). Six men were needed to haul this monster out of the woods to a vehicle. Field-dressed, but with the heart, lungs, and liver left in the carcass, it scaled 361 pounds (164.7 kg). Wayne Porath, a deer biologist with the Missouri Game Department, later estimated its live weight at 423 pounds (191.9 kg). In recent years, several other gigantic bucks weighing well over 350 pounds (158.8 kg) have been registered by Missouri's Show-Me-Big-Bucks Club.

Even a state as far south as Louisiana has produced some massive whitetails. The biggest, killed in 1964 in the Russell Sage Wildlife Management Area, had a live weight of 321 pounds (145.6

kg). And at the Thistlethwaite Wildlife Management Area, a buck killed in 1968 weighed 314 pounds (142.4 kg) and one taken in 1969 weighed 312 pounds (141.5 kg).

While photographing whitetails in the Louisiana bayous, contrary to what I had been led to expect, I saw good, big-bodied bucks. Even more amazing, some of those bucks had racks as large as I have ever seen on any bucks, anywhere.

Still another huge southern whitetail was taken in Worth County, Georgia, in 1972 by Boyd Jones. Hog-dressed, it weighed 355 pounds (161 kg). The estimated live weight was 443 pounds (200.9 kg).

Mule deer tend to be larger than whitetails, and this is especially true of Rocky Mountain mule deer *(O. h. hemionus)*, so it is hardly surprising that records of very heavy Rocky Mountain muleys have come to light. Dr. Ian McTaggart Cowan, writing in *Safari* magazine in 1975, commented on their size:

> The Rocky Mountain mule deer is the largest of its species. I have known autumn bucks to have a live weight of 475 pounds (215 kg)—but this would be most unusual. Only 2 percent of 360 Modoc County, California, bucks weighed over 250 pounds (113.4 kg) dressed and only 1 percent over 300 pounds (136 kg).
>
> This is also the deer with the great difference in size between the sexes. Adult does on good range probably average about 140–150 pounds (63.5–68 kg). This size difference appears to be related to winter cold together with the poor energy status of the bucks as they enter the winter. The larger animal is more efficient in conserving heat than the smaller one of similar shape. Thus there has been a selection pressure for bigger bucks ... whereas in the more southerly ranging races of mule deer the bucks are considerably smaller and the bucks and does more nearly the same size.

In the 1930s, the following measurements for a Rocky Mountain buck were reported by Joseph Dixon of the California Fish and Game Commission: overall length, 70 inches (177.8 cm); tail, 8 inches (20.3 cm); height at shoulder, 44 inches (111.8 cm); ear from crown, 9 inches (22.8 cm); hind foot, 20 inches (50.8 cm). His measurements for an adult doe were as follows: overall length, 61 inches (154.9 cm); tail 7 inches (17.8 cm); height at shoulder, 32 inches (81.2 cm); ear from crown, 8½ inches (21.6 cm); hind foot,

19 inches (48.3 cm). He did not give the weights for these deer but quoted the records of a co-worker, J. S. Hunter, who reported in the January 1924 issue of *California Fish and Game* that a Dr. Trusman of Adin could vouch for a Rocky Mountain muley buck from Modoc County that weighed 380 pounds (172.4 kg) hog-dressed. The live-weight estimate was 456 pounds (206.8 kg). Another hog-dressed buck weighed 350 pounds (158.7 kg) for an estimated live weight of 420 pounds (190.5 kg).

The largest Rocky Mountain mule deer that Dixon could personally verify was shot on September 30, 1930, by Arthur Oliver in Shasta County, California. This buck weighed 308 pounds (139.7 kg) dressed, for an estimated live weight of nearly 400 pounds (181.4 kg).

John Russo recorded that the heaviest muley buck from the Kaibab Plateau in Arizona in 1963 weighed 225 pounds (102.1 kg) hog-dressed, for an estimated live weight of 281 pounds (127.5 kg). Obviously, the biggest, heaviest mule deer are generally found farther north, even though some record-book antlers have come from Arizona and New Mexico.

Gilbert Hunter, a state game manager of Colorado, has verified the scale weight of the two heaviest mule deer on record. A California hunter, O. I. Ranch of Long Beach, shot a monster muley near Meeker, Colorado. The dressed scale-weight was 360 pounds (163 kg) for an estimated 459 pounds (208.3 kg). He did not give the year the buck was taken.

However, in 1938, Hunter weighed a huge mule deer buck taken by Laurence Rowe near Ellens Park, Colorado, at a hog-dressed weight of 410 pounds (186.1 kg). Using the .1275 figure that many westerners prefer, the live weight of that buck would have been 522.75 pounds (237.3 kg).

Paradoxically, the Sitka blacktails up in Alaska tend to be smaller and lighter than other mule deer. This, as I indicated earlier, is chiefly a result of their forage and habitat, but the heredity of the race also has an effect. If Sitka blacktails were transplanted to very lush habitat, genetics probably would still prevent them from growing to the size of a big Rocky Mountain muley. The record Sitka blacktail, killed on one of the islands of Prince William Sound, weighed 212 pounds (96.2 kg) field-dressed. That would be an estimated 253 pounds (114.8 kg) live weight. Columbian blacktails occasionally grow a bit larger, given ideal forage.

In 1939 a severe forest fire in Tillamook County, Oregon, burned over 305,000 acres (123,432.5 ha) of land inhabited by Columbian blacktails. The entire area grew back with sprouts, berry bushes, ferns, forbs, and other excellent deer food. This area was closed to hunting for four years because the Oregon Game Commission had found that four years will allow the food to grow back and will also allow the deer to reach their greatest individual and population potential. Before the burn, the deer harvest averaged two deer per section, with an average live weight of about 120 pounds (54.4 kg). Four years after the burn, when the season was opened, there were 14 deer per section and their weight had increased 30 to 50 percent. Only seven deer weighed less than 180 pounds (81.6 kg). Twenty-four deer weighed 180 to 200 pounds (81.6 to 90.7 kg), 19 weighed over 200 pounds, and one buck weighed 310 pounds (140.6 kg). This last weight is the heaviest record I can find for a Columbian blacktail.

10

The Digestive System

Deer are ruminants, equipped with a four-chambered stomach. Ruminants—that is, the deer family, the giraffe, the antelope, and sheep, goats, and cattle—are cud-chewers. A four-chambered stomach is needed to process the large quantities of low-nutrient food that these animals eat. The ruminant's stomach is also a very important survival feature; it allows the animal to gather and swallow a lot of food in the shortest possible time and then chew it later, at its leisure, while relaxing in a safe place.

When a deer is feeding, it is at a tremendous disadvantage, because it is concentrating on locating and obtaining food, rather than concentrating totally on avoiding predators. The nose and the eyes are used in locating the food. Moreover, the gathering of the food makes considerable noise, which is conducted via jaw bones to the auditory nerves. So engrossed, deer are vulnerable to predation, and they know it. This is one of the reasons why deer prefer to feed facing into the wind; the scent of danger ahead of them will be borne down to them. It is also one of the reasons why cud-chewers do not feed steadily in one spot. Deer take a bite of food and then take a couple of steps and nibble again. This prevents over-utilization of the food plant while also keeping deer moving away from any stalking predator. Deer lack the reasoning ability to calculate where safety lies, but natural selection favors protective behavioral patterns, establishing and reinforcing these patterns through evolution. Almost all predators hunt into the wind, whether they hunt by sight or scent, so that their own scent will not be carried to their intended prey. Any predator stalking a feeding deer has to make a longer stalk because the deer are constantly moving away into the wind. The longer the stalk, the more chance there is for the predator accidentally to betray its presence.

Depending on the type and abundance of the food that the deer is eating, it can fill its paunch in one or two hours. Naturally, if food is scarce, it takes much longer, or the deer may not even be able to fill its paunch.

In most areas of the country, deer vary their diet seasonally, feeding upon whatever is most easily obtained. Deer turn aside from woody browse to feed on the first succulent grasses that sprout in the spring. As the season progresses the broadleaf *forbs*—nongrass plants—may become the most important food plants. The forbs grow in superabundance at the time they are needed most, the period of giving birth and lactation for the doe and of antler growth for the buck. Later in the summer, when the heat has slowed the growth of the forbs, deer supplement their diets with the broad leaves of various trees and bushes. During this period mushrooms and berries are avidly sought and eaten by the deer. In early fall apples begin to drop, and the last part of September brings the nut mast crops. In areas where they are available, the deer will forsake almost all other foods to concentrate on the acorns.

The amount of acorns available to the deer varies greatly from year to year. In good years some acorns will be on the ground all winter. As the acorns disappear, the deer again begin to feed on grasses, hardwood browse, and especially on the newly fallen leaves. Deer continue on this diet until there is a snow depth of about 8 to 10 inches (20.3 to 25.4 cm). After that, the deer are forced to feed almost exclusively upon browse, but will feed more on evergreen browse because they will be seeking shelter among the conifers from the winter wind.

Except for those living in true wilderness, deer also feed on farm crops to a greater or lesser degree. In the early spring they seek out the green winter wheat and the first shoots of alfalfa and clover. They taper off on the alfalfa as it matures until it is cut and resprouts. Deer also feed heavily on the sprouts of most legumes, such as peas and beans. Corn plants may be eaten up until the tassel ripens. Many times deer eat the center, tassel-bearing stalk, preventing the corn ear from maturing and drastically reducing corn production. The corn ear itself may be eaten from the time the kernels begin to fill out all through the fall right up until the last vestige of corn is gone. Newly sprouted winter wheat, planted in the fall, is a major food at that time of year.

Strawberries, a favorite deer food, are followed shortly by the peach crop, and apples are taken whenever and wherever available. In some areas orchardists suffer tremendous losses due to deer depredation. The deer damage the trees, and regeneration is sometimes difficult because of the size of the deer population and the paucity of other foods.

Deer are primarily browsers, feeding on the tips of twigs, branches, shoots, and leaves. In many areas, due to a lack of browse, the deer also obtain much of their food by grazing on grasses, forbs, and similar types of vegetation.

When a deer is feeding on browse, it prefers the tenderest, newest shoots and tips. Such tips are more palatable, more flavorful, easier to bite off, and most nutritious.

As forage plants mature their cell walls thicken. The center cell solubles are up to 98 percent digestible. The walls contain such components as cellulose, hemicellulose, lignin, cutin, pectin, and tannin. Some of the cellulose can easily be broken down by the microorganisms in the stomach; other substances are tougher. Lignin, a noncarbohydrate polymer that binds the cell together, is indigestible. The older the material ingested, the higher the proportion of indigestible material consumed, and the lower its nutritional value. This is why many deer die of starvation even though

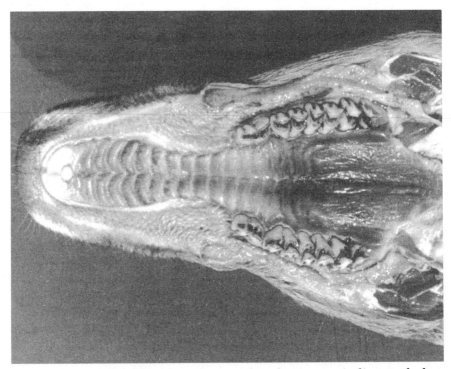

This is the roof of a deer's mouth: note that there are grinding teeth, but no incisors. Lacking upper front teeth, deer use the rough pad at the front of the mouth to grip plants. Browse is nipped off by the lower incisors bearing against the pad.

The Digestive System

their paunches may be filled. The foods that deer are often forced to consume in the winter are the least nutritious.

Tips of twigs up to about the size of a wooden kitchen match are nipped off by the deer's front lower incisors against the pad in the front of the top of its mouth. Lacking upper front teeth, the deer has to tear the twigs loose, always leaving a ragged edge. Where deer and hares or rabbits are competing for low-growing vegetation, it is easy to tell which twigs were eaten by deer. Hares and rabbits use their upper and lower incisors to clip twigs off as if with pruning shears.

When deer are feeding on larger twigs, those up to the size of a pencil—which they do only in times of food shortage—they take the twigs in the side of their mouths and bite them off, using the third premolars and first molars. That is why the greatest amount of wear shows on these teeth.

When deer are grazing, they use their lips to maneuver the grass into their mouths, hold it with their incisor teeth against the mouth pads and tear it loose by moving their heads forward. I have never seen deer use their tongues to tear grass loose as cattle and bison often do. The food is moved by the tongue to the back of the mouth, where it is chewed just three or four times before being swallowed. The food passes down the gullet into the first part of the stomach.

The four sections of a deer's stomach are the *rumen*, the *reticulum*, the *omasum*, and the *abomasum*. The food initially goes into the rumen, which can hold 8 to 9 quarts (7.6 to 8.5 liters) and has the combined functions of storing the unchewed food and acting as a fermentation vat.

In the mammal world, males are usually larger and sometimes many times larger than females. The male deer's larger body size accommodates a larger paunch, which allows the bucks to process more food of lower nutritional value, and maintain a lower metabolic rate, than can does and still survive. This is of particular importance in areas where deer yard up, or congregate in one protected spot, in the winter. Researchers have found that in the winter yard areas the bucks tend to stay out on the fringes of the yard, where they are exposed to more danger and where the food may be of lesser quality.

After the deer has filled its paunch, it retires to some spot of comparative safety to chew its cud or bolus. A deer may chew its cud while standing up but it usually lies down. The deer can rest or even doze while doing this, but the head is held upright. Cud-

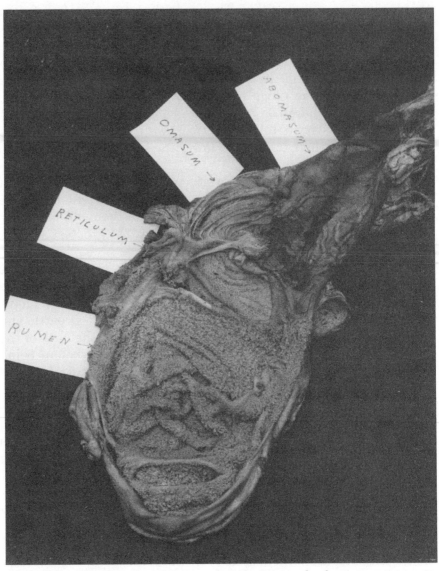

The labels alongside this deer stomach designate the four compartments: the coarsest food passes through the rumen twice, once when it is eaten, and then again after it has been regurgitated and chewed as a cud.

chewing is almost automatic, progressing without the deer consciously starting it, but if the deer becomes alarmed or alerted the process is stopped instantly.

To break down the fibers, cellulose, and other basic plant components, and convert them to materials that can be assimilated

The Digestive System

and used by the body's metabolic system, the deer depends on a symbiotic relationship with billions upon billions of microorganisms that live in its stomach. Among these microorganisms are ciliate bacteria, and bacilli, cocci, and spirilla. Through digestion by these organisms and by fermentation, the food produces metabolites known as short-chain volatile fatty acids, such as acetic, propionic, valeric, and butyric acids. The lining of the rumen has small, spaghetti-like papillae, varying in length from ⅜- to ½-inch (9.5 mm to 12.7 mm). There are 1,600 of these papillae to the square inch (6.45 sq. cm), and I know because I have counted them. Over 40 percent of a deer's essential energy is derived from the acids absorbed through the papillae and the walls of the rumen. Minimal digestion starts to take place in the deer's mouth, but it is the organisms, acids, and heat of the stomach that really do the job.

The cud is vegetative material that is regurgitated from the rumen and comes back up the gullet. It is about the size of a big lemon, and its passage up to the mouth can easily be seen. In masticating its cud, the deer chews with a sideward motion of the lower jaw. A whitetail gives the cud an average of 40 chews before reswallowing it. Muleys and blacktails give it an average of 56 chews. I know how thoroughly deer have to chew the cud not just from reading biological monographs but from personal observation. In the course of taking over 20,000 photographs of deer, I

Hunters and naturalists can easily tell whether a twig has been browsed by deer or rabbits. Rabbits clip plants neatly. Deer rip them and leave ragged edges as seen here.

have often had to wait for an animal to come within camera range, or to get up, or to just stop chewing its cud. With not much else to do while waiting, I have counted the number of chews.

The purpose of chewing a cud is to break down foodstuffs into smaller pieces so that each piece has a much greater surface area for microorganisms to work on. This allows for more active microbial fermentation and decreases the length of time the foodstuffs stay in the rumen. The finer rechewed particles then bypass the rumen and are floated into the reticulum and omasum.

As the food is reswallowed, it does not make a noticeable lump. It is passed on from the rumen to the second portion of the stomach, the reticulum. The reticulum has a lining that looks like a honeycomb, with hexagonal ridges about ⅜-inch (9.5 mm) apart. Each ridge is about 1/16-inch (1.6 mm) high. The reticulum can hold foodstuffs up to about the size of a standard softball. Some digestion takes place in the reticulum, but one of its main functions is to filter out foreign material.

Like other ruminants, small children, and Indian fakirs, deer habitually swallow indigestible material. I have found stones, a

A young whitetail doe chewing her cud, visible as the bulge in her left jaw. Deer move constantly while they are feeding, but usually chew their cud at rest and in relative safety.

The Digestive System

.22 caliber cartridge case, and even a piece of melted glass in the deer stomachs I have examined.

Occasionally a "hair ball" forms in the stomach. Deer frequently lick themselves to help remove their old coats when they are shedding, and they often lick hair from one another. Usually this hair passes through with the food being digested. Sometimes a mass of it does not. The action of the stomach, combined with food material, puts a coating on the hair mass, creating a ball that lodges in the stomach if it becomes too large to be passed on.

On very rare occasions a "bezoar stone" is found in a deer's stomach. I have never seen one in the many deer I have examined, but I own a bezoar given to me by a masqi hunter who took it from an African antelope. A bezoar—also called a *calculus* or *madstone*—is formed by layers of calcium or resinous material, built up around some indigestible object in much the same way that a pearl is formed in an oyster. Some bezoar stones have reached a length of more than 2 inches (50.8 mm). At one time they were highly treasured, as they were supposed to neutralize poison. These foreign objects are usually found in the reticulum.

There is a lapse of 14 to 18 hours from the time a deer's food is eaten until it has been passed up again as cud, chewed, reswallowed, and passed through the reticulum. The food then goes into the omasum, where intensive digestion and absorption take place. The omasum's lining has forty flaps of varying heights, from 1/8-inch (3.18 mm) to 1 1/8 inch (28.58 mm) high.

The last stomach compartment, the abomasum, has a very smooth, slippery lining with about a dozen elongated folds. It is interesting to open a deer's stomach carefully and observe how digestion has progressed in each of the four compartments. From the jumbled mass of recognizable vegetation in the rumen, the food becomes finer, more liquefied, and less identifiable as it goes on. It enters the rumen as a coarse mass and passes through the abomasum as a liquid slurry, or soup. The deer's rumen holds about 80 percent of the stomach contents; the reticulum, 5 percent; the omasum, 7 to 8 percent; and the abomasum, 7 to 8 percent.

The deer has 65 feet (19.8 m) of intestines. Of that, the small intestine is 49 feet (14.9 m) in length and 2/5-inch (1 cm) in diameter. In addition to the acids and enzymes in the food, bile from the liver is added here to help break down and absorb the remaining long-chain fatty acids and the fat-soluble vitamins. Most of the nutrition that the deer gets from its food is absorbed into the bloodstream in the small intestine.

Characteristics of droppings sometimes indicate what the animal was eating. The pellets at the upper left are from a deer that was eating browse; the pellets at right are from a rabbit. The masses at the bottom left were dropped by a deer eating in a grassy forage area; the pellets at the right are from a porcupine.

The large intestine is about 16 feet (4.9 m) long, which is about two and a half times the deer's total body length, ³⁄₂₅- to ³⁄₂₀-inch (0.33 to 0.5 cm) in diameter. It is here that most of whatever water is left is absorbed, leaving an impacted mass of undigested particles. This mass is, of course, passed out as excrement. A deer defecates an average of 13 times every 24 hours. The type of food eaten determines the shape and consistency of the excrement. This can be of interest not only to biologists but to hunters, since there are times when the excrement can furnish a clue to where deer have been browsing or grazing.

If a deer has been feeding on grasses, forbs, or fruit, its feces are usually a loose, formless mass. When the deer is eating drier material or feeding on browse, the feces will form elongated pellets. A deer's feces can easily be told from a rabbit's, because the

The Digestive System

deer's pellets are always elongated, sometimes misshapen, and often pointed, while a rabbit's or hare's feces are round. There may be anywhere from just a few pellets to well over 300 in one defecation. Ordinarily, food will pass through the deer in 26 to 30 hours. Very rough, fibrous food may take a bit longer.

The deer's food utilization depends on the food that is eaten, the age of the deer, and the condition of its teeth. Averaging out these variables, biologists figure that of every 100 pounds (45.4 kg) of food, 65 percent will be used for maintenance, growth, and heat production. Five percent is likely to be lost as methane gas, 5 percent as urine, and 25 percent as feces.

Deer and some of the other herbivores lack a gall bladder. (The musk deer of Asia is the only deer with a gall bladder.) Acorns are a very high-fat food, yet they are easily assimilated by the deer without the help of gall, a digestive fluid that aids other animals to neutralize acids and emulsify fats. It is thought that the lack of gall enables a deer to eat the leaves of the rhododendron, which are poisonous to sheep and cattle.

Mule deer bucks frequently urinate in this squatting position, like a doe, but I have never seen a whitetail buck do this.

11

Deer Hair—
Summer Red to Winter Blue

A deer's color changes somewhat from summer to winter. To describe the difference in appearance, hunters and naturalists sometimes refer to deer as being in their "red" (summer) coats or "blue" (winter) coats. There are two complete hair molts per year. The exact time of the molt depends on altitude and latitude, as well as on the subspecies, physical condition, and age of the deer. Deer in the North shed their winter coats later and their summer coats earlier than do deer in the South. Adult deer shed their coats before the young ones do, and those in good physical condition shed earlier than those in poor shape.

My friend Manny Barrone, of Lincoln Park, New Jersey, is a taxidermist. He constantly supplies me with bits of information and specimens that he encounters. His wife, Elaine, has also been of enormous help to me. On the night of December 15, 1976, Elaine did not get to bed until 4:00 A.M. because she sat up counting the number of hairs in a square inch of a deer's summer coat and on a square inch of winter coat for me. To my knowledge, no one has published such figures, and her findings were amazing.

On a single square inch (6.45 sq cm) of skin from the top of the neck of an adult buck, taken in late November, there were 2,664 hairs. These hairs were 2 to 2¼ inches (50.8 to 57.15 mm) long and had a diameter of 0.007 inch (0.1778 mm).

An average-size adult deer has about 1,200 square inches (7,741.5 sq cm) of skin covering. Using the above figures, we find that the deer has about 3,196,800 hairs in its winter coat. By applying the same calculations to Elaine's count of hairs on skin taken from a buck "in the red," killed in late August, I discovered that the summer coat actually has almost twice as many—5,176 hairs to the square inch—or about 6,211,200 hairs. These counts explain why so many birds' nests are lined with deer hair. There

These swatches of hair show, from left to right, the short dense hair of summer coat; the long, hollow insulating hairs of the winter coat; and rare, a wooly base winter coat.

must be gobs of it in any woods with a good deer population.

The winter hairs are kinky, brittle, hollow, and filled with air, and they can easily be pulled out of the skin. Beneath these long guard hairs is a very soft, fine, kinky undercoat like cashmere wool. Elaine tried to count these undercoat hairs also but found it impossible. It is very difficult to see the undercoat, and when the wool is pulled out, it is so buoyant that it practically floats away. The long, hollow hairs combined with the undercoat give the deer terrific insulation, or protection against the cold. Because the guard hairs are hollow, a deer in its winter coat floats much higher in water than it does in its summer coat. I have often seen a deer bedded down during a snowstorm, with several inches of snow covering its body. None of the snow melts because so little body heat is lost.

Many hunters like to have a deerskin tanned with the hair on it, for use as a rug or throwpiece. This is a waste of time and money, unless the skin is to be used only as a wall hanging. Because the winter hairs are hollow, they are very brittle. If they are walked on, they soon break off.

There is one record of a Pennsylvania doe that had no guard hairs on her coat at all. Her entire body was covered with just the silky undercoat, although the undercoat hairs were longer than on a normal deer. This doe undoubtedly would not have survived the winter because the undercoat is not waterproof, as the guard hairs are. In a freezing rain, this doe's hair would have become soaked, matted, and frozen.

There are records of three deer, one each from Pennsylvania, New York, and Michigan, that had their long winter guard hairs but also had unusually thick, dense undercoats. Now there is a fourth record. Manny Barrone has sent me such a hide from a deer taken in New Jersey. These deer must have been exceptionally warm with all that insulation. Nevertheless, it is obvious that such additional dense coats are not needed or nature would have provided them for all deer.

Generally speaking and regardless of species, the deer's winter hair is a grayish brown at the tip. The base of the hair is gray, shading to almost black. The deer of the Far North have the darkest coats, while those in the South are quite a bit lighter. Deer of the desert or open areas are the palest of all. The difference in the color between the winter and summer coats is much more noticeable in the North than in the South.

Because a deer's winter coat is so much heavier, I had always assumed that it contained more hairs; Elaine's count was a great surprise. The "red" of the summer coat extends about 80 percent of the hair length, while the basal 20 percent is grayish. The summer hair is about an inch (25.4 mm) long and only 0.003 inch (0.0762 mm) in diameter. The hairs are thin, straight, solid, and without an undercoat. A summer deerskin could be used for a rug because the solid hairs are not brittle. Being solid, they provide little insulation, but they offer another kind of protection. The increased number of hairs shields a deer somewhat against biting insects.

Hairs grow through the skin. If they have not grown out to their full length, part of their dark roots show on the inside of the skin when it is removed from the animal. The roots make the inside of the skin dark, and the skin is then said to be "blue," or unprime. Once the hairs have grown all the way through, the inside of the skin is pale, and when tanned it is creamy white.

A whitetail deer has whitish circles around its eyes, white inside its ears, a white band around its muzzle and chin, and a white throat patch. It has a white belly, a narrow white rump, white beneath its tail, and white running down the inside of each leg. The amount of white around the individual deer's eyes, on its muzzle, and in its throat patch is very variable and genetic. Some deer have two white throat patches; others have variable amounts of white around their hoofs.

For unknown reasons, the hair on the brisket, or chest, of all deer points forward, while all the other hair points to the rear or downward.

Mule deer and blacktail deer have white over the nose, a gray face, white inside the ears, and a white throat patch. Mule deer

The hair on a deer's brisket points forward—against the grain of all the rest of its body hair, sometimes creating a fringe at the meeting points as seen on this deer.

A "double-bibbed" buck shows genetically determined two patches of white throat hair. photo courtesy Joni Norris, Cleveland Metroparks.

The mule deer face has a whiter muzzle, white inside the ears, and the characteristic dark lines descending from the antler base.

have a very heavy patch of dark, coarse hair on their foreheads that comes down the face beyond the eyes. Both types of deer have much less white on their bellies than does a whitetail, and a smaller amount of white on the inside upper portion of the legs. They have big white circular rump patches, the size depending on the subspecies. And, of course, the top of a blacktail's tail is black, while that of a mule deer is white.

Whether the deer is in its dark winter coat or its reddish summer coat, the color tends to fade in the sun. The winter coat is more subject to bleaching because it is worn for a much longer period—seven to eight months—and it is exposed to the sun more. In the winter, deer lie in the sun for warmth. They also feed more in the daylight hours, and with the leaves off the trees, the sun penetrates to the forest floor.

In the spring, while a deer is shedding its winter coat, it has a ragged, moth-eaten appearance. The hair sloughs off in patches. The old hair evidently itches, because the deer spend quite a bit of time licking away swatches of it. Two deer may team up and, through mutual grooming, remove the hair from each other's coats. It is chiefly this licking of the long winter hair that forms the hairballs sometimes found in a deer's stomach.

The deer's winter coat first starts to shed out in late March, although the deer seldom loses big patches of hair until May. The summer coat is worn for a little over four months, and it is complete only in June, July, and part of August. By mid-August the summer coat starts to be replaced by the winter coat. Though it takes about three months for the deer to shed its winter coat, its summer coat is usually replaced in about one month.

Even in their thin summer coats the deer are very susceptible to heat, feeding only in the late evening, at night, or in the early mornings. When the temperature gets about 65° F (18.3° C), deer seek shade.

The deer are usually in their complete winter coats by mid-September, although the hairs have not grown out to their full length by that time. We often have periods of "Indian summer" as late as the end of October. Some of those glorious fall days get downright warm. The bowhunting season opens in many states in middle to late September. When we have periods of Indian summer during the bow season, hunters have little chance of seeing deer.

Deer undergo a molt twice a year; shedding begins at the muzzle and progresses towards the tail. This doe has shed her summer coat back to her shoulders. The shedding of winter coat in spring is more noticeable than the molt of the summer coat.

The winter-coated deer just can't stand the heat and will not move until darkness brings the temperature down.

Any exertion by the deer before the weather is actually cold causes stress. The deer stand with their mouths wide open, pumping air in and out of their lungs at a fast rate in an effort to expel as much heat as possible.

Eccrine glands are sweat glands that produce a thin, watery sweat such as we humans give off when we perspire. *Apocrine* glands are heavy-duty sweat glands that give off a much thicker liquid, usually in the form of sebum. These glands are usually scent-producing glands, frequently exiting through hair follicles.

All of the literature on deer says they do not have sweat glands. I can't prove that one way or the other. Nevertheless, I have seen many bucks with their body hair so wet that it was plastered and would fold into layers instead of lying flat. I have seen this when bucks have run hard. I have seen it particularly in the rutting season, when the dominant buck is chasing the does and chasing off lesser bucks at the same time. I have seen it when two bucks fight.

When I asked wildlife biologists about this, one suggested that the body heat confronting cold external air caused condensation and that made the hair wet. Another suggested that perhaps the frost, which often covers a deer's hair on a cold night, wet the hair when it was melted by the sun. But I have seen this condition on

Biologists say that deer do not sweat, yet I have seen numerous deer, elk, and mountain sheep soaking wet from their exertions during the rut, as this buck is.

warm days in late October and early November when there had been no frost and the air was not cold.

Horses sweat; everyone who has been around horses has seen horses lathered with sweat after they have been ridden or worked hard. I have seen mountain sheep, during their breeding season, with their coats so thoroughly soaked by exertion that I don't see how they were able to stay warm. Both bucks and rams were so wet that their coats of hair must have lost all insulating value.

Heat is ordinarily lost through four methods. The deer can lose heat through conduction when it is lying on the cold ground. Then the heat is transferred directly from the animal's body to the cold earth. Deer lose heat in this fashion; they will melt the snow beneath their bodies. Most kinds of insulation lose their ability to retain heat when compressed, and the deer's hair is compressed when the deer is lying down. (This is the reason you need twice as much insulation beneath your body as you do on top of it when you camp out.)

Heat is lost through convection, which occurs when air is flowing over the body. Deer lose very little body heat when the air is still. Frost on the deer's hair usually melts off in the sun, and even snow will remain dry and fall off when the deer moves. It is because of convection that deer will forsake food to seek out hollows, swamps, and evergreen thickets when the wind blows in winter. A deer can stand all the cold in the world; it cannot, because of convection, withstand the wind.

Heat is lost through radiation. Deer lose very little heat in this manner. To prevent radiation, a deer's winter coat is a darker color than its summer coat. Dark attracts and retains heat. That's why deer seek sunny hillsides or areas to feed and bed on in the winter; their coats absorb the heat from the sun's rays, warming their bodies with very little caloric expenditure.

Heat is lost through evaporation. We humans sweat and are cooled as the moisture evaporates. A fan does not make the air cooler, it just moves air past our moist bodies, and it is the evaporation that cools us. When deer and sheep become wet, as I have described, they must lose a tremendous amount of body heat. I've often wondered why this does not prove fatal to the sheep.

If not from sweat glands, I just can't explain where the moisture I've seen comes from. I hope my remarks here will stimulate research or evoke an answer if there is one at this time.

People often report seeing "albino" deer. Of course, some deer are albinos, but most of the white deer seen are not true albinos:

they are mutations. These mutations are becoming more prevalent with the help of man. When the wildlife population was controlled by natural predators, any deer that had the deficient genes to result in a white coat would have been killed, because it would have been so easily seen. Ordinarily, the other deer will shun the mutations, even going so far as to drive them away.

I have also found that these white deer usually have poor hearing. During the many years when I was chief gamekeeper for the Coventry Hunt Club, we had a number of them from time to time on our club lands. They were almost always loners and they were easy to creep up on because they could not hear well. All of these white deer happened to be females, although this mutation can occur in either sex. These deer appeared at our club before New Jersey allowed does to be hunted. With the large predators gone and man protecting the mutant does by law, the deficient genes were passed on. So white deer are more common today than ever before.

Around 1928, when white deer were a rarity, they suddenly became very common near Mifflinburg, Pennsylvania. It was thought that overcrowding had something to do with the prevalance of this condition, but even today we do not know the reason.

In 1983 I did a series of deer seminars in Illinois. In the evening paper, I read that Governor Thompson had just signed a bill that provided full protection to all albino deer within the state. What the article failed to explain was just what the new law was protecting. If the law was protecting true albinos, then perhaps it had merit; if it was protecting all of the mutations, then the law was doing the Illinois deer herd a disservice. Mutant deer are definitely inferior, degrading the deer they breed with by passing on harmful recessive genes.

True albinism is due to lack of pigment. Ordinarily the production of pigment is controlled by the pituitary gland. In human beings, albinism is caused by the absence of the enzyme tyrosine, which is needed to produce the darker pigments. A true albino deer will have all-white hair, grayish hoofs, and pink eyes. The eyes appear pink because, in the absence of pigment, the blood can be seen coursing through the blood vessels.

The Seneca Army Depot in New York State has an entire herd of white deer. The first white fawns were seen in 1956, born to a normal brown doe. The commanding officer issued orders for protection of the white deer, and the boom began. In 1958 one white fawn was born, in 1959 two white fawns were born, and in 1960 two more white fawns were born. The white herd went from seven deer in 1960 to 150 in 1968 and has continued to increase in size.

The New York State Department of Conservation has conducted breeding studies with these deer. The biologists have found that, contrary to the usual laws of genetics, in this case white is the dominant color. Except for being white, these deer are as hearty in all ways as the other deer at the Seneca base. Although their coats are white, their eyes are brown and not pink, as in true albinism.

Albino whitetails are much more common than albino mule deer or blacktails. In October of 1963, Pete Peterson shot an albino muley doe in Ekalaka, Montana. It was a very rare specimen, for biologists think that albinism in mule deer may be as rare as 1 in 500,000 deer.

Occasionally a *melanistic* deer will be seen—one that is totally black. This is a rarer occurrence than albinism, but it has been reported in a number of regions. Melanism is caused by overproduction of the skin pigment melanin.

Archibald Rutledge, the late dean of South Carolina outdoor writers, reported two melanistic deer. One was a small black buck killed near Georgetown, South Carolina. Rutledge did not see this animal, but the reports were reliable. The second was described as "a huge melanistic buck—as black as coal." He saw this deer on four occasions. It had the most magnificent rack of any buck he had ever seen in the Santee River area of South Carolina.

In 1965 Frances Arens of Lansing, Michigan, shot a black spike buck near Roscommon. On the opening day of California's deer season in 1965, Barney Patten shot a melanistic forkhorn mule deer in Modoc County. The same county produced another melanistic mule deer on opening day in 1966—a four-point buck shot by Kim Sigler. It is most unusual for two such deer to be taken in two years in one area.

Yet another oddity of coat was displayed by a whitetail buck killed in 1967 in Dutchess County, New York. This buck, shot by Rudy Selvaggio, was normal in all respects except that it had a short, stiff mane of hair running from its forehead back and down the top of its neck to the shoulders. The ridge of hair looked just like the short-clipped mane of a showhorse.

I have since received many reports and have seen a number of photographs of both bucks and does having manes. This can not be called a common occurrence, but one taxidermist I talked with in Indiana had mounted either five or six such deer that were

Maned deer are not common, but this genetic trait is being found in different parts of the country. photo courtesy Richard P. Smith

taken in his area in that state. A taxidermist in Tennessee also reported mounting a number of maned deer. Having a mane would also be a genetic characteristic passed on by a dominant maned buck. The hair of the manes is usually less than 2 inches (50.8 mm) long. Some of the manes may be only 6 inches (15.2 cm) in length, but the buck killed by Mike Jaskaniec near Blue Mounds, Wisconsin on November 11, 1983, had a mane like a horse that extended from the top of its head to its shoulders.

12

The Five Senses

Constant vigilance is the price of life. Wild creatures are always alert, always monitoring the air for particles of scent, listening for the slightest unnatural sound, watching for the movements of anything within their range of vision. Adult woodchucks, for example, stand upright an average of seven times per minute just to watch for danger over the tops of the vegetation they are feeding on.

All of a deer's five senses are important to its survival, but by far the most important is its sense of smell. This is certainly true of the whitetail and the blacktail deer; it may be less true of the mule deer. Whitetails and blacktails usually inhabit heavy, brushy cover, whereas the mule deer is a creature of the more open spaces. This is not to say that the sense of smell is less important to the mule deer than to whitetails and blacktails; it is to accept that perhaps the mule deer's huge ears and vision are of equal importance.

SMELL

A world of odors encircles and informs most of the creatures that share this earth with us. It is a world we can only guess at, occasionally venture into, and never fully know—not even with the artificial "noses," or scent-detecting devices, that are now being devised. Birds are the exception, in that few of them have a well-developed sense of smell.

The sense of smell is the response to chemoreception by the limbic system, found on the base of the cerebrum, the front portion of the brain. Most odors in the natural world are of organic compounds and are released as molecules of gas. For gases to be smelled, they must be mixed or dissolved with moisture. With the exception of moose and caribou, all members of the deer family have moist, hairless muzzles. The moisture both inside and outside

This deer is tense and alert as he searches for scent, but he is not yet fully alarmed, since the tail hair is still relaxed.

of the nostrils traps scent particles. Deer frequently lick their muzzles, and this provides additional moisture.

The nostrils are lined by the epithelium, containing mucous membranes and sensory nerve endings. Scent molecules are inhaled and dissolved on the moist surface of the epithelium. Responses to the chemical action are carried by the sensory nerves to one of the two olfactory bulbs. A deer's olfactory bulbs are much larger than a human's. The olfactory bulb in turn sends electrical impulses to the brain stem, where the odor is classified. This area of the brain also controls appetite, digestion, and emotions, linking the sense of smell closely to all three.

The average human being can detect skunk odor, or mercaptan, when it is dissolved to $\frac{1}{25,000,000}$ part of 1 milligram. Most of us can identify hundreds of odors, and professional perfumers have committed thousands to memory. What the most sensitive human nose can do pales beside the achievements of dogs. Bloodhounds can, under ideal conditions, follow a track that is two weeks old. But dogs have far more epithelium than humans have. A dog's epithelium equals $\frac{1}{80}$ of its surface skin area, while a human's equals about $\frac{1}{8000}$. A deer's epithelium covers a much larger area than a human's and may be equal to a dog's. Unfortunately, it is difficult to test animals such as deer for their ability to detect odors because deer cannot be trained to respond as dogs do.

The Five Senses

Many variables, including wind, temperature, and moisture, affect scent. The reception and perception of olfactory stimulation is heightened by a fairly warm, moist, still atmosphere. But there is a limit to the amount of humidity that is beneficial. Both rain and falling snow carry the molecules of scent to the earth and dilute and dissipate them. Light mist or heavy fog also blocks scent from traveling great distances. High humidity—that is, 50 to 70 percent—is about ideal for scenting purposes. High humidity makes deer very nervous because more scent is carried to them, and they become doubly alert.

In November 1976, I attempted to photograph deer from a permanent blind that I have on a large estate. The humidity was high, and there was not a breath of air to dissipate my scent, which slowly suffused to a large circle around the blind. The blind is located where the deer are accustomed to feed, and they were hungry, yet they would not venture into my circle of scent.

Low humidity—10 to 20 percent—works against wildlife because the nasal passages have a tendency to dry out, and this hampers the ability to capture and register the scent molecules.

High temperatures cause air convection, and the rising thermals carry the scent molecules upward before they can reach a deer. In the Southwest, on a hot day, I have been able to approach a deer very closely, even with the wind behind me. Extremely low temperatures also handicap a deer because the scent molecules are usually pushed downward.

Air movements play a big role in the deer's ability to use its sense of smell. In most regions, breezes usually move at 3 to 15 miles per hour (4.8 to 24.1 km per hour). Ideal scenting conditions would mean a humidity of 20 to 80 percent, with temperatures ranging from 40° to 90° F (4.4° to 32.2° C), coupled with breezes up to 15 or 20 miles per hour (24.1 or 32.1 km per hour). Personal observations lead me to believe that under such circumstances a deer might detect danger half a mile away and perhaps farther.

Human scent can, however, be masked so that deer are not alarmed by it. As a farm boy, I always rode my horse bareback. I would often bend forward, lie along my horse's neck, and ride very close to groups of deer without their being alarmed.

Deer use their sense of smell not only to locate danger but also to locate other deer. Deer cannot recognize their own fawns by sight but depend upon scent. This can be seen in late summer, when the does and their spotted fawns mingle in groups as they feed. A fawn will try to nurse from any doe. But a doe will turn and smell the fawn. Her own fawns will be allowed to nurse. Any other fawn is usually driven away with hard kicks.

During the rutting season, a buck will track a doe with all the skill of a hound following a game trail. As I noted in the chapter on external glands, deer leave a glandular musk wherever they walk, and other deer recognize the odor.

The sense of smell is also important to deer in locating food. Some foods are rejected because of their odors. Deer can locate such foods as apples, acorns, and corn under a foot or more of snow if there is no crust to lock the scent beneath the surface.

The winter of 1960–61 was an extremely hard one in northwestern New Jersey, and we had deep snow on the ground. In February, when we had our first thaw, the deer came out of the mountains in droves. On Coventry Hunt Club lands, we had planted many game-food patches, and the patches for deer had a lot of rape, a relative of the cabbage plant. The fields were covered with 24 inches (61 cm) of snow, and the deer had to leap to get around at first. They came out of the woods to feed on the rape. They located it under the snow and dug down for it. Within a few days, deer trails crisscrossed all of the patches, and within weeks every rape plant had been located and consumed.

I am a naturalist today because I was a trapper as a kid. I was raised on a farm at the tail end of the Depression, when kids just had no money. We lived well, but there were no extras. To get spending money, I combined my love of being in the outdoors with my ever-increasing knowledge of wildlife and trapped all kinds of fur-bearing animals, especially fox. The use of scent was the most important aid in drawing wildlife to the exact spot where the trap had been placed. I still use all types of scent both in my photography work and for hunting. Many people are amazed that I use scent for photography, but when I am taking photographs, I am simply hunting with a camera instead of a gun. The reason to use scent is the same: to attract the deer.

Hunters, particularly bowhunters, who do not use scents while hunting will not be as successful as they should, or could, be. I am so firmly convinced of this that I am a co-owner of the Helping Hunters Scent Company.

There are basically two kinds of scents: attractants and masking scents. An attractant may be a basic deer odor, a sex scent, or a food scent. Each of these works best at specific times of the year, or several may be chosen carefully and used in combination.

Over most of the eastern half of the United States and on the

West Coast, acorns are probably the preferred deer food. In early bow season, September and October, this is one of the best scents to use because the season coincides with the dropping of the acorns. Oak trees do not bear a good crop every year, yet the deer know where every oak tree is located and check out every one of those trees whether it is a good mast year or not. Acorn scent is always good, but it is even better when acorns are scarce. White oak acorns are preferred over all others where they are available. (My company is the only one that produces a white oak acorn scent.) Apple scents also work well.

The basic deer-odor scent can be used year-round and in conjunction with a food scent. It does not alarm the deer at any time, because it just smells as if another deer were feeding in the area.

I would not use a sex scent except during the rutting season, because this scent, used too early, may scare the deer off. It won't seem natural to the deer if the rutting season hasn't begun.

The human scent cannot be eliminated, but there are many things that you can do to nullify it. Personal cleanliness is the chief way. I recommend that you use an unperfumed soap, such as Ivory. Do not use any lotions, colognes, or perfumed powders. Avoid onions, garlic, and highly spiced foods; they leave an odor that is not only on your breath but will ooze out through your pores. Keep your hunting clothes clean and away from gas, oil, and household odors. Place your hunting clothes in a clean plastic bag and hang the bag out on your back porch, not in your garage. Put your clothing and a clean pair of rubber boots on after you have arrived at your hunting area and are away from your vehicle.

Do not put deer scent, or any other scent, on your boots, foot pads, or your person. You can make a cottonball drag on a string to lay down scent trails. My company offers a sheepskin scent drag that works quite well. I often squirt deer lure on a tree in the area where I want the deer to stop. I *don't* want a deer tracking me to my blind or stand.

It is imperative that your stand be downwind of the deer's trail. For photography, you also have to have the sun behind you or off to one side.

Fox scent placed around your blind or stand helps, because fox are always scenting the boundaries of their territories. The deer are used to the odor of fox and have nothing to fear from them. I do not recommend the use of skunk scent: the only time a skunk discharges its mercaptan is when it is in danger, and all wildlife associate this odor with danger.

On a warm day the molecules that make up a hunter's scent will dissipate by being lifted up and away within an hour after the hunter has walked to the stand. A tree stand has many advantages: it allows the hunter to see farther, to look over intervening brush, and to make slight movements without being seen. On warm days, the height prevents the hunter's scent from getting down to where the deer can smell it. But on a foggy day, the hunter's scent will emanate out of the tree directly to the ground and saturate the area, eliminating all scent advantages.

Most hunters use far too much scent. Just a few drops of scent will attract deer for as long as three or four days under regular fall conditions. Because the deer lure is concentrated, deer will be able to smell it long after your passage through the woods can be detected.

Deer do not panic when they detect human scent; they encounter it too often. Almost all deer come in contact with human scent a number of times each day. But human scent does alert all deer. How they react to it is based on their individual experience with humans.

TASTE

Deer have a well-developed sense of taste, as shown by their decided food preferences. They may also acquire a taste for food to which they do not ordinarily have access. Humans (and perhaps deer, as well) can only taste things that are sweet, sour, bitter, or salty. Our sense of smell is 10,000 times more discerning than our sense of taste. Most of what we *think* we taste, we actually smell. Food flavors depend on a combination of taste and smell. There is no evidence that deer are endowed with any more tasting ability than we have, so undoubtedly their "taste" for foods depends on their sense of smell, too. In northwestern New Jersey, there are more than 650 foods that deer will eat, but there are also many that they will not touch. For example, in different parts of the country, at certain times of year, mushrooms are important deer foods. The deer do not eat the poisonous ones, such as the amanitas, and no one has yet discovered how they can distinguish between them. Smell is the logical explanation.

When I did my basic research for this book, I read a listing of foods eaten by deer in Massachusetts, one list among dozens. I was amazed to find that the researcher had listed spice bush. Spice bush, *Lindera benzoin,* has a very aromatic oil that gives the plant a pungent odor. Early settlers used it to spice up their foods, to cover up the odors of meat that might be tainted in those days of no refrigeration. I had never seen deer eat spice bush. Joe Taylor had never seen deer eat spice bush, and neither of us had ever seen the bushes, which abound in our area, browsed upon. Deer usually avoid this shrub because the volatile oil inhibits bacterial action in their paunches. I first saw deer in our area eat spice bush only in 1988.

HEARING

Deer also have an extremely keen sense of hearing. Like radar antennae, their large ears constantly sweep back and forth, turning, twisting, swiveling to locate the source of the slightest sound. No sound is too slight to attract their attention.

Sound is a form of energy that reaches the ear as cyclic vibrations. With low-pitched sounds, the waves are fairly shallow and wide-spaced. High-pitched sounds compress the width of the waves, forcing them to high peaks, or *frequencies*. The adult human ear can hear in the range of 40 to 16,000 cycles per second. For some people, this range is extended on either end—as low as 20 or up to 20,000. Children can hear much higher frequencies, but they lose this ability with age.

Personal observations lead me to believe that deer have an upper range extending to 30,000 cycles and perhaps beyond. I

An alerted muley doe tries to catch sounds from all directions. Note that one of her huge ears is swiveled rearward, and the other is aimed forward.

A whitetail doe is focused—ears, eyes, and nose—on the direction of a strange sound; these ears actually have 24 sq. inches of reception area per ear, compared to 3½ sq. inches on a human ear. A mule deer's ears have almost twice as much reception area as a whitetail's.

carry a "silent" dog whistle that I often use to get an animal's attention. The human ear cannot hear this very high-pitched whistle but dogs and deer respond to it readily.

The external ear of the deer, the *pinna,* is a skin-covered flap of cartilage and muscle. The ear on an average whitetail is about 7 inches (17.8 cm) long and about 3 inches (7.6 cm) across. It has a cupped structure. The actual surface, if flattened out, is 4½ inches (11.4 cm) across at its widest. This gives the whitetail approximately 24 square inches (154.8 sq cm) of receptive surface per ear. The length I have given is from the skull to the ear tip. This measurement should not be confused with the scientific measurement of the deer's ear, which is taken from the tip to the notch near the base on the inside of the ear. A whitetail's ear is a complete tube for the first inch, so the scientific measurement states the ear to be only 6 inches (15.2 cm) long. Naturally, the various subspecies of the whitetail have different body sizes and correspondingly different ear sizes, but the figures given are an average for the species.

The mule deer got its name from its exceedingly long ears. Particularly on a fawn, the ears seem enormous. The average mule deer's ear measures 9 inches (22.9 cm) long and 6½ inches (16.5 cm) at the widest point when the ear is flattened out. Cupped, the ear is about 4½ inches (11.4 cm) across. A muley's ear measures out to about 42 square inches (271 sq cm) of receptive surface. The average mule deer is larger than the average whitetail, but the ears are disproportionately large. This is because the mule deer inhabits more open country where the need to hear danger at a great distance is more important due to the scarcity of cover.

The exceptionally large ears of the muley may also be a thermoregulator, as on some of the other large-eared mammals. In the summer, the hair on the ears of all deer is exceptionally short and thin. If the ear is viewed with the sun behind it, the blood vessels can usually be seen. Even in the desert, where the sun temperature may be high, the shade areas where the deer bed are much cooler, and there is almost always a breeze blowing. The warm blood, after passing through the deer's ears, would be cooler than its body temperature and would thus help to cool the animal. Conversely, in winter the ears are amply protected with a thick coat of hair.

The average blacktail deer is slightly smaller than the average whitetail, but it has the characteristic ears of the mule deer. The measurements fall about halfway between those of the whitetail and the mule deer. They average 6¾ to 7 inches (17.2 to 17.8 cm) in length and 5½ inches (14 cm) in width, with the ear flattened, giving an average receptive surface of 32 to 33 square inches (206.5 to 212.9 sq cm).

These sizes seem all the more impressive when compared to a human ear, which has only about 3½ square inches (22.6 sq cm) of receptive surface and cannot be swiveled about in the manner of a deer's ear. It is no wonder that deer can catch very small sounds at great distances.

Humans can use numerous devices to hear better, further. If there is something that I have to hear, or can locate only by sound, I use a large parabolic reflector with an amplifying microphone that goes to a headset. I take out my two hearing aids and use just the amplified headset. One of the latest gadgets is a hat that looks like the ones the "Mouseketeers" wore, except that the large ears are so placed as to funnel sound into the wearer's ears. I'm not sure I want to be seen in such a hat, but I am sure it will work. We cup our hands behind our ears to hear better. The deer's ears are naturally cupped.

Regardless of the ear's external size, deer, and humans as well, have an auditory-canal opening about ⅓-inch (8.7 mm) in diameter. Sound waves entering the auditory canal are compressed and directed to the tympanic membrane—the eardrum —

causing the membrane to vibrate. The vibrations activate the three tiny bones of the inner ear, which in turn amplify the incoming sound as much as 90 times. These vibrations cause the hairs in a fluid called *endolymph* to be stimulated, activating impulses of the auditory nerve. The nerve impulses go to the temporal lobe of the brain, which deciphers what is heard.

The inner ear is also important because it helps to establish equilibrium, allowing animals to stand erect without falling over.

Hearing in deer also seems to lead to the "bump of curiosity." A deer becomes curious about the origin of any unidentified sound, though a strange sight or odor seldom makes it yearn to investigate. Times without number, I have watched deer catch a sound and then cautiously advance, trying to locate and identify its source. When deer do this, they are extremely nervous and alert. They stalk forward stiff-legged, with head outstretched and usually lowered to or below the level of their backs, and their big ears are held straight out at right angles to the head. Sometimes the deer advance silently; at other times they stamp each forefoot as it steps. I think this stamping warns other deer of suspected danger and is also an effort to startle whatever made the original sound into betraying its presence. If the deer cannot identify the

A whitetail doe nervously stamps the ground as she advances. Stamping may startle a predator into revealing itself, and serve to warn other deer.

source of the sound, it usually makes a circle so that eventually it will catch any scent of the source.

The deer's brain has a file of stored sounds to which it has become accustomed. Deer are not panicked by common sounds that bombard them constantly. They recognize the scampering of a squirrel in dry leaves, the hopping of a rabbit, the rubbing of one wind-tossed branch against another, the sound of falling fruits and nuts. And they are not alarmed. But they associate danger with the crunching of a human's heavy footfall, the snapping of a twig, or even the whisper of twigs rubbing against hard-surfaced clothing. For this reason, wool is still the best—because quietest—clothing for hunting.

Deer also pay close attention to the warning signals made by the birds and animals in their area. The raucous alarm call of the crow, the strident screaming of the jay, the whirring take off of a grouse, the chattering scolding of the squirrel, the slap of a beaver's tail against the water are all sounds with a purpose. Something or somebody caused these alarm signals to be given. Sometimes the greatest clue to danger is the total *lack* of sound. When all the wild creatures stop eating, stop moving, and become silent, they are listening and watching and smelling for danger. They have detected something that worries them, and their silence is a danger signal to all animals in the vicinity, deer included.

Deer can become accustomed to almost any sounds. When they make a habit of feeding along a highway, they seldom lift their heads no matter how much noise the traffic makes. Deer populations are often high on military reservations that have vast areas of good habitat on which there is no hunting. If these reservations are used for artillery practice, the deer become accustomed to the booming of the big guns and pay scant attention to the noise. Similarly, they may become accustomed to the noise of a power saw at a rural home site or a lumber camp, and have been known to approach rather closely to investigate that sound. Yet they remain alert to other alien sounds, which may portend danger.

Deer do not associate the sound of a rifle shot with danger. The sharp report of a gun may startle them, but because of the short duration of the sound, they may be unable to locate its source. Because of this, many deer that have been shot at and missed may stand long enough to allow the hunter to get off a second shot. Deer almost always try to locate the source of danger before running off. They instinctively know that to run blindly may cause them to blunder right into danger.

High winds make deer extremely nervous, because the crashing of the wind-lashed trees and branches covers almost all sounds, and the eddying and reversal of air currents makes scenting almost impossible. With the whole world in motion, deer are almost panicked. To attempt to hunt in a high wind is a waste of time.

SIGHT

A deer's vision is geared to detect motion. No movement, including the blinking of an eye, seems to be too slight to be noticed. But motionless objects, even if they do not blend with their surroundings, are seldom seen and if seen are seldom recognized. Many times I have been caught out in the open by deer but I escaped detection by remaining absolutely motionless. I am convinced that a deer—unlike a fox or a wolf—does not recognize a human as a human if it cannot detect motion or scent. I am not suggesting that hunters should allow themselves to be silhouetted. Skittish deer may become alarmed just because something new and strange is in their area, even though they do not recognize that something as a person. And I am firmly convinced that deer know their areas so intimately that they are instantly aware when something new has been added. For this reason, anyone photographing or hunting

A deer's eyes are adapted to night vision, which means that the light shined at them in darkness is reflected from the tapetum in the rear of their eyes, almost like a mirror. photo courtesy Len Rue, Jr.

deer will be wise to blend into whatever bushes, trees, or big rocks are in the area. I wear Trebark camouflage.

Humans and the other primates and some diurnal squirrels are the only mammals capable of seeing a spectrum of colors consistently. Tests have proved that most mammals, probably including deer, see the world in shades of gray. Deer do not see the brilliance of the blaze-orange that hunters are now required to wear in most states. The use of blaze-orange has reduced hunting fatalities, because it reduces the chance that a hunter will be mistaken for a deer. Although deer see blaze-orange as a light shade of gray, rather than as a bright color, the deer can see it more easily than the blended shades of camouflage cloth. Moreover, a blaze-orange jacket or jumpsuit is a large expanse of one smooth, solid, uninterrupted shade, unlike almost anything else in the woods, and it therefore looks suspicious. A hunter or photographer wearing such clothing is well advised to pick a stand where vegetation or rocks will break up the expanse of a silhouette, or a stand well above a deer's normal line of sight. The difference between blaze-orange and camouflage is almost nullified if the wearer remains motionless, but how many people can remain motionless?

It is important to remember that when you have been spotted by a deer and are trying to remain motionless, the deer will do everything in its power to get a better look at you or to startle you into motion. Most commonly, the deer will keep its eyes on you but will move its head as far as it can to one side of its body and then to the opposite side. This increases dimensional perception, so that an object stands out from its surroundings. Flying squirrels do this to gauge distance before launching themselves from one tree to another.

If a distant deer spots you, do not try to get out of that deer's sight. If you drop to the ground or disappear down a hollow, when you look again for the deer it will be gone. Almost all animals feel extremely threatened when an object seen suddenly disappears. They feel they are being stalked and they will not stay around to find out if the assumption is correct.

A deer's eyes are located on the sides of the skull but are angled at about 25 degrees toward the nose. Eye placement on the side of the skull is typical of most creatures that are preyed upon. This gives them primarily monocular vision but allows them to see a much greater portion of a full circle so that they can perceive danger from all sides. Most prey animals also have big, bulging eyes to increase their ability to see behind the head. A deer's eyes protrude about ¼-inch (6.4 mm) beyond the skull. The deer cannot see completely behind its head, but it can view at least a 310-degree arc of its surroundings—and probably more due to the eye bulge and curvature. Also important is the fact that the deer's eye placement also allows it to have 50 degrees of binocular vision, which makes three-dimensional perception possible.

An alarmed deer constantly flicks its ears back and forth, often laying one ear back along its neck. Perhaps it is not only trying to locate the sound of whatever has alarmed it, but also getting an ear out of the way for a better view of any danger to the rear.

Ordinarily, it is safe to move closer to a deer when its head is down because then the deer is focused on feeding. You can tell when to "freeze" because all deer invariably jerk their tails before bringing their heads up. However, a deer that is particularly wary may only pretend to feed. The deer will put its head down and then almost immediately jerk it back up to see if the object it was studying has moved. Or the deer may put its head down and pretend to feed but will keep its eye on you. There is no way you can tell when this is taking place beforehand; you will know it only as the deer wheels around and dashes off.

Basically, a deer's eye is like that of a human in that light passes through the cornea and the amount of light admitted to the lens is controlled by the opening or closing of the oblong pupil. The image is then projected onto the retina, which is connected to the optic nerve. There the nerve impulses are transmitted to the occipital lobe of the brain. In humans, more of the total area of the brain is devoted to seeing. Ours is a world of sight, while in the deer's world, sounds and smells dominate.

In proportion to body size, deer have much larger eyes than humans have, and deer can see very well at night. The retina, the receptive surface at the back of the eye, is composed of rod cells and cone cells. Sharpness of vision and sensitivity to color depend on the number of cone cells, or cones, in the retina. The *fovea* of a human eye is a small circle of the retina directly in line with the center of the eye's lens. The sharpness of sight, or *accuity,* depends on the number of cones in the fovea.

Surrounding the fovea are the rod cells, or rods, which are used primarily in night vision and register sight in black and white. That is why in extremely poor light a human can see objects better by looking above, below, or to either side of the object being viewed. This brings into play the rods, which are ordinarily used for peripheral vision.

The Five Senses

To a deer's monochromatic eyes, this hunter wearing a Trebark-pattern coat must look almost as he appears in the black and white photograph here; even his blaze orange hat appears as a shade of gray.

The deer's retina is composed almost entirely of rods. This is why, it is supposed, the deer cannot see color and has poor vision except for movement, yet is able to see at night. The rods also act as a mirror, the tapetum reflecting and doubling the amount of light available. At night some light is always present, although the amount may be too small to be recorded by the human eye. With mammals and birds that have good night vision, the available light goes through the eye and registers on the retina, then strikes the cones and is bounced back through the retina—thus doubling impulses received. This is why a deer's eyes shine when a bright light is directed at them at night. If the deer is looking directly at the light, the eyeshine is a silvery-white. If the deer is not looking directly at the light, the eye acts as a prism and the reflective color may be red, orange, yellow, or green. In the daytime, a deer's pupil is usually a dark bluish brown. About half an hour after death, the pupil turns green.

Recent research on elk using the electron microscope showed that elk have more cones in their eyes than was previously acknowledged. This led the researchers to claim that perhaps elk could see more color than was previously thought. By extrapolation it was theorized that, if elk could see some color, which was not a proven point, then perhaps deer, too, could see color. The consensus today seems to be that perhaps deer can see some color in dim-light situations. Deer cannot see colors at night, because color is a reflection or a refraction of light, and there is infinitely less light available on a dark night than in bright sunshine. On the other hand, deer cannot see color in the bright light of day because the light intensity nullifies the action of their rod cells. Only in dim light is there a chance that deer can see color, when their rods and the small number of cones they do have can work in conjunction.

I myself think that deer, and most mammals, see their entire world as shades of gray at all times. I base this conclusion on the fact that there are no brightly colored mammals. There are brightly colored birds, reptiles, insects, and fish, all of which are known to see color.

If deer can see some color, they are probably dichromatic, that is, can see two colors. The retinas of dichromatic creatures compare short and medium wavelengths of the color spectrum, not utilizing the long wavelengths that are so important to diurnal creatures such as humans. Yellow is the lightest color the human eye perceives (white is the absence of color). The brightest color nocturnal creatures such as deer can perceive is green, which registers as a very light shade of grey. The most prominent color in the deer's world in late spring and summer is green; in late fall, winter, and early spring, it is gray.

While doing my wildlife photography, I always wear camouflage; I have found Trebark to be the best pattern. The camouflage was originally designed for spring turkey-hunting, but it has now become the most popular camo for the bowhunters as well. It allows me to blend in with any woodland habitat.

Deer do not usually look for danger lurking above them. Mule deer and blacktails tend to look up more than whitetails because they are hunted by the cougar, also called the mountain lion. A cougar may be in a tree or on a rocky ledge above them.

Until recently, the whitetail seldom, if ever, looked up, and a tree stand well above the ground had several advantages. All of that has changed. Today the whitetail looks up more often than either the mule deer or the blacktail, because it has now discovered predators, human predators, lurking in just about every other tree. As I have said so many times, the whitetail is one of the most adaptable of creatures. As hunters took to tree stands in droves, the whitetail took to looking up in the trees to watch for them.

Whereas I once could state that a tree stand 10 feet (3 m) above the ground would probably escape detection, today I have to say that the minimum distance would have to be at least 20 feet (6.1 m). In some areas where deer are hunted extensively for three or four months at a time, hunters are finding that their stands have to be 30 feet (9.1 m) or more off the ground to be effective. Nevertheless, hunting from a stand is probably still the best method a bowhunter can use on whitetails.

TOUCH

Relatively little has been written about the fifth sense—touch. We are only beginning to realize how important the sense of touch is to deer. From the moment of birth, a deer is exposed to tactile stim-

A doe washing her new fawn; this scent-bonding is important since does cannot recognize their fawns by sight.

Bucks engaging in mutual grooming.

ulations. The fawn is licked clean by its mother's tongue, and touch, coupled with sight and scent, helps the doe recognize her fawn, linking the two to each other. The bond is continuously strengthened as the fawn and the doe touch noses or lick each other in gestures of affection. (I make no apologies for using the anthropomorphic term "affection.") Sexual stimulation is increased prior to mating when the buck and doe rub against each other, and when the buck licks the doe's vulva. Strong ties are established and maintained between bucks by their mutual grooming of one another. I will show in the next chapter how deer use a combination of senses—including touch—to communicate with one another.

13

Communication, Instinct, and Intelligence

I noted earlier that when a deer becomes alarmed but cannot identify the object of its suspicion, it stamps a forefoot. It may stand in one spot and repeatedly stamp while doing all it can to discover what alarmed it by using its nose, ears, and eyes. Or the deer may advance cautiously, stamping one forefoot and then the other. The tremors set up by the stamping travel a long way through the ground, alerting all deer in the area. Some of the deer may be feeding or screened by vegetation from the alerted deer, and they cannot see a visual signal. But even if they cannot hear the stamping, they can feel the ground tremors. Communication has been achieved through the sense of touch.

An olfactory warning may also be involved. Each time the deer stamps, there is a probability that scent from the interdigital gland is placed on the ground. I have often noticed that deer put their noses down to smell the spot where a previous deer has stamped its foot. They do not seem unduly alarmed, but they take note of it.

Caribou definitely use the scent of the interdigital gland as a warning, and it is heeded. When caribou are moving along and one is alarmed, it rears up on its hind feet, whirls about, and dashes off. Dr. William Pruitt, Jr., an Alaskan researcher, calls this the "excitation jump." When a caribou does this, scent from the interdigital gland is deposited on the ground. I have witnessed this jump many times while photographing caribou, and I know that every caribou that comes to the spot will sniff the scent. In my experience, every caribou then became excited and none ever advanced beyond that marked spot. It was as effective as if a fence had been erected.

Deer that are suspicious of danger also flare their tarsal glands into rosettes, releasing both the gland's scent and the concentrated odor of urine with which these glands are saturated.

The biological definition for territory, according to my *American Heritage* dictionary, is "an area inhabited by an individual animal or a mated pair, or group of animals, and often vigorously defended against intruders." In the strict sense of this definition, the whitetail cannot be described as having a territory. The one exception to this is during the doe's birthing period, and we will go into that later.

Whitetail deer do have a home range of about 1 to 2 square miles (2.6 to 5.2 sq km). During the rutting season, the bucks extend their travels, in search of receptive does, by five to six times, so that they cover 10 to 12 square miles (25.9 to 31.1 sq km). Mule deer, not counting their migratory route, have home ranges almost twice the size of a whitetail's, while the blacktails have slightly smaller ranges.

Scrapes and Rubs. Prior to, and all during the rutting season, whitetails, both bucks and does, depend heavily upon what might be called advertising. We have already discussed both buck rubs and the deer's forehead scent glands. The bucks make the first rubs to peel the velvet from their antlers. They often make rubs when fighting with the bushes, strengthening themselves. The rubs are also made to advertise their presence, because the inner white wood can be seen for a long distance in the woods, where there is nothing as bright to compete with it. While making these rubs, the bucks inadvertently, or perhaps deliberately, deposit scent there from their forehead glands.

Twenty years ago, almost nothing was said or written about deer scrapes. Deer have probably always made scrapes; why was no significance attached to them? I first became aware of scrapes after reading William O. Pruitt's account of how a buck made scrapes on the E. S. George Reserve in Michigan in 1952. Even when I wrote the first edition of this book in 1977, I devoted only two pages to scrapes.

My awareness of scrapes was undoubtedly delayed because of the fact that, in my area, the northeastern United States, the bucks don't make as many scrapes as they do farther south. The bucks in my area don't have to advertise their presence as much as do the southern deer, because they don't encounter as much competition for breeding does as do the southern deer. And make no mistake about it, bucks make scrapes to advertise their presence and availability to the does.

Communication, Instinct, and Intelligence

In no part of the United States is the deer-hunting pressure more intense than it is in the Northeast, simply because no other area has such a high concentration of human population and correspondingly large number of deer hunters. Most of the northeastern states allow hunting basically on a "bucks only" system, though most also hold special days to reduce the doe population, according to the dictates of management. Even when either sex can legally be taken, as in the bow and arrow season, most hunters want a trophy, not just the meat. The "bucks only" system combined with some hunters' "macho" image and, in some cases, misinformation, results in deer herds that are skewed dramatically in favor of the females. The ratio is often three or four to one, and, in extreme cases, may be as high as six does or more to one adult buck. In the south, and particularly in Texas, where the deer are heavily managed, the sex ratio is closer to 50/50.

In my home area of northwestern New Jersey, I am lucky if I encounter five major scrapes in a day's scouting. In South Carolina I saw as many as ten primary scrapes within a mile. The scrapes were in a row, along the edge of planted fields adjoining forests. I also saw many secondary, or "pause," scrapes.

A pause scrape is made when a buck suddenly stops or pauses momentarily, then scrapes, rakes, or paws at the ground with either or perhaps alternate front feet. He may or may not rub-urinate (rub his tarsal glands together and then urinate on them) over the pawed area.

Almost every primary scrape I have seen has been beneath an overhanging branch, usually a branch 5 feet (1.5 m) off the ground,

The primary scrape is always made under a tree with an overhanging branch that is low enough to be reached by the bucks, so it can be used as a licking branch. The twig here has been broken many times by different deer hooking it with their antlers and chewing on it.

low enough for the buck to chew on it. Some hunters overlook the overhanging branch, especially when it is 7 feet (2.2 m) above the scrape and the buck has to stand on his hind legs to reach it. Occasionally the branch is so high that the buck can't chew on it at all, but he can reach it with his tongue to lick it. The significance of that overhanging branch is that it is the main depository for the individual scent from the saliva of the buck that made the scrape; it's his calling card.

Some scrapes become traditional: the sites are used year after year by succeeding generations of deer. Some are used for just one season.

Primary scrapes are usually made on, or near, the main deer paths. Deer, like humans, tend to do things the easy way. While feeding upon acorns or browsing, the deer wander about through the woodlands in what appears to be an aimless fashion. When they are moving from their bedding areas to their feeding areas, they follow trails, because it is easier for them to do so. These trails will invariably be the paths of least resistance, although not necessarily the shortest distance between the two locations. Most mammals have the built-in instincts of a highway engineer when it comes to laying out a trail with the easiest gradient to negotiate elevation. (In fact, many of our modern highways are laid out on ancient game trails.) The bucks make their scrapes on, or near, the main trails because most of the deer in that area will be using those trails. No one puts billboards on rutted side roads; they are built on major highways for everyone to see.

When a buck makes a scrape, or when he reuses it, his actions usually follow the same sequence. The buck often starts by hooking the overhanging branch with his antlers, especially if the branch is more than 5 feet (1.5 m) off the ground. This hooking is done to break some of the overhead branch to get part of it to hang down. If it is low enough to be reached easily, the buck takes the branch in his mouth and chews on it gently. He is not attempting to cut the branch off; he crushes its twigs with his premolars and molars, covering them with his saliva. If the branch is a bushy one, he then rubs both his preorbital and forehead scent glands all over it.

I realize that I will be accused of anthropomorphism for saying that the buck is in ecstasy while he rubs his forehead scent glands on the branch, but he is. Depositing scent from these glands is sexual, it is sensual, it is a means of gratification.

After several minutes of rubbing with his head glands, the

Communication, Instinct, and Intelligence

The ritual visit at a scrape usually involves all three steps seen in this series of photos: first the buck "signs in" by chewing on the overhanging branch. He paws in the scrape, and then steps into the scrape and rub-urinates over his hocks and tarsal glands. photos courtesy Len Rue, Jr.

A buck rubbing his forehead glands onto the branch above a scrape.

buck begins to paw the scrape with his forefeet. I find it interesting that most bucks paw first with their right forefeet. The reason for this is that about 85 percent of wild creatures, like 85 percent of humans, are right-handed.

If the scrape is a new one, it will take considerable pawing to open up the ground directly beneath the overhanging branch. Usually the buck will paw three or four times with one foot and then three or four times with the other. On a new scrape, the buck may turn around on the spot while scraping. The actual scrape may be anywhere from 18 to 36 inches (45.7 to 91.4 cm) in diameter, forming a very shallow saucerlike depression in the earth. Usually the earth is simply pawed back from the center, but I have also seen the earth thrown backward for a distance of 5 to 6 feet (1.5 to 1.8 m). A primary scrape will be regularly "cleaned" by pawing.

The pawing completed, the buck will stand with all four feet in the center of the scrape, then balance himself on his two front feet, rub the hock or tarsal glands of his hind feet together, and urinate on them. The urine and the scent from the tarsals is thus washed down his feet to the scrape beneath. The more dominant a buck, the more frequently he rub-urinates, in some cases staining his feet from the tarsal glands to his hoofs. The buck may also defecate, and occasionally even ejaculates by masturbation. So in the rut, a primary scrape has a well-turned "muddy" look.

Most scrapes are made by the dominant bucks, but they do not get to use them exclusively. Almost every buck that comes by will also stop to urinate in the scrapes, although the lesser bucks may not go through the entire procedure. It's a little like dogs urinating on car tires. The dogs urinate on the tires to include the car in their personal territory. But all passing dogs, not just the dominant territorial dog, also urinate on the tires, making those tires an olfactory bulletin board where every dog can post notices.

Lesser bucks will approach a dominant buck's scrape cautiously, and usually from the downwind side. They don't want to get caught adding their urine calling card if the dominant buck is in the area. Several bucks from the same bachelor group may use the same scrape, with the dominant buck using it first.

There are several records of does making scrapes and several of does pawing and urinating in a buck's scrapes. I have never seen a doe make or paw in a scrape, but I have on two occasions seen a doe urinate in the dominant buck's scrape. Both occasions were during the breeding season, and the does were advertising the state of their estrus, looking for a mate. A doe in estrus is as anxious to be bred as the buck is to breed her.

As I have already mentioned, the bucks usually check out their scrapes each time they pass by. Do they go through the entire procedure, as described, each time they check out a scrape? In the Northeast they may, because they don't have all that many scrapes to check out. In the South a buck cannot possibly paw and urinate in each of his scrapes each time he passes; he just wouldn't have enough time and probably not enough urine. When urinating in a scrape, a buck does not empty his bladder, as he is prone to do after arising from his bed; he urinates just enough to leave his scent. Dogs do the same thing, just a squirt here and a squirt there. The lesson here for hunters is to put out bottled lure sparingly. You don't have to saturate the area, just leave a definite trace of the scent.

If you want to see whether the buck is using his scrapes on a regular basis, just cover a scrape with dead leaves. Most of the time, the buck will paw the leaves back out. But don't give up on the scrape even if the leaves are not removed at first, because the buck will often check out the scrape, to see if a buck has been there, without stopping to go through the entire procedure.

Many times during the hunting season, hunters will hunt a particular scrape for a week and never catch sight of the buck. It may be that the buck is coming to the scrape only under cover of darkness. I want to make a point here that I think is of great importance. I am firmly convinced that the really big trophy bucks, where there is heavy hunting pressure, move only at night, even in the peak of the rut. It's how they get to be big trophy bucks. When they are seen moving in the daylight hours, it is because they have been pushed out of their places of concealment by some hunter or other disturbance.

It must also be remembered that during the rut the bucks have increased the area that they cover by five or six times. Covering so much additional territory may keep the buck from getting back to his home range immediately.

I am often asked whether bucks have a specific breeding area, and I have to say that I don't think they do. The scrape is going to be located in the general area containing the largest number of does. The does in any given area are going to be in the vicinity of the greatest amount of available food at that particular time. So it is the food source and not some specific preferred area that determines where the bulk of the breeding will take place.

There is probably no hard-and-fast time frame within which a buck will visit his scrapes, but I think that unless the buck is killed he will come back to his scrapes eventually. It may, however, take longer than the hunter can devote to watching one particular scrape.

I have discovered that bucks will rub their scents in two quite similar circumstances that I have never seen described in print. I

This buck is depositing scent from his forehead glands onto a horizontal low branch of an evergreen.

This buck is rubbing the licking stick between his ears and antlers, depositing his forehead scent on it. In the second photo, he is licking it off. He did this repeatedly while I photographed him. Other deer, including does, will be equally stimulated to this behavior.

discovered both while taking deer photos, and if I had not witnessed bucks making use of them, I probably would never have been aware of them myself. (It must be remembered that in doing photography, I spend countless hours sitting in blinds looking for deer, or following deer.) I call these situations rubbing sticks and licking sticks. They are not to be confused with the regular buck rubs on the branches overhanging scrapes. There is no evidence of scraping or other disturbance to the ground or to the leaves in the vicinity of these marking sticks. Their sole purpose seems to be the placement of scent from the forehead and saliva glands.

In the first situation, the deer selects some tree branch, usually one horizontal to the ground about 4½ to 5 feet (1.4 to 1.5 m) above the ground. Most of those I have seen have been on some evergreen, because evergreens usually have a number of radial branches growing horizontal to the ground. The end of the branch is usually broken off and the bark removed. The buck rubs his head back and forth on the end of this branch, rubbing the glands between his ear and his antler.

In the second case, a small sapling about ½- to ¾-inch (12.7 to 19.1 mm) thick has been broken off by the buck's antlers about 30 to 36 inches (76.2 to 91.4 cm) above the ground. The top 6 inches (15.2 cm) of the sapling stub will be bright where the bark has been rubbed off by the buck's antlers. As on the horizontal branch, the buck rubs the glands located behind his antlers and in front of his ears on the sapling stub. Although I have not seen bucks lick the horizontal branch after depositing scent on it, they always lick the sapling after depositing the scent.

I watched one buck deposit his scent, then lick it off, deposit more scent, and lick it off, for a period of at least fifteen minutes. The significance of the scent marking on the sticks is that it stimulates other bucks to do the same thing. On one occasion I saw a doe lick at one of these sticks, although I did not see her deposit any scent there except that which was in her saliva.

I have found only a dozen or so of these scent saplings, but know I must have walked past many more without recognizing what they were. Having never heard anyone else tell or write about them, I have to assume that everyone else is just passing them by, too. I predict that once we know more about these horizontal branches and broken-off saplings and their role in marking communication, we will find that they are as good a spot over which to hunt as are the scrapes.

Body Signals. A deer that is alarmed often stands with its body tensed and rigid, leaning forward. The head and neck are extended and held below the level of the back, and the head is bobbed up and down. This behavior instantly alerts every deer witnessing it to potential danger. It is an effective visual signal.

A whitetail investigates danger with characteristic body language: head extended below the level of the back, head bobbing up and down. photo courtesy Irene Vandermolen

Here a blacktail deer trotting with his tail held in the characteristic just-under-horizontal position. Note that this mule deer subspecies' antlers are hard to distinguish from the whitetail's.

The best-known visual signal is the flashing tail of whitetail deer. Mule deer do not raise their tails when they run, nor do they signal with them. Blacktails usually run with their tails held below the horizontal line of their backs. When they raise the tail above the horizontal, I have noticed that they often cant the tail forward much more than the whitetail is able to. The blacktail deer does not wag its tail when it runs, nor does it flare the hairs of its tail.

Both the whitetail's tail and the white hairs of its rump are used for signaling. The pronghorn antelope is often called the "heliographer" because it is able to erect outwardly its big white rump patch. This large rosette reflects a lot of light, and the raising and lowering of the hair causes flashes. On the open plains, home to the pronghorn, reflected light can be seen at a distance over which the unaided human eye cannot see the animal. And so can the light reflected from a whitetail's rump and tail.

Most artists portray a whitetail buck running with his tail held jauntily aloft. That bucks often do run like this can be proved by many photographs. I have found, however, that bucks will run with their tails down as often as with them up. Does almost always run with their tails up, and they usually wag them loosely from

A pair of whitetails tensed and ready to flee: note the buck's flared white tail hair. When they finally broke, the buck lowered his tail while the doe kept hers up.

Communications, Instinct, and Intelligence

side to side. I think that the doe's bouncing white tail is to guide her fawns as they follow her.

In the chapter describing the tails of the three types of American deer, I mentioned that behaviorists have recently come up with the theory that the whitetail's "flagging" is also a message to a predator that its whereabouts have been discovered. Predators that stalk to get close to their prey before attempting to capture it will frequently abandon a hunt if they find they are discovered, because the chance of success is then too poor. So the whitetail's flashing tail may not only warn other deer but also inform the predator that all of the deer in the area know about its presence.

In the mid-1800s, when Charles Darwin first wrote his book *The Expression of the Emotions in Man and Animals,* the term *behavioral scientist* had not yet been coined. Since that time, William Hornaday, Konrad Lorenz, Nikko Timbergen, Karl von Frisch, Robert Audrey, Desmond Morris, John Alcock, John Scott, David McFarland, and many, many others have documented the visible expression of emotions of wildlife. Fast, Nerenberg and Calero, Pease, and others have documented the visible expression of emotions in man. All creatures, including humans, are constantly giving off visible clues to their emotions. In the case of humans, it is well known that these visible clues may directly contradict what is being said. This so-called "body language" is of extreme importance because it is the more truthful expression. It is almost always produced instinctively, without conscious effort, and accurately depicts the emotions felt. Humans can subvert or modify this "language" if they are conscious of it, but with wildlife, it means what it says. Nevertheless, even with animals there may be some confusion among the recipients of the message, although only one message may be intended. You have undoubtedly encountered a dog that was barking, even snarling, but that was, at the same time, wagging its tail. Your quandary was in knowing which end of the dog to believe; it was sending out mixed signals.

The book by Darwin is the first one written on the subject that I know of. I'm sure that extensive research would show that Aristotle or others of the ancient Greeks, who were astute observers of wildlife, wrote on the subject. Humans have always been aware of the "body language" of both man and animal, probably even more so eons ago than we are now, because early humans lived closer to nature and had to be forewarned of the actions of wild animals in order to survive. As our spoken language increased, our dependency on the visual language deteriorated. In my book *How I Pho-*

tograph Wildlife and Nature, I spelled out the most common danger signals of the various types of big game in North America, including the deer. Because the wild creatures have not increased their vocalizations, they are as dependent as ever, or perhaps even more so, on visual signals.

People have always commented that deer seem to know when they are being hunted and when people are just observing them. I myself have found that all types of wildlife avoid me as much when I have a camera as when I have a gun. This all has to do with how a prey species perceives a predator. When we are hunting with a gun, a bow, or a camera, we all walk the way a predator walks when it is hunting its prey. Our entire mien, our attitude, our very posture, is more tense; we are more alert, we move more cautiously, more deliberately, and we send out all these signals to the prey species. Predators don't just saunter through the woods, we don't leisurely stroll: we move with determination, and all prey species are instantly on guard.

Olfactory signals, the various scents, are sent out on a "to whom it may concern" basis. These are extrinsic signals; they serve as a surrogate for seeing the sender. The sender may have a definite receptor in mind, but whether or not the receptor is present, the signal goes out to whatever creature encounters it.

Visual signals are usually sent directly to the receptor. They are usually, but not always, given when the receptor is in sight. These are intrinsic signals, given with the body. A whitetail buck, during the rutting season, may erect the hair on his body when he encounters the odor of a rival buck, even though the rival is not in sight. You know how varied your own emotions may be when you respond to a letter, a phone call, a photograph, a song, and so on, even though the person stirring these emotions is not present.

The response to any given signal will vary, depending upon the receptor and its interpretation (or misinterpretation). How many times have people responded entirely differently than you thought they would to what you said or wrote? How often does a whitetail buck send out signals, intending to dominate another buck, only to be met with an act of aggression showing that the other buck has no intention of submitting? Many signals are given by a sender anticipating a certain response from the receptor; but that does not guarantee that the response given is the one the sender wants to receive.

It is very important to remember that most of the signals given by wildlife, particularly deer, are given to convey the message that establishes dominance, but with the greatest conservation of energy and the least likelihood of a physical encounter. Most wild animals do not want to fight; fighting can cause injury and perhaps death. Every creature wants to be dominant, to be top dog, the boss, and wants the world to know and to recognize this. The dominant male among the different species of wildlife has the first chance at available food resources, the choice of the best habitat, and the opportunity to pass on his genes by breeding—and sometimes the opportunity to prevent the lesser males from breeding at all. Among humans, the most powerful urge is survival, and then dominance; these two may, at times, be reversed. At times survival depends upon dominance. Dominance, that is, power, is much more important to humans than is either sex or money, because usually these two are prerogatives of power.

The most obvious sign of dominance among many of the wild creatures is simply an attribute it possesses: size. With bears, for example, it is actual body size that determines which bear is boss. With horned and antlered animals, the animal with the largest body usually also has the largest set of horns or antlers. If two animals have approximately the same body size, the one with the largest horns or antlers will be dominant. About the only time this

A dominant buck scatters the does with a "hard look": the fact that his ears are flat back against his neck suggests there's a rival in his view.

rule is broken is when the larger animal is past his prime and a mature but younger rival with smaller horns or antlers challenges the older animal on the basis of strength.

Both moose and elk rock their huge racks of antlers from side to side as they approach a rival, while deer tuck in their chins and tilt their antlers forward. Both of these gestures are made to ensure that the challenger sees the size of the dominant animal's horns or antlers and has the opportunity to withdraw while he is still able.

The most common visible sign of aggression among whitetails is called the "hard look." This is a low-intensity threat given by both does and bucks to any deer they intend to dominate. The dominant deer stares hard directly at the lesser deer. At the same time, the dominant deer lays its ears back along its neck. If the hard look does not achieve the desired result, the dominant deer will walk purposefully toward the lesser deer. The subordinate responds by turning or backing away with averted eyes, so as not to look at the dominant deer. The lesser deer's ears are also put back, but up. If the hard look is ignored or not responded to immediately, the dominant deer will often drop its head well below the level of its back, and extend its head toward the subordinate. This posture is usually followed by a charge.

There are two variations on this behavior. Any dominant deer uses it just as described against a lesser deer; a buck uses it just as described against a rival. But it is also used by a buck approaching a doe to check her for estrus; the only difference being, in this latter case, that the buck's ears are held level and forward instead of back against his neck. When approached by a buck like this, most does run off. Whitetail bucks are known for chasing the does. If the does weren't so used to being chased off with this aggressive approach by every other dominant deer of either sex, perhaps the bucks wouldn't have to chase as much or as far. Chasing, either aggressively or sexually, is done with the head held low in the extended-neck threat position.

The hard look is also used in conjunction with the head held high, the ears laid back along the neck, and the chin tucked in. When does use this approach to a subordinate, it usually precedes their striking out with one forefoot at the lesser animal or their rising up and flailing out with both front feet. A buck without antlers will follow through exactly as does a dominant doe. If the buck has antlers, he will tuck his chin in so that it projects his antler tips forward. He then usually launches an attack.

Both these bucks are employing the "hard look" as a challenge to dominance. Since neither one shows signs of submission, a fight will probably ensue. Note that both buck's tails are tightly clamped.

The buck on the left is dominating the other with the "hard look," backed up by other signals of aggression: lowered antlers and laid-back ears. The lesser buck is acknowledges submission by averting eye contact as he shows a humped back and backs away from the aggressor.

A whitetail trailing an estrous doe with lowered head and his ears low, but cocked forward. Note the doe's ears are trained to the rear.

When two equal bucks challenge one another, they use a combination of visual, auditory, and olfactory signals. As they approach one another, they give each other the hard look with their ears laid back along their necks. They tuck their chins in and project their antlers forward. They use muscles called *erector pilli* muscles, which lie just beneath the skin, to stand all the hairs on their body on end. This makes them appear much larger than they actually are, an important consideration when size alone often means dominance. They sidle up to one another, walking in a small circle of about 30 feet (9.1 m). They walk with a shortened, stiff-legged stride, almost as if they were crippled. This circling is to make sure that their opponent has every opportunity to see how big and dangerous they are, in a last effort to intimidate the other. While circling, they either make a low grunting sound or emit a high-pitched whine. If neither buck gives way, they crack head on and the battle is joined. We will discuss deer fighting in a later chapter; here we are discussing the signals leading up to a fight and the ones used to prevent one if possible. These signals are occasionally seen among blacktails and mule deer, as well.

When a whitetail is in hiding, the white hair on its rump is turned inward and the tail is pressed down so that only the dorsal

Communication, Instinct, and Intelligence

brown hairs are visible. This position almost completely covers the white that is usually seen when a deer is viewed from the rear.

The most widely known sound of all three deer is their explosive, blasting snort. They make this snort with their mouths closed, expelling air forcefully through the nasal passageway and thus causing the closed nostrils to flutter. Deer frequently snort when they are surprised or startled. And their snorting has probably surprised or startled everyone who has ever spent any time in deer areas.

When a deer snorts nasally, all the other deer take instant notice, and most of them prepare to flee the area. Deer have another, higher-pitched, whistling snort that is made with the mouth open and is combined with a vocal sound. Upon hearing this higher, whistled snort, every deer in the area explodes into action. There is no hesitancy. That snort means danger, and in seconds the deer are all gone.

Indians have been using calls for deer hunting for eons, and a large number of good commercial calls are now available to hunters. The first commercial calls were patterned after the ones used by the Indians and imitated the raspy, low-pitched blatting call of

In his body language, this mule deer buck is showing many signs of extreme aggression: all of his body hair is standing on end; his chin is tucked in projecting his antlers forward, he is walking stiff-legged, and his tongue is flicking out of his mouth. This is a very dangerous deer.

This buck exhibits the unusual sideling gait of a very aggressive buck: all his hair is standing erect, and his head and ears are lowered. He is offering his opponent the last chance to back away from a fight—which could ensue with a quick lunge if he does not.

the bucks. Anyone who has ever heard an old domestic ram blat will know exactly what a buck, and these calls, sound like. Does also blat, but use a much higher tone. Both bucks and does will respond to the calls used for each sex. There are also commercial calls that imitate the regular snorting of the deer. Whereas snorting is often used to signify danger, it also invokes curiosity. Many times deer snort when they don't know what it is that frightened them. Upon hearing the snorting call, deer will come toward the caller to see if they can locate what it was that frightened the deer they think made the snort.

Some of the latest deer calls imitate the deer's grunting. Does have a soft grunt that they use when conversing with their fawns. The buck's grunt is much lower in tone and is given with much more volume. When a buck is tracking a doe about to come into estrus, he follows her trail with his nose to her tracks, his tail raised straight up, grunting every couple of steps.

Recently, in Pennsylvania, I watched a dominant buck simultaneously pursue a doe in estrus and chase away two lesser bucks. He really had his work cut out for him, because the young bucks constantly tried to cut the doe away from him. The big buck was steadily on the run, and with every jump he grunted loudly in both

Communication, Instinct, and Intelligence

rage and frustration. I observed the deer for about 15 minutes, and he grunted continuously the entire time.

Using a variety of these calls, I have been able to lure both bucks and does within range of my cameras. Calls should be used sparingly and the factors of wind and concealment must also be right, but if there are deer in the area, they are likely to come to investigate the sound.

The doe makes a very soft, catlike, mewing sound when she calls to a hidden fawn. I first heard this sound while I was in a blind I had placed in the woods so I could photograph red-eyed vireos. I did not know a fawn was in the area. As soon as it heard the call, the fawn responded by jumping to its feet and going to the doe to nurse.

Fawns bleat, making a sound more like that of a calf than a lamb. It is higher-pitched than the sound a calf makes and lacks the volume. It does not vibrate the way a lamb's call does. Fawns and young deer that have been injured just bawl loudly in terror.

Although it cannot really be labeled communication, the sound of deer's antlers being clashed together, either in a real fight or by a hunter simulating a fight, is a sound that attracts deer. We will discuss rattling in the chapter on the autumn cycle of a deer's life later in this book.

I suspect that the means of communication I have described above are only the most easily detected ones. We have only begun to know the means of animal communication. The communicators I've mentioned are the obvious ones, the ones I have seen, heard, or smelled. There must be many less obvious signs and factors that I am missing. I am sure we have far more to learn about wildlife communication.

One often hears about the sixth sense of wildlife, a feeling that warns animals of impending danger without any of the obvious clues. This exists. I have experienced it myself, though I cannot fully explain what it is. Even with a hearing aid I miss some of the sounds in the outdoors. While photographing wildlife, I am often exposed to potential danger. At such times, with no sound to warn me, my body evidently sets up some kind of electrical field that is sensitive to vibrations in the area around me. My body hair stands on end and I tingle. I feel more "alive" than at other times. A number of other people have reported similar sensations when exposed to danger that they could not have detected by ordinary means.

I have seen sleeping cats awaken when a mouse showed itself in our barn. I once watched a fox, sleeping in a washed-out gully in Alaska, awaken because a wolf trotted over the horizon a third of a mile away. The fox could not see over the gully, the wind was quartering so that the wolf's scent was being blown away, the distance made it impossible to hear the wolf move on the soft tundra or to feel any vibrations. Yet the fox awoke like a spring uncoiling, and was gone before it could even verify the wolf's presence. That fox was warned, it heeded the warning and lived.

Many tests, particularly in the Soviet Union, have indicated that animals have some psychic means of perception and communication. Does this sound incredible? New fields of knowledge usually do. I am convinced that as our knowledge of psychic phenomena increases, our knowledge of wildlife communication will also increase. We must learn more about the fabled sixth sense, intuition.

When confronted with danger, an animal usually has five options. It may hide or remain hidden. It may flee. It may cry out as a warning to others of its kind or to let the "danger" know it has been detected. If young are involved, a parent may either try to lead the enemy away or perform a distraction display. Or it may threaten and bluff or actually attack.

Considerable research has been done on "flight distance" and "attack distance," especially with the larger, dangerous animals such as bears, lions, and tigers. A person can approach a dangerous animal to a certain distance before that animal feels acutely threatened. Up to that distance, the animal feels instinctively that it still has the option of running away. When the distance is shortened, the animal feels it can no longer count on flight to save itself. It feels "cornered," even if it is out on an open plain. That is when the animal attacks.

The deer's adrenal glands are astride each of its kidneys. When a deer is frightened, threatened, or enraged, the inner area of these glands, known as the *adrenal medulla,* secretes a hormone known as *epinephrin,* or *adrenaline,* into the bloodstream. This hormone gets the deer's body ready for "flight or fight" situations. In the deer (and in humans), the pouring of those hormones into the bloodstream causes the heart rate to increase. Increased heart rate, with the constriction of the arteries, causes the blood pressure to go up. The hormones cause the liver to mobilize glycogen and fat stores, which, in turn, releases the glucose and fatty acids needed for energy and increased oxygen consumption and heat production, so that the body can respond to whatever situation arises. The deer becomes extremely high-strung, with increased

Communication, Instinct, and Intelligence

awareness. This same hormone will increase blood coagulation so that, if the deer is wounded, it will suffer less blood loss. Peristalsis, the muscular contraction that moves food through the intestines, is inhibited while the stomach muscles are tightened.

A deer unsuspecting of danger when shot will bleed out faster and drop sooner. Its meat will taste better and be more tender. A deer that has been frightened and is running with epinephrin coursing throughout its system may run on for half a mile even after being shot through the heart.

Deer seldom attack, but they too have a flight distance that varies with the pressure the individual deer has been subjected to. Most hunters and all wildlife photographers become very aware of the flight distance of animals. I can "feel" when I've approached as close to a creature as it will allow. Most wild animals have become so used to automobiles that it is possible to get closer in a car than on foot. The animals don't feel threatened by the car.

Argument goes on endlessly as to whether all of a deer's actions are instinctive responses to situations. It is the argument of "purposive actions" versus "reflexive actions." Behaviorists have found deer to be very poor subjects in a laboratory. The deer are too wild to cooperate, and they are not interested in being rewarded for something they accomplish. Removing a deer from its natural environment to the artificiality of the testing laboratory just hasn't worked. Conclusions about the mental processes of deer must be made from observations in the wild.

From the thousands upon thousands of hours that I have spent working with deer, studying, observing, and photographing them, I believe they often show a response to new situations that is akin to reasoning. Of course, most of their actions are instinctive. Yet I have often seen deer react to an unusual situation by coming up with a solution that could not be instinctive. Deer do have memories: they can remember exactly where a hole in a fence is; they can remember a new source of food; they can remember many things. Memory is stored data retained for future use by means of mental processes. And a good memory must indicate some degree of intelligence. I have seen deer do smart things and I have seen them do many "dumb" things, but then I've done "dumb" things myself. I would not claim that deer are as intelligent as many of the other creatures. I am convinced that the coyote is the "smartest" animal in North America today. A deer's intelligence does not begin to compare to a coyote's. But I also think that some deer have more reasoning ability than we give them credit for.

II
HOW THE YEAR
GOES

14

Life in Spring

T he foregoing chapters have made it obvious that each of the three types of American deer differs from the others in many ways, and even local populations of a single subspecies may have their own peculiarities. But—with exceptions that I will note—all have roughly the same habits and life histories. So in this part of the book, I will deal with all types of deer simultaneously, singling out a particular species or subspecies as appropriate. I will divide the complex story not by subspecies or region, but by the four seasons that govern their lives, allotting April, May, and June for spring; July, August, and September for summer; October, November, and December for fall; and January, February, and March for winter.

As Ecclesiastes tells us, for everything there is a season, and a time for every purpose. The characteristics of the seasons dictate the annual events in the life cycle. It is true that these events do not occur for all deer in all regions at the same time. But in previous chapters I have sketched the times of several major events for all types of deer in all regions—the time of the rut, of birthing, of growing antlers, of peeling the velvet, and of shedding the antlers. Those earlier chapters will enable you to adjust the timing of the happenings I describe hereafter to the deer in your region. I will discuss additional events, and their times, as they occur in their seasons. I will also explain differences in food and habitat from region to region and from one variety of deer to another, so the simultaneous treatment of their life histories should cause no confusion.

Let us begin with spring, the time of the vernal equinox, when the sun crosses the equator and day and night are of equal length. In the South and in desert areas, the temperature is high. In the North, the sun's rays are still weak, but stronger than they have been, with enough warmth to make the snow compact. The warming is the key that unlocks the small streams from winter's bondage. The streams gather strength with each passing day, although their activities are curtailed each night. Bare banks are beginning to be seen, and they are bare only briefly because vegetation quickly surges into growth. It is along the stream beds that the deer find the first green shoots of spring. For deer in many areas, spring cannot come too soon, and for many deer it comes too late. Their decomposing corpses are often found along the stream banks where their body nutrients, that were plants in previous years, are leaching back into the soil to produce the plants of tomorrow.

As soon as the snow has settled enough to allow the whitetail deer of the North to move, those that have "yarded up" leave the yards—the relatively protected pockets where they gather in winter—not to return until they are forced to do so by the rigors of the following winter.

MIGRATION

Mule deer and blacktails move to the lowlands rather than yarding up in winter. Now, at a leisurely pace, they begin to wend their way back to the mountains. Whereas their winter migration had been sudden, direct, and hurried, their spring migration is accomplished in easy steps. The does and their yearling fawns are the first to leave the wintering areas. The bucks act as if they are reluctant to go, as if they know their journey will be longer because they move to higher elevations. Most of these deer wintered at elevations of 1,000 to 3,000 feet (304.8 to 914.4 m) above sea level but will spend the summer at elevations from 4,000 to 12,000 feet (1,219.2 to 3,657.6 m). The journey may take weeks because some of the deer migrate as far as 100 miles (160.9 km). The rate of travel is determined by how fast the snow melts at the higher elevations and the availability of food. The migration is usually under way by the middle of April.

This is no forced march. The deer feed as they make their way upward. Those that have lost the most weight in the winter are the ones to gain weight most rapidly in the spring when food is again available. The does and yearlings move directly to their summer grounds, while the bucks tend to wander about.

Although the whitetail is not considered to be a migratory animal, some make seasonal shifts between their winter and summer ranges that are long enough to classify as migration. Carlsen and Farmes documented a seasonal shift of some of the whitetails of northern Minnesota at 55 miles (88.5 km). Verme noted that some

"Bachelor" groups of mule deer are common in the spring, and even white-tail bucks may keep each others' company in the spring and summer, until the rut begins.

of the whitetails in Michigan's upper peninsula moved up to 32 miles (51.5 km). Sparrow and Springer noted that some South Dakota whitetails traveled up to 34 miles (54.7 km) between their winter and summer ranges.

FOODS

The sprouting vegetation has more nutrition in it than it will have at any other time of year. Its protein content soars. Deer throughout the continent just about forsake all browse and hungrily turn to grasses and forbs, although the new shoots on the browse species are also highly nutritious. For a list of the common foods preferred by the three types of deer during the spring, see Appendix I.

Throughout all of North America, wherever it is grown (and over 30,000,000 acres [12,140,000 hectares] are grown in the United States), alfalfa is a prime deer food. Deer can be seen feeding in the fields from the early spring until late fall. Deer usage peaks on the new growth and tapers off as the alfalfa plants mature and blossom. Following each cutting, the deer are back feeding on the new sprouts. The new sprouts of alfalfa are as much as 24 percent protein. (The protein content drops to as low as 15 per-

Life in Spring

cent when the plant matures.) Alfalfa plants are also a source of nitrogen. The new growth is easier to digest than the mature plants. These factors explain the deer's preference for new sprouts.

The deer often bed in alfalfa fields at night but have to bed elsewhere during the day because the plants do not grow high enough to provide cover.

Deer feeding on alfalfa can represent a substantial loss to the farmers. In some areas as much as 20 percent of the crop may be lost to deer. Researchers in Pennsylvania found that, in that state, alfalfa losses averaged about $600 for each farm. The Wisconsin Department of Natural Resources concluded that the average deer eats 92 pounds (41.7 kg) of alfalfa each year. The heaviest losses occurred in the spring and early summer, when as much as 5 pounds (2.3 kg), green weight, may be consumed each day. Deer will also feed heavily upon the resprouting winter wheat because it is about 13½ percent protein.

The size of the home range for the different species of deer depends on the available food. Under ordinary circumstances, does of all three types have a spring home range of ¾-square miles to 3 square miles (1.9 to 7.7 sq km). The buck will have a slightly larger range, from 2 to 5 square miles (5.2 to 12.9 sq km).

ADULT HABITS AND BEHAVIOR PATTERNS

Bucks of all three varieties, in all parts of the country and at all times of the year, bed down earlier in the morning, seek heavier cover, are generally less active in the daytime, come out to feed later in the evening, and are more active after dark than the does and fawns. Bucks are almost always more wary and suspicious than the does, even where both sexes are hunted.

In the springtime, whitetail bucks are often solitary, or sometimes a big buck is followed by several younger bucks. Occasionally there may be a small group of mixed-age bucks, but the largest such whitetail group I have ever seen consisted of five bucks. In the more open grassland areas of Texas, the buck groups are larger.

The grouping of blacktail bucks is very similar to that of whitetails, whereas muley bucks, being more sociable all year long, will often be found in larger concentrations. The largest such mule deer grouping I have found consisted of 11 bucks. Some of the really big muleys are found by themselves. This is a trait found among many mammals. With advanced age, the males tend to be loners. Most frequently this is by choice, although occasionally they are driven out by the younger, dominant males that have supplanted them in the hierarchy. Either way, the exile of the aged has survival value to the general population. The presence of the old and infirm would attract predators to the herd.

The bucks are active for two or three hours before dawn and seek a sheltered vantage point within an hour after dawn. The whitetail bucks usually bed up on the tops of ridges. When possible, muleys like to bed up against some rim rock, and blacktails slink off into impenetrable thickets. If the bucks have to bed down on level ground, they usually walk with the wind for some distance before bedding. Then, when they lie down, they can watch downwind with their eyes, and their noses will tell them if anything is following their trail.

Their antlers are growing rapidly at this time and are soft and tender, so the bucks take great care not to hit anything with them. The bucks are gaining weight rapidly now on the new, nutritious food.

The bucks are up, milling around and feeding lightly in their bedding areas, around noon. Then they bed down again until about an hour before dark. The evening feeding is the heaviest of the day: the bucks may feed for as long as three hours. Under the cover of darkness, they bed down in the fields and open areas where they have been feeding. Another period of activity takes place around 10–11 P.M., and the bucks again bed in the open. Around 2:00 A.M. they feed for about an hour. They are again active before dawn. By the time the sun comes up, they will be heading back into heavy cover to spend the day.

A doe's daily pattern is quite different. Does usually feed from dawn until about 9:00 or 10:00 A.M. Starting in late winter and on through the spring, does do a lot of feeding in the morning. The demands made on the doe's body by the developing fetus and by the fawns after their birth require longer feeding periods to cram in all the food possible. Although the new grasses, forbs, and browse are high in nutrition, they are also high in water content, and more pounds must be eaten to meet the demands on a doe. Around 10:00 A.M. the does usually bed down at the edges of the fields and clearings where they have been feeding. They are usually up feeding between 12 noon and 1 P.M. Around 4:00 P.M. the does consume their greatest amount of food of the day, and feeding may last up until darkness. Nighttime feeding periods correspond to those of the bucks, with one around 10–11 P.M. and the other

Deer find an abundance of food to graze in the spring, and they choose the most nutritious grasses and forbs as they become available. Here a white-tail feeds on new shoots in the everglades.

around 2:00 A.M. They, too, bed down in the fields and clearings at night. In the morning, the does start to feed later than the bucks and stay active longer. Once they have fawns, they usually nurse them before each of their own feeding periods except for the noon feeding period.

In the springtime, just before dropping their fawns, does of all three varieties become extremely territorial. This is the only time of the year that this is true. The exact size of each doe's territory depends upon the habitat, the availability of food, and the total population of the deer herd in that particular area.

About one month before giving birth, the does become restless and very antagonistic toward each other and even toward their own yearlings. This leads, first of all, to the breakup of the females' wintering concentration and, later, to the dispersal of the young. It is this traumatic change in the yearlings' lives, when they are no longer led and guided by the older does, that results in so many of the yearlings being killed by automobiles. The two peaks in deer highway fatalities occur first in May, during this period of territorialism, and then in November at the height of the rut.

David Hirth did a comparison study on the social behavior of whitetails in two diverse habitat areas. He studied deer on the E. S. George Reserve near Pinckney, Michigan, which has the old farm-woodland habitat so common across most of our northern states. He also studied deer on the Wilder Wildlife Refuge near Sinton, Texas. This latter refuge is in southeastern Texas, and has a mixture of dense brush and flat, wide-open grasslands. I will refer back to this study a number of times, because the behavior of the deer was similar in some instances but diametrically opposed in a many more. Hirth found that the whitetail does in Texas did not establish birthing territories, and that they did not chase their yearlings away. I have observed that whitetail does in the north do not allow the fawns to follow after them for the first 2–3 weeks after their birth; rather the does keep their fawns hidden. Between the third and fourth weeks the doe's yearlings are allowed to rejoin her. Hirth found this pattern among the whitetails in Michigan.

The does' territorialism scatters the deer as widely and as uniformly over their entire range as is possible. This pattern offers each doe and her fawn or fawns the greatest chance of survival. It prevents the fawns from following the wrong doe, which would probably reject them. It prevents other deer from attracting predators to the territorial doe's fawns. It may prevent predators from finding both of the fawns if they do invade the doe's territory. Territorialism lasts for about two months, starting two to four weeks before parturition and extending four to six weeks afterward.

By this time, deer throughout this continent have changed their daily habits as they themselves are being changed. As automatically as plants respond to the increased sunlight, the autonomic nervous system of the deer also responds. The pituitary gland, which has been functioning at a greatly reduced rate since December, becomes more active in late March under the stimulus of increased sunlight. The pituitary, in addition to releasing the somatotrophic hormones that start the buck's antlers developing, also stimulates the adrenal and thyroid glands. The activated endocrine system provides the impetus for greatly increased growth of the fetuses the does are carrying.

The does were bred in the late fall or early winter, during the 24 to 28 hours when they were in estrus. In the pre-estrus period, the new ovarian follicles grew, then burst, discharging the fully developed egg into the fallopian tubes. Countless thousands of motile sperm from the semen of the male swam up the fallopian tube where one penetrated the egg, impregnating the doe some 14–20 hours after copulation took place.

Life in Spring

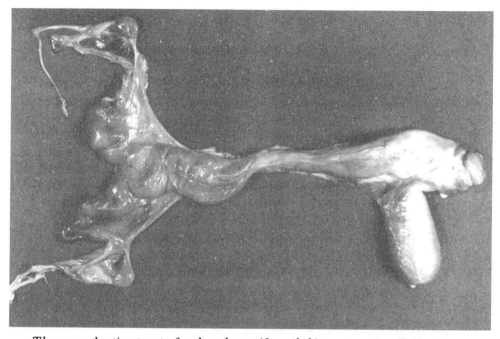

The reproductive tract of a doe shows (from left) two ovaries, Fallopian tubes, uterus, vaginal tract, bladder and vulva. In many states, hunters are encouraged to remove the tract and send it to the game department so that biologists can sample the herd's reproductive health.

The fertilized egg, or *blastocyst,* floats loose in the doe's uterus for about 30 days before it becomes implanted in the uterine wall and begins to develop. The development and multiplication of the egg cells is a phenomenal mathematical progression. It is not until 37 days after implantation of the blastocyst that the embryo is considered to be a fetus.

In deer, as in humans, identical twins result when a fertilized egg splits and produces two fetuses. Identical twins are always of the same sex. When two different eggs are fertilized, the resulting fetuses are known as fraternal twins, and they may be of one or both sexes. Triplets or quadruplets are produced in the same manner. Fraternal twins are more common than identical twins, and they are also much more common than triplets or quadruplets.

E. L. Cheatum, G. H. Morton, and R. A. Armstrong did the work that forms the basis for the following discussion. Their specimens were pregnant does that were killed on the highway or shot illegally. These does were examined at New York's Delmar Wildlife Research Laboratory near Albany.

The researchers worked up tables that allow the age of the fetus to be determined by measurements, which are made by two methods. Everyone is aware of the term *fetal position,* a position in which a human, or any mammal, lies on its side, curls the body and head forward, and draws the legs up toward the chest.

The deer's fetus is curled into such a position until the forty-eighth day, and until then measurements are taken in a straight line, not following the curve of the body, from the crown of the head to the rump. No attempt is made to straighten out the curvature of the fetus. At 37 days the crown-rump measurement is about 11/16-inch (17.5 mm), and the fetus weighs ½- gram. At 48 days the fetus is 1½ inches (38.1 mm) in length and weighs 2.2 grams.

After that, the fetus begins to straighten out, and the measurements are taken from the fetus's forehead to its rump at the base of the tail. At about 65 days, the fetus will have a forehead-rump measurement of 3⁵/₁₆ inches (84.1 mm) and weight of about ½- ounce (14.1 g). The eyelids are formed, the mouth is closed, and the vibrissae follicles are present that will develop as the whiskers of the eyebrow, mouth, and chin.

I well remember the hysteria that arose when New Jersey first proposed having a doe day. The anti-hunters and many hunters joined forces to denounce the state's actions, and they obtained court orders to stop the proposed harvest. The hunt was planned for December but was delayed by court action and finally took place in February.

The hue and cry against the hunt was that does are mothers. Of course they are. Many hunters loudly proclaimed that they would never shoot a pregnant doe. The fetuses at that time of year average 60 to 90 days in development. At 90 days a deer fetus measures 6½ inches (165.1 mm) in length and weighs about 90 grams, or roughly 3 ounces. At that time I took pride in having excellent eyesight, and I couldn't detect that the does were carrying fetuses. Since I was also working as chief gamekeeper at Coventry Hunt Club, I gathered many of our does' reproductive tracts. In not one instance did I find twin fawns, though twins would have been normal if the food supply had not been depleted. New Jersey, like an increasing number of other states, now has a doe season, for a doe harvest is a legitimate tool of herd management. The number of does to be taken during this season is based on the estimated deer population in each of the state's management zones.

At about 120 days, the fetus weighs 22 ounces (623.7 g) and measures 10½ inches (266.7 mm). At this stage it has incomplete

Life in Spring

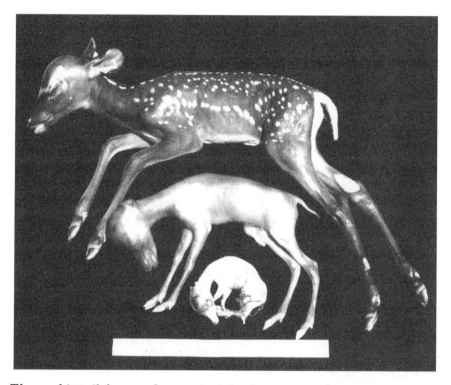

These whitetail fetuses show typical development at about 80 days, 120 days, and 180 days, which is considered a full-term fawn. Most fawn development is accomplished in the last 30 days of the term, when the doe has nourishing spring forage available.

hair covering with some pigmentation showing in the hair, although the spots are clear. The tops of the hoofs are dark with white tips.

At 125 days the fetus weighs 24.2 ounces (687.2 g), and at 135 days it weighs 36.3 ounces (1,029.3 g). This 50 percent spurt in growth in just 10 days is followed by a gain of more than 100 percent in the next 10 days. Initially slow development, changing to greatly accelerated growth, is nature's way of preventing an undue drain on the doe during the period of food shortages. The doe's metabolism slows down each winter, and the fawn develops very slowly during the first four months. Then the development spurts ahead so that the young will be born at the optimal time of year.

From 180 days on, the fetus is considered full-term. Almost completely developed, it has a length of about 18 inches (457.2 mm) and weighs about 2,600 grams, or 5¾ pounds. I provided detailed birth weights in an earlier chapter, so it suffices here to say that the fawns of all three types of deer average 5 to 8 pounds (2.3 to 3.6 kg) at birth, after a gestation period of 187 to 212 days. (A period of 200 to 210 days is considered normal.)

BIRTH OF FAWNS

When a fawn is born, its dimensions are far out of proportion to those it will have as an adult. The legs are much longer at birth in relation to the body than they will be later in life. A newborn fawn seems to be all legs, none of them steady. This is true of many large herbivores, including horses and gazelles. The long legs soon become steady, for they are needed so that the young can be moved from the place of birth—a place to which predators may be attracted.

The fawn's muzzle is short, the length from eyes to nose being somewhat out of proportion to the section from the eyes to the back of the skull. The jaws gradually grow longer for almost a year to provide room for the permanent teeth. The short muzzle is particularly apparent when the fawn has its first winter coat. The long hair makes the muzzle look even shorter. The fawn's body also lengthens proportionately with age.

At birth, the pupils of the fawn's eyes are brown and the surrounding area is bluish gray. After nine or ten days, the blue portion of the eyes begins to turn light brown as in the adult.

As birthing time approaches, the doe wants solitude. Her fawns of last year are puzzled by her behavior. For almost a year, almost every minute of every day, the doe's main concern was for the yearlings she now seeks to elude. She may even drive them off by striking at them with her front feet. Although puzzled, at last the yearlings leave or are left. In the case of a young buck, this separation may be permanent. The young does will almost surely rejoin their mother within a week or two.

If the young does are pregnant, they, too, are seeking solitude. Often the yearling does and their fawns will rejoin the matriarchal

A doe with twin fawns born minutes apart; as she washes the second, the first one already looks alert. Twins are common and triplets are not rare.

doe within a month or two. Sometimes the little family unit is broken up only long enough for the older doe to give birth to her new fawns.

It becomes very obvious when a doe is in the last stages of pregnancy. Her swollen abdomen emphasizes her condition and so do her actions. Being heavy, she does not run and bounce around as she usually does; her manner is that of a sedate matron.

This is particularly noticeable with blacktail deer. Blacktails often run with what appears to be a very stiff, bone-jolting gait, particularly when they are nervous or frightened. This gait is abandoned by blacktail does in the last month of their pregnancy.

Another indication of imminent birth is the swelling of the doe's udder. About two weeks prior to giving birth, her udder swells noticeably and the skin there turns pinkish with body heat. This swelling and coloration of the udder can be seen on wild deer if checked from the rear. We have found with captive deer that, within 24 to 30 hours after milk can be hand-stripped from the doe's teats, she will give birth.

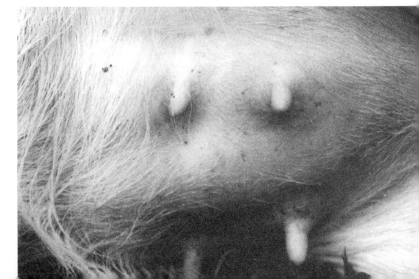

About two weeks before a doe gives birth, her udder swells noticeably as it fills with milk.

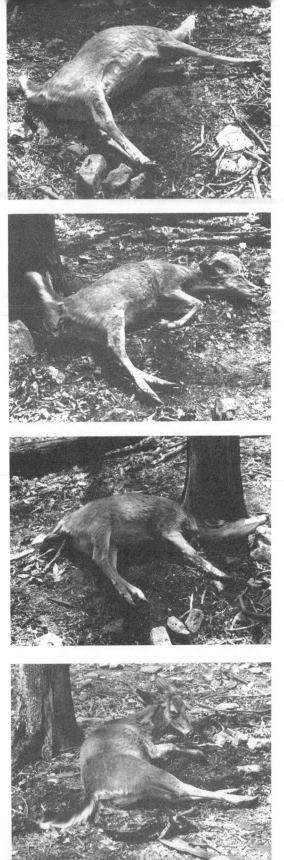

Birthing

1. *A whitetail doe lies down during the throes of labor.*

2. *As the doe strains, the fawn's front feet emerge.*

3. *After additional spasms from the doe, the fawn's head appears.*

4. *With the fawn almost completely expelled, the doe rests.*

5. As the fawn moves towards the udder, the doe begins cleaning her new arrival.

6. The doe thoroughly cleans the fawn as it nurses.

7. Thirteen minutes after birth, this fawn tries to stand.

8. The eighteen minute-old fawn takes its first wobbly steps.

Ordinarily, a doe does not seem to look for any special spot for birthing, other than seeking a little privacy and some sheltering cover. In most cases, the young seem to be dropped almost anywhere when it is time for them to be born. Sometimes, however, a doe makes her way to an island where she drops her fawns, using the water as a protective barrier. Where predation is heavy, does seem to take more care in selecting a spot for birthing.

I have been most fortunate in being able to witness the birth of fawns on two different occasions. Both does were semitame, living on a deer preserve owned by Joe Taylor. The first occasion was in 1960, while I was gathering data for my book on whitetail deer. Joe's deer at that time had the run of an entire mountain. Their only link with humans was the feed that was regularly put out for them. I was building up a series of photographs on antler development. While walking through Joe's woods looking for the bucks, I located a doe that had part of her water bag, the amniotic sac, protruding from her vaginal tract. I forgot about the bucks. Although I had seen many newborn fawns, I had never witnessed a birth, and not many other people had, either.

The doe was standing when I first discovered her. When she was sure I had seen her, she walked several hundred feet up a hillside and went into a fairly open thicket of alders. Apparently, the doe had been in labor. Since I did not want her to abort the young by moving any farther, I did not immediately follow her. She lay down, and in a few minutes I could see the feet of the fawn protrude from her vagina, then the head and then foreshoulders.

At this point, the doe stood up and the fawn slipped out of her body. Knowing that the doe was now anchored to this spot by the birth of that first fawn, I cautiously approached. She showed absolutely no concern about my presence; her only interest was in licking her new baby dry. The little fellow (for I soon afterward had a chance to examine it and found it was a buck) attempted to get on his feet in less than 5 minutes and could stagger a few steps in less than 10 minutes. His attempts to get on his feet were not helped at all by the doe's licking. Her washing of the fawn was so vigorous that she often knocked him over, and he kept getting tangled in part of the placental sac. At one time he became completely enveloped in the sac. He nursed when he was 8 minutes old, while the doe was lying down. All the while, she continued to wash him.

After about 20 minutes, she stopped her ministrations and began to strain—a strain that sent a rippling shudder through her body. She strained and pressed and was rewarded by the protru-

This doe had dropped one fawn and will soon give birth to a second, while her yearling fawn, now a button buck, approaches to investigate. She will tolerate but ignore the yearling at this stage.

This fawn is still entangled in the placental sac, which the doe is pulling away and eating. Consuming the sac is instinctive, defensive behavior that keeps the scent of birth away from predators.

sion of the feet of her second fawn. The straining continued until about two thirds of the fawn's body protruded. Then she stood up and the little one slid out as easily as a diver knifing into the water. The fawn was actually in a diving position with its head stretched out between its forelegs. I do not know how long the doe may have been in labor before I first saw her, but I do not think it was long, considering the ease with which the fawns were born and the short time that had elapsed. This doe had given birth twice before. Those births had stretched her reproductive tract. Thus, her tract was considerably larger than that of a doe that had never had young.

While she was giving birth to the second fawn, a little doe, her first fawn lay curled up beside her. Shortly after giving birth to her new fawns, the doe's yearling buck came over to investigate. The doe paid no attention to him, just as she paid no attention to me. The second little fawn was washed and licked dry until the hair on its coat stood on end. The actions and timing of the second fawn closely paralleled those of the first.

About 20 minutes after the second fawn was born, the doe began to eat the afterbirth, or placental sac. Most of the sac had come out of her body with the two fawns, and she pulled the remaining string of tissue out of her body with her teeth.

Almost all wild animals and most domestic ones eat the afterbirth as soon as they can. The primary reason, I think, is to get rid of it so that it will not attract predators. Consuming it is a defensive instinct. Canine scavengers throughout the world are attracted to herd animals when they give birth, for the placental sacs provide an easily scavenged feast. No doubt another reason why the mother consumes the afterbirth is that pregnancy and birthing have drained her of substantial nutrients. Eating the sac returns some of these nutrients to her system. A doe giving birth loses from 14 to 25 pounds (6.4 to 11.3 kg) in overall weight, depending on the weight of her fawns. It has also been suggested that eating the afterbirth is a stimulus to lactation, although I can find no research reports on this.

After the doe had eaten the afterbirth and had nursed both fawns, she was evidently exhausted. For a while she was content to lie still and rest. The fawns had curled up alongside their mother. The doe, being comparatively tame, had accepted my presence and was not concerned as I cautiously moved about taking photographs from different angles. The fawns struggled to their feet and tottered over to me. I gently put them back beside the doe, but they instinctively came right back to me because I was moving. It would have been very easy for me to imprint these fawns—that is, have them accept me as their mother. This was something I wanted to avoid, so I put them both back beside the doe and quickly withdrew from the immediate area. They were then content to lie there, snuggled against their mother.

The fawns had been attracted by my moving because, in the wild, all the hoofed mammals move their offspring from the place of birth as soon as they are able to walk. Although the doe has eaten the afterbirth, her lost body fluids have seeped into the earth and that odor can attract danger.

I had not hesitated to touch these fawns because the doe was accustomed to human beings and would not be alarmed by my odor. I never touch any of the wild fawns that I locate in the woods and fields.

Every year game departments across the country are deluged with reports of and by people who have picked up "orphan" fawns. Unless the doe has been killed, these fawns are not orphans, but if a fawn is touched and petted, there is a chance that it may become an orphan. It may be rejected by its mother if she is afraid of human scent. At one time it was thought that touching a fawn practically guaranteed its abandonment.

Research at the Wilder Wildlife Refuge, near Sinton, Texas, where fawn-tagging has been carried on for many years, has proved that in only about 5 percent of the cases will the doe abandon her fawn because of the human scent. These statistics have been substantiated by the New Jersey Game Department, which has also been doing a lot of fawn-tagging. Nevertheless, I still constantly stress to everyone: *Don't touch the fawns.* The fawns you touch may be among that 5 percent, and no one should ever deliberately cause a fawn's abandonment.

In New Jersey, the middle of May brings dogwood blossoms, Mother's Day, and rattlesnakes. The second birthing I photographed took place on a day when Joe Taylor and I were coming off the mountain after hunting for rattlesnakes. It was about May 15, a week to 10 days early for deer to drop their fawns in our area.

The doe, another semi-tame one on Joe's preserve, was a yearling that had not been bred as a fawn. This was her first parturition and it was a hard, long period of labor for her. Her inexperience made it even more difficult.

We first saw the doe about 10 A.M. and she was in the throes of hard labor. This was evident because of the large area that she had trampled and cut up in her pain. Biologists recognize that most does thrash around and trample quite an area while giving birth. On fawn-tagging expeditions, they look for such areas in the grasslands, hay fields, and woodlands that pregnant does are known to frequent.

The doe moaned and groaned piteously and strained mightily but to little avail. She had not dilated sufficiently to allow for the fawn's passage. There was also no doubt that her body was feverish. Several times she got up and walked over to a spot that was wet from a small seepage. She would lie on the damp earth for a while, strain some more, then rise to go back to the spot where we had found her originally. When she strained, she put her whole body into the effort. She kicked out with her hind legs, thrashed around with her front legs, and twisted and turned her head and neck from side to side. Her eyes were closed, her mouth open. After about an hour, her vagina gradually dilated and the fawn's front feet emerged, wrapped in a placental sac.

Now the doe's inexperience multiplied her problems. If her actions had not caused her pain, the situation would have been humorous. She strained until it seemed as if her body would be split asunder. The fawn's front feet would emerge a little farther with each effort, but after every massive push the doe climbed stiffly to her feet and turned around to search the area for her fawn. Every time she got to her feet, the fawn would slide back into her body and the feet would disappear from view. When the doe could find no fawn, she would lie down again until the next spasm drove her to greater exertions. Then the water bag burst and the amniotic fluid poured out. Six times she got up to see if the fawn was born and six times the fawn slid back into her body. Her vagina was dilating now and at each subsequent spasm the fawn slid back out much more easily than before.

We were happy to see that the fawn's head was in the proper position between its front legs. Since the labor was so difficult, we thought that perhaps the fawn's head was turned back over its shoulder. This can happen, and when it does the fawn is not in the streamlined, least-resistance position best for a natural birth. Fawns that are turned that way sometimes die in the birth process, and occasionally they cause the death of the doe. These complications are similar to the breech-births of human babies, except that in the case of wild animals, there is no attending doctor to turn the infant to the proper position.

Finally the fawn's head and shoulders were out. The doe's labor was still intense, and she still thrashed and moaned. Fortunately, she did not get up again. With more effort, the fawn was pushed out a little farther. When it was halfway out, the doe rested; the worst was over. A little more effort and the fawn was almost completely expelled. The doe then lay quietly for a while with the fawn's hind feet still inside her body. The fawn shook its head and moved its floppy, sopping ears. Noticing the movement behind her, the doe immediately turned and started to clean the fawn with her tongue.

In four minutes the fawn started to crawl up between the doe's hind leg and nuzzled her udder and nipples. It continued to crawl forward while she continued to wash it vigorously. Between six and seven minutes after emerging, the fawn turned around and again searched down along the doe's belly for her nipples. The fawn had difficulty getting a nipple into its mouth but instinctively knew what to do and kept on trying. The fawn was successful in nursing when it was 10 minutes old. It nursed for two minutes, then turned around again, lying against the doe's front legs. The doe lovingly lavished all of her attention on her new baby. She missed no part of the fawn in her cleaning efforts. It was as if she had decided to give it a thorough examination and was pleased with what she found.

At 13 minutes of age, the fawn heaved up onto its front knees, raised its hind quarters into the air, and attempted to stand. This was a tremendous effort, and the fawn trembled violently with the exertion. It fell. It tried again. It continued to fall or to be pushed over by its mother's ministrations, but it went on trying.

When the fawn was just 18 minutes old, it took its first few wobbly, spraddling steps. Then, thoroughly tired, it sank to the earth and curled up into a ball. But it had walked.

I imagine that most does have more trouble giving birth for the first time than they have with subsequent births. Whether this was an exceptionally hard labor, I cannot say. The total time elapsed for this birthing is not known because I did not see the very start of the doe's labor, but I was there for over two hours before the fawn was born. At the other extreme was a birthing described by Edwin Michael at the Wilder Wildlife Refuge.

Doing fawn research early one morning, Michael was watching a doe when he suddenly realized she was about to drop a fawn. And here the word *drop* is most appropriate, because the doe did not lie down. In fact she acted as if she were not aware of what was taking place. The doe was walking along browsing on little tidbits here and there. At 8:25 A.M. the fawn's feet were protruding from the doe's vagina, but she leisurely continued with her feeding. At 8:55 A.M. the fawn was almost halfway out of the doe's body, but she still showed no concern. She went on feeding and at no time did she even turn around to see what was happening. About 9:00 A.M., Michael lost sight of the doe when she entered high grass where she might have dropped her fawn. Michael measured the distance and found that the doe had walked about 275 yards (251.5 m) in the period that he watched her.

C. W. Severinghaus reported a birthing that took a long time, but he did not say whether the labor was a difficult one. A whitetail doe that was in captivity at the Delmar Wildlife Research Laboratory gave birth to a normal, brown fawn during the night of July 7, 1968. It was the first of two, but its twin did not appear until early on the morning of July 9. The second fawn was a white one.

When fawns are born, the bottoms and tips of their hoofs are covered with a soft, gelatinlike substance about ⅛-inch (3.2 mm) thick. It is similar to the cartilage on the tip of the breastbone of a young frying chicken. The padding prevents the hoofs from puncturing the placental sac while the fawns are being carried. This soft material is usually worn off or shredded the first day or two of life. The hoofs are dark gray at birth and get darker as they harden in a couple of days.

The placental, or navel, cord usually tears loose when the fawn is born. The portion attached to the fawn dries up after the third day. Usually the dried cord drops off in about two weeks and a scab forms over the spot.

A doe giving birth for the first time generally has a single fawn whether she is bred as a six- to seven-month-old fawn or at 1½ years of age. Thereafter, if she has access to an ample supply of nutritious food, she will usually have twins. We frequently hear about barren does. If the forage is so poor that a doe is starving, she may fail to conceive because her body is not functioning properly or because she herself needs all the nutrients that would be consumed by the embryos. Old age, however, very rarely renders a doe barren.

A study of 1,322 mule deer does in five states showed an average fetal rate of 1.44 fawns per doe. That is a good average, because 1.44 fawns per doe means a herd increase of about 50 percent per year. The does examined were all at least 18 months old at the time of breeding. A single fawn was being carried by 29 percent of them; 56 percent carried twins, and 1 percent had triplets. Those 13 sets of triplets had to come from areas where the does had a superabundance of food. Fourteen of the does were barren.

Some of the 14 may have been 1½-year-old does that were not yet sexually mature. Quite frequently, blacktail does on poor land do not breed until they are 2½ years old. Some of them could also have been suffering from some ailment or physical impairment that was not primarily sexual. The study showed that the fertilization rate for the 1,322 does was about 95 percent. Rarely do herds achieve a better average than that.

Most mammals produce more male than female young, and deer, under normal conditions, are no exception. Females, throughout life, are more resistant to disease, less affected by stress, less adventuresome than males, and live longer. Individual states, and individual areas within states (or provinces), may vary, but throughout their ranges the ratio for all three types of deer starts at 106 males for every 100 females born.

A number of factors will skew this normal ratio. The fecundity of all of the deer depends upon the availability of nutritious food, which is the end product of good habitat. High-protein consumption is needed if the deer are to realize their maximum breeding potential. The protein level should be at least 13½ percent, though it had been found that a 7 percent protein diet drops the production ratio only from 1.9 fawns per doe to 1.7. The latter figure is well within the normal ratio for healthy deer.

Good nutrition for fawn production is important at all times of the year, but it is of somewhat less importance in the winter than at other times. Although this sounds like heresy, it must be remembered that a deer's metabolism declines in December and remains low until March. In effect, the deer's engine is idling. This is the period when the least amount of food is available and, as already discussed, the period of very little fetus growth. The deer's metabolism goes back up in March in response to photoperiodism, at a time when new green plants are beginning to sprout and the growth of the fetus is beginning to spurt. During the winter the deer survive primarily on their stored fat reserves, and these reserves are the result of an abundance of good food in autumn.

Summer forage is very important to the does because of the heavy drain of lactation. The doe is attempting to regain her own body weight and to replace the minerals depleted from her bones by the developing fetuses. On good habitat the doe can do this and still have the surplus needed for milk production. On a badly deteriorated range, or if a severe drought causes the existing vegetation on good habitat to dry up, the fawns will probably be lost.

A doe of an expanding deer herd that has not yet reached the carrying capacity of the land will usually have access to all the good nutritious food that she needs. Under these conditions the doe is able to nurse her fawns, regain her body weight and minerals, and go into the fall breeding season in excellent condition. Surprisingly, such does will give birth to a preponderance of female fawns. Louis Verme, in Michigan, found that well-fed does gave birth to 57 percent females and 43 percent males, while does on a starvation diet produced 28 percent females and 72 percent males. W. L. Robinette had very similar findings in a study of mule deer in Utah. There the does on the best range produced 58 percent females and 42 percent males, while the does on the poor range produced 35 percent females and 65 percent males.

I have stated that, under average habitat conditions, more male than female fawns are produced. The apparent contradiction in these higher-female ratios from does on the best range comes about because the carrying capacities of the ranges have not been reached; more females will increase the rate of population growth.

Likewise, does on poor range often lose their fawns and are thus freed from the drain of lactation. The food, even though limited, that the doe eats can then go entirely to her own needs. Such does usually go into the breeding season in as good condition as are the does on average habitat that had the drain of nursing fawns. Apparently the doe that lost her fawns because of poor range conditions, yet regained her own body reserves, is as apt to produce as high a ratio of male fawns as is her better-fed counterpart. This relationship of habitat to the sex of fawns seems to be one way in which deer populations are self regulating, although the research is not conclusive.

It is thought that the high ratio of males born to females on a deteriorated range is nature's way of limiting the size of the deer herd. With fewer females being born, the reproductive potential of the herd is cut back drastically, and this condition continues until the habitat has had a chance to recover.

The actual timing of conception in the deer's estrus period also has a bearing on the sex of the fawns. This has long been known to be true of humans. With humans, the secretions of the womb change from being acidic to being alkaline, and acidic conditions favor the XX, or female, chromosomes, while the alkaline solution favors the XY, or male, chromosomes. Whether or not this explanation is true of deer as well, does impregnated in the early part of their estrus period produce mainly female fawns, while those impregnated in the last stages of estrus produce males. The general effect is to increase the number of males in the herd if there is a shortage—and there must be a shortage of bucks if the doe is not bred as soon as she is able.

Gary Connolly, doing fetal research on the blacktail deer in California, recorded a much higher male-to-female ratio among does giving birth for the first time and among old does. Middle-aged does had the higher female-to-male ratio. These findings have been collaborated by several researchers working with white-tails.

Nutrition and Fawn Mortality. Study after study has confirmed the all-important link between the nutrition of the doe and the survival of her fawns. Malnourished does lose fetuses and fawns at a rate of anywhere from 70 percent to 93 percent. Well-fed does drop fawns in the 6 to 8 pounds (2.7 to 3.6 kg) range, while those on a starvation diet give birth to fawns weighing about 4 pounds (1.8 kg). Any fawn weighing less than 5 pounds (2.3 kg) at birth has little chance of surviving. The underweight fawns are often too weak to stand, are unable to nurse because they can't reach the doe's nipples, have no resistance to adverse climatic conditions, and are unable to escape from predators.

Malnourished does are often unable to produce milk, are too run-down to establish the doe-fawn bonding, and are indifferent to their fawns. B. A. Schulte, working with deer in Michigan, concluded that this maternal negligence by the doe was the result of an insufficient pituitary secretion of the hormone prolactin, due to malnutrition. It is prolactin that stimulates the secretion of milk and promotes the maternal instinct.

Fawns of does giving birth for the first time also have a much higher natal mortality. The young does are not experienced and, being on the lowest rung of the social hierarchy, are forced to accept the less favorable birthing places. The older, larger does monopolize the best birthing territories, forcing the lesser does onto marginal areas.

The high male-to-female ratio at birth quickly reverses itself once the fawns are out in the world. Mortality among the males is greater. Probably this is because the males tend to be more active, more aggressive, and more independent, with the result that they get into more trouble than do the females. At any rate, the male-to-female ratio continues to widen throughout the deer's life. Hunting greatly accentuates the process.

When deer are on a poor diet, the gestation period is usually four to six days longer to give the fetuses the maximum opportunity for development. Usually male fawns are carried longer than are female fawns, because it takes a little longer for the males to develop. (In human pregnancies, too, males are usually carried longer than are females.) Among both sexes, fawn mortality is high. Studies have shown that 6 to 7 percent of all whitetail fawns born in the wild die within the first 48 hours.

A Missouri study of whitetail fawns investigated the relationship of doe nutrition and postnatal mortality. Forty-two percent of the fawns whose dams were kept on a 7 percent protein diet died. Twenty-seven percent of the fawns from does that were on a 10 percent protein diet also died. But there was no mortality at all among fawns dropped by does on a 13 percent protein diet.

Colorado conducted a study in which muley does were given increased protein in order to find the effects on birth and fawn mortality. The does were fed a diet containing 16 to 18 percent protein. The 172 fawns born to these does during the program had an average weight of 8.13 pounds (3.7 kg), which is substantially higher than the average weight of wild-born fawns. Usually, when only a single fawn is born, it weighs more than either of twin fawns, but twins may also be large if the mother has been well fed.

The does on the high-protein diet came into rut five to seven days earlier than did the control deer. They also dropped their fawns earlier; their fawns, as noted, were above average in weight, and fawn mortality dropped. Four sets of triplets were born during this program. Again and again research proves that food is the key controlling most facets of life (and not only for deer).

Some interesting figures appeared in a New Jersey study of multiple deer births from 1952 to 1962. On some of the best land in central New Jersey, three does that bred at six to seven months of age produced twins. The national breeding rate for six- to seven-month-old fawns on good land runs between 40 and 60 percent. Those that do breed usually have just one fawn, but these three, because of their good diet, each produced twins. In that study 30 does had triplets and one had quadruplets.

In 1968 the New Jersey Game Department collected a large number of bucks and does from estates near Harbourton. This is rich land with a superabundance of food. The 138 adult does collected were carrying, in addition to the expected sets of twins, eight sets of triplets, three sets of quadruplets, and one set of quin-tuplets. That is what adequate nutrition will do.

A doe killed on a highway near Espanola, Ontario, in June, 1961, evidently had access to good food. She contained four full-term fetuses—three bucks that weighed 7.9 pounds (3.6 kg), 5.8 pounds (2.6 kg), and 5.3 pounds (2.4 kg), plus a doe that weighed 4 pounds (1.81 kg). That's a total fawn weight of 23 pounds (10.4 kg). Another doe, killed by a car in Cambria County, Pennsylvania, in 1958, was carrying full-term quadruplets—three does and a buck.

Henry Patha, near the town of Chautauqua, New York, killed a doe with his automobile on April 26, 1949, that contained four perfectly formed fawns. The sex of the fawns was not recorded. On March 27, 1980, a motorist killed a doe on Highway 45 in Birnam-wood, Wisconsin. Tom Meueriden dressed out the doe and found she was carrying four perfectly formed fawns, two does and two bucks. Fortunately not all of the known quadruplets were killed. Abe Miller of Baltic, Ohio, had a doe give birth in captivity, on May 28, 1984, to a set of quadruplets, and she was able to raise them all. In addition to the New Jersey quintuplets I mentioned, there have been other records of quintuplets, but I have not been able to ascertain whether the fawns lived.

When it comes to records, a semitame doe by the name of Di-ana stands out. Diana lived in the woods near the Tomhegan Camps in Maine. She produced 31 fawns in 14 years. She had six sets of twins and five sets of triplets. And she produced one of those sets of triplets when she was 15 years old. From the age of 12 on, she became very docile, and even her offspring "bossed" her around, but she never became barren. She was killed by an overly amorous buck just before the rutting season when she was 18½ years old.

A whitetail doe in a commercial herd at Hamburg, Pennsylvania, also set some kind of a record. This doe gave birth to triplets on Memorial Day for four consecutive years. I think the results of research on the punctuality of the birthing of deer would be inter-esting. Mink breed on the same day and give birth on the same day year after year, and this precise timing is passed along to the daughters of each female.

One other fawn record of interest involved a doe injured by dogs at Higgins Lake, Michigan, on February 25, 1957. The doe, because of her injuries, was killed by a game warden. When her reproductive tract was checked, it was found that she was carrying twin fawns. One was normal, but the other had two heads that were joined at the ear. This fawn had two outside ears but only one in the center at the joint.

Albert Strausbaugh of McSherrystown, Pennsylvania, shot an odd-looking deer during Pennsylvania's 1962 anterless deer season. It was a six-month-old fawn with three eyes, one on the right side and two on the left. It also had two noses, one where it belonged and another a third of the way between that nose and the eyes on the left side. Everyone has read of two-headed calves, five-legged sheep, and similar anomalies. They are rare, but there is nothing especially strange about their occurrence in any mammalian species, including deer. It is rather strange, however, that a three-eyed, two-nosed fawn lived for six months, because few such unusual youngsters survive for long.

Nursing. The milk that the fawns get for the first couple of days is a particularly thick, rich, sticky, yellow milk known as *colostrum*. It is vitally important to the fawns because it contains immunoproteins—antibodies against disease.

The fawns grow rapidly on their mother's rich milk. A good Jersey cow has about 5 to 6 percent butterfat in her milk, but a deer's milk is 11 to 12 percent butterfat. The fawns gain about 10 percent of their birth weight per day for the first week and then slow down to about a 5 percent weight gain per week. The doe's milk's butterfat content drops down to about 8 percent after the first three weeks. Her diet will affect the *amount* of milk she can give but it has little effect on its *quality*.

The fawns are nursed three to four times per day, usually under cover of darkness. They will drink 6–8 ounces (170.1–226.8 g) of milk per nursing for the first week. As they grow, their consumption goes up. The fawns will have about doubled their birth weight at two weeks of age. During their third and fourth weeks, they will be nursing about 9 ounces (255.1 g) four times a day. Their body weight will have tripled since birth. Single fawns will gain weight faster than will twins because they have access to the doe's total supply of milk. Single fawns usually maintain their head start throughout their lives.

The volume of lactation puts a tremendous drain upon the doe's body, and many lose weight during the fawn's first three weeks of life. If the doe was in good condition at birthing time, this weight loss is not appreciable. If not, she may not be able to nurse one or both fawns, a major cause of fawn mortality. A doe in good

Life in Spring

By early summer, fawns move about freely with their mothers, sampling vegetation. Fawns make this transition—from remaining hidden to traveling with their does—during the period from four to ten weeks of age.

shape, on good habitat, will be a successful mother.

For the first two weeks of its life, the fawn is not a ruminant. Although the rumen and reticulum are present, they are not used. Fawns have what is known as an *esophageal groove,* a tightly folded segment across the top of the rumen and reticulum that extends from the fawn's esophagus to the omasum, taking the milk directly there. The omasum and abomasum are fully developed and functioning.

Between two and three weeks of age, fawns begin to eat their first greenstuffs. They are often forced to, because the doe is producing less milk and the fawn's nutritional needs are not being met. Once the fawns start to eat vegetation, their paunches and reticulum begin to expand and become functional. When this occurs, microbial fermentation begins. It has been found that newly weaned fawns and adult deer have the same microbes, and in the same proportion, according to the amount of their rumen content. But up until the fifth week, fawns deprived of their mother in the wild cannot survive on their own because their rumens are not developed enough to allow them to digest sufficient vegetation.

Over the years I have seen a number of newborn and young fawns eat dirt in an apparent effort to get needed minerals that they were not getting in their dam's milk.

Early Life. Newborn fawns spend up to 96 percent of their time curled up in their beds. They do, however, get up and change the location of their beds four to six times each day. They usually do not go more than 10 to 20 feet (3 to 6.1 m) before bedding again. They walk from one place to another with their heads down for the first week.

When the fawns are bedded, they hold their heads up and are extremely alert to all that goes on around them. The scampering of a squirrel, the hammering of a woodpecker, a plane passing overhead, are all noticed and located. If anything, except their dam, approaches, they drop their heads and curl tightly in a circle; their ears are held back along their necks, and they make themselves as small as possible. Their eyes remain open.

When the fawn is not disturbed, it holds its head high and is alert to all sounds; when alarmed, it drops its head and lies curled. In the forest, the spotted coat is good camouflage, since it looks like dappled sunlight on dead leaves. Out in open fields, it is not as effective.

Nadine Jacobsen, of Cornell University, did some very interesting research that has added greatly to our knowledge of fawns and their physiological activities during the first week of life. She put radio telemetry equipment on five fawns that monitored their heartbeats and breath rates. A newborn fawn, at rest, has a heart rate of about 177 beats per minute. This rate drops to about 130 beats per minute at two weeks and is down to 102 beats per minute at one month. An adult deer at rest averages about 37 beats per minute. The fawns had unrestricted use of a large pen with natural vegetation and were observed constantly, both visually and by telemetry. Their reactions to various stimuli were recorded.

Phase I was considered to be a fawn, undisturbed, resting. Its head was up, its ears were forward. It averaged 177 heartbeats and 21 breaths per minute.

At phase II, an unfamiliar observer approached the fawn. The fawn put its ears back, dropped its head, curled into the prone position, and remained motionless. Instantly, within just one beat, the heart rate dropped from 177 beats per minute to 60 beats per minute. In 53 out of 74 tests, all breathing was suspended.

At phase III, as the observer sat quietly about 1 yard (1 m) from the fawn for a short period of time, the fawn's heart rate crept back up to 102 beats per minute and it began to breathe normally.

At phase IV, when the unfamiliar observer again moved, the fawn's heart rate instantly dropped again, to 82 beats per minute, and breathing was again suspended.

At phase V, after the observer left the fawn, its heart rate went back up to 183 beats per minute and its breathing went up to 30 breaths per minute. The higher rate is undoubtedly an effort to repay the oxygen debt incurred during the alarm period.

As the fawns grew older, their normal "resting" heart rates decreased when tested, and conversely, the abnormally low heart rate at phase II increased. After five weeks the fawns responded to an alarm stimulus with increased breaths per minute as they prepared to run to safety.

The survival benefit of the slowed heart rate is tremendous. The drop was so sudden that in a split second the fawn literally went into suspended animation or pseudo-hibernation. All body functions slowed down so that the brain was not deprived of oxygen, nor could the slowed heartbeat be heard. During the period of suspended breathing, the breaths could not be heard nor was there any exhaled carbon dioxide to be smelled.

At birth, a fawn's coat is the reddish color typical of the summer coat of all deer. There are two rows of white spots, one on either side of the spine, from the fawn's neck to the base of its tail. There are usually 60 to 80 spots in these two rows. On either side of the body, 100 or more irregularly sized white spots are scattered rather randomly. The fawn's coats I have counted have had a total spot count averaging between 294 and 306. The hairs of these spots are not white to their roots but only terminal white tufts on regular reddish brown hair. As the fawn grows older, the white is gradually worn away, so that the fawn begins to lose its spots before it molts its summer hair for its winter coat.

The spotted coat provides excellent camouflage when the fawn is lying among the brown leaves and forest litter. The sunlight filtering through the leaves of the trees casts a dappled shadow pattern that matches the fawn's coat and masks its presence.

For the first three or four days, fawns usually lie curled in a small circle with their heads resting on their legs or on their bodies. If frightened, the fawn may stay in this position or, more frequently, it will lie almost straightened out because it can get its head lower in this position. The fawn holds its ears backward and flattened out along its neck.

As more of our woodlands are being converted to agricultural use, does spend more and more time grazing than they do browsing. This has led to an increased number of fawns being dropped in the grasslands. The fawn's camouflaged coat and its ability to curl tightly and remain motionless make it hard to discover them even in the grasslands. Unfortunately, this may work to the fawn's disadvantage. The birthing time of the fawns coincides, in many areas, with the first cutting of alfalfa hay. Many fawns are killed or maimed by mowing machines simply because a farmer could not see them.

After washing her fawns thoroughly, thus removing odor, the doe soon leads them away from the birthing spot. The washing not only serves to clean the fawns but deposits on them the doe's individual odor, from her saliva glands. This helps her to locate the fawns and to tell her own from any others. The washing helps the female to establish the maternal bond to her fawn. But fawns do not imprint as readily to their mothers as do many animals who spend most of their time with their young. This lack of a close bond can be seen by how often a fawn will attempt to nurse any doe, or even a buck, that it encounters. When a fawn is hungry, it doesn't care whom its next meal comes from, it just wants to be fed. When a fawn approaches the doe, she will sniff its coat carefully. If she

The doe consumes everything the fawn vacates while it nurses; by licking its anus, she stimulates the fawn to defecate and thereby removes the scent that could attract predators.

does not identify the fawn as her own by its odor, or with her own salivary odor, she may simply walk off or, more commonly, hit the fawn with her head or forefoot.

The doe does not bed twin fawns together but hides them in different locations, sometimes within 25 feet (7.6 m) of each other, sometimes as much as 250 feet (76.2 m) apart. Then she deliberately stays out of the immediate area. The doe never goes far from her fawns, and she always knows about where they are. She in-

When they heard the doe's mewing call, each of these twin fawns emerged from its separate hiding place. Does never put two fawns in the same place.

stinctively stays away from them so that her body odor will not attract predators to the young.

I have often seen dogs run unawares right past fawns that were curled up in the forest. Many biologists in their reports have claimed that there is no discernible odor to very young fawns. One reason is that they do not move around much during the first week. Without movement, there is little odor. Another, probably more important, reason is that there is no excrement.

While a doe nurses a fawn she licks it continuously, and she concentrates on the anal region more than anywhere else. This constant licking stimulates the fawn's bowels, and the doe consumes the excrement as it is voided. A captive fawn that is not having regular bowel movements should have its anus rubbed with a warm, moist rag.

People trying to raise fawns seldom have as much trouble with constipation as with scours or diarrhea. At the first sign of runny excrement, the milk should be heated almost to boiling before it is fed to the fawns. That is a simple old remedy that works for man or beast. Medicine is available that will stop this looseness of the bowels, such as Kaopectate.

One of the first indications that a newborn fawn is not in good health is that the tips of its ears become slightly twisted. As soon as the physical problem—malnutrition or diarrhea, for example— is corrected, the ear tips will straighten out.

Getting enough proper nutrition into a captive fawn is another problem. I don't know how the practice started, but frequently a fawn is fed ordinary cow's milk cut with water. This mix is less than one fourth as rich in butterfat and protein as a doe's milk, so the fawn is literally being starved to death. Evaporated milk is better: it has about twice the nutrition of whole milk with at least 7 percent butterfat. Today, calf-starters with a fat content of about 10 percent and protein content of about 22 percent are being used with better success. But nothing can take the place of the doe. There is no way to improve on the care provided by the natural mother, and there are only two exceptions to the rule that a fawn should never be taken out of the wild. One is in the event of a known emergency—that is, when there is absolutely no doubt that the fawn is orphaned. The other is when fawns are needed for scientific research, either as fawns or later in life. Fawns reared by their mothers are always wilder and more unmanageable than those that are bottle-fed. This is true of all wild animals. If they are taken from the mother, bottle-fed, and handled frequently, they become tame. In the absence of this handling, it is as if they take in wildness with their mother's milk.

A doe usually nurses her fawns about four to six times in 24 hours, although some does nurse more frequently. Fawns at first need about 2 to 4 ounces (59.2 to 118.3 ml) of milk every four hours. At one week of age, they are taking about 30 (887.2 ml) ounces a day. At three weeks they will take a little more, but by then they are already starting to eat green vegetation. Young does with their first fawns cannot produce as much milk as older does. Researchers, by milking tame does and by calculating how much bottle-fed fawns consume, figure that most does produce 1½ to 2 quarts (1.4 to 1.9) of milk per day.

Like cows, deer do not all give the same amount of milk, and as the quantity of milk diminishes, the butterfat content goes up. The doe's milk has 18 percent butterfat just before the fawns are completely weaned at five months of age.

In the beginning, the doe probably nurses the fawn as much because of the discomfort of her swollen udder as from maternal instinct. Because newborn fawns do not empty the udder at each nursing, the udder is swollen and often hot to the touch. Since the udder is not emptied, it quickly fills up again, and the sooner the udder is filled, the more often the doe is prompted to feed the fawns. As their intake increases, the pressure on the udder lessens and the time span between feedings lengthens.

For the first week or so, the fawn is content to drop to the ground and sleep as soon as it has finished nursing. Up to four days, fawns will lie still if discovered. Between five and six days, at least half the fawns will run off, and at this age it is almost impossible for a human to catch one. It is very difficult to hold a five- or six-day-old whitetail fawn because it kicks, jumps, and thrashes, bawling all the while at the top of its voice. A mule deer or blacktail fawn at this age is still docile.

At seven days of age, almost all fawns will run if they feel they are discovered. They will also often get up and wander about by themselves while the mother is away. Often if a fawn is in an area that becomes too warm, it will move into the shade. When the doe comes back to nurse the fawn, she may have to hunt for it. The doe usually goes to the precise spot where she left it after the last nursing. If she cannot find it there, she will try to track it by the odor left by its hoofs. If the scent is too weak to follow, she will then call—uttering a soft, mewing, catlike sound or a soft grunt as she searches. Upon hearing this sound, the fawn immediately jumps to its feet and runs to the mother. Quite frequently both fawns hear this sound and run to her. She then nurses both at the same time. She has four nipples, and they generally have no difficulty in finding and reaching one.

The fawns may nurse from the side or the rear. If the doe lies down, they will nurse her in that position. Usually, after the first three or four days, the doe stands up while the fawns nurse.

Occasionally, if a doe is killed, another doe will adopt one or both of the orphans. More frequently, a wild doe will allow only her own fawns to nurse. Does in captivity often allow others to nurse. The actions of deer in captivity should not be interpreted as normal behavior. In captivity, deer are forced to be in one another's company constantly; they are deprived of the space that would be normal and desirable for them in the wild.

Joe Taylor had one doe that was an excellent mother and produced a lot of milk. Whenever her fawns ran up to nurse, it was as

A doe nursing her two fawns; twins are common—almost the rule after the doe's first birth year.

if someone had sounded a dinner gong. All the fawns in the pen ran over and started to nurse, too. Of course, as the doe had only four nipples, only four fawns could nurse at a time. She welcomed them all. When young hoofed mammals nurse, they pull on the nipples and then butt upward into the bag with considerable force. There were times when four good-sized fawns were nursing this doe and actually lifted her hind quarters off the ground as they all butted and pushed.

The fawns nurse from 4 to 10 minutes at a time. While they are nursing, a milky froth appears around their mouths. Either the fawns will lick this milk off their muzzles or the doe will do it.

Between the first and second week, the doe begins to have a problem leaving her fawn after nursing it. By now the fawn is strong enough to travel and wants to stay with its mother. The doe usually does not want the fawn to accompany her yet, and she will make the fawn lie down. Usually she does this by pushing the fawn down with her head—or with a foot, if necessary.

Years ago, I had a bird blind in the woods where I was photographing the eggs and young of the red-eyed vireo. I had entered the blind very early in the morning and made no noise because the action at the nest was very slow. The blind was made of burlap and canvas over a wooden frame. I could see out through the burlap mesh but nothing could see me sitting on the inside. I was unaware that a whitetail fawn was curled up about 75 feet (22.9 m) from my blind until a doe came into sight, calling softly. The little fawn bounced to its feet, ran over, and began to nurse while the mother licked it thoroughly. After about six minutes, the fawn had finished nursing. The doe turned to leave, but the fawn followed. Using her head, the doe pushed the little one back, but it persisted in tagging along. Finally, the doe lifted her forefoot, placed it on the fawn's back, and pushed the fawn to the ground. This time the fawn got the message. It lay still and the doe left the area.

If a predator ventures into the area where the fawn is hidden, the doe may bolt for safety while the fawn drops to the earth and remains hidden. The doe's running off may divert the predator so that it focuses all of its attention on her. At times the doe will remain hidden with her fawn, hoping the predator will not discover either of them. And on rare occasions a doe will actually attack a smaller predator.

Research at the Wilder Wildlife Refuge in Texas shows that in some years the fawn mortality is as high as 70 percent by the time the fawns are a month old. Predation by coyotes counts for half of all the fawns killed. The coyotes usually kill by biting through a fawn's head, neck, or spine.

An alert four-day old fawn: now it will run as often as it will hide from danger. Since it was born with long legs, at this stage the fawn can easily outrun a human.

15

Life in Summer

I t is summer. It is hot. In the Midwest the golden wheatfields are ripening. In corn country the leaves of the corn are turning slightly upward and inward, reducing their exposed surface area in order to cut back the sun's evaporation of vital juices. The leaves will continue their gradual inward curling until the next hard rain makes such conservation measures unnecessary.

In the Southwest, the air is redolent with the odor of pine pitch, which is slowly oozing from countless thousands of pines. Even the fragrance of the pine needles is masked by the sharp, raw smell of the pitch.

The temperatures soar in the Southwest until even the lizards do not venture out into the sun. Yet it is not unbearable in the shade. The rising heat thermals create a breeze, and rapid evaporation of sweat cools the body considerably. In the shade of the rim rock, the deer are not uncomfortable.

At the higher elevations of the western mountains, the days are hot but the nights are cool. These mountains are probably the most pleasant place to be at this time of year, and most big-game animals forsake the lowlands for the high country.

In the North Woods, across the continent, life seems to stand still in summer. The growth taking place on all sides is discernible only in one's mind. The leaves hang apparently listless, yet they are all factories working at top speed and on double shifts in the long summer days. They are converting the sunshine into something wildlife can eat.

Heat makes the living easy. It is the one time of year when deer, even the does with their young, can take their ease. In the noonday heat, there are no thoughts of the frigid winter to come, yet all of this growth is in preparation for the long, cold nights that will follow.

Daily, countless millions of living things die so that countless millions can live. In the summer the number of living things is so vast that those eaten are but a fraction of the total. Nature is lavish with seed and young.

INFANCY

Fawns are not ruminants at birth because they are on a milk diet. All that is needed for them to become ruminants is to consume vegetation; the proper symbiotic bacteria are already in their stomachs before they are born.

About ten years ago I was returning to my home from nearby Belvidere, New Jersey. It was about 11:00 A.M. on the last day of May, and the highway traffic was heavy. Four miles out of town, a doe dashed out of a roadside cornfield and tried to cross Route 46. A car ahead of me hit the deer broadside at about 50 miles per hour (80.5 km per hour). Glass and chrome flew and anti-freeze sprayed in all directions as the fan chewed through the radiator. The doe was ricocheted down the highway for perhaps 100 feet (30.5 m). As a deputy game warden, I stopped to be of assistance. Fortunately, the driver was uninjured, though the car was a wreck. The impact not only killed the doe, it burst her abdomen apart, spilling out her paunch, intestines, and two full-term fawns. I was driving a pickup truck, so I put the dead doe in the back. When I gathered up the two fawns, one of them moved. To keep the little fellow warm until I could dry him off, I put him back into his mother's body cavity. I made a brief stop at a garage to send back help for the motorist, then dashed home to care for the fawn.

When I got home I toweled the fawn dry and got some warm evaporated milk into him. I had always heard that fawns get their needed stomach bacteria from their mothers via the milk or by mouth contact. This fawn never saw another deer until long after he was a functioning ruminant, thriving on natural vegetation. That laid to rest another old myth.

When fawns are about three weeks old, they begin to follow their mother. At this time they begin to sample all kinds of vegetation, either as a trial-and-error experiment or in imitation of the doe. At the age of four to five weeks, they are quite selective in what they eat, having already developed taste preferences, and they seek the plants they like. Fawns usually prefer broadleaf forbs. At the age of three weeks, captive fawns begin, voluntarily, to take dry feed. At two weeks they also begin to drink water, although they still depend on milk for most of their liquid and food requirements. One researcher, watching a day-old fawn as it

Fawns often engage in play sessions, both alone and together after their does form family groups. The improvised running, jumping, and bucking are good practice for later life, but sometimes their behavior seems to be simply curiosity.

changed its bed site, saw it step into a small, shallow brook. The fawn lowered his head and drank for about four minutes.

Fawns are usually weaned when they are four to five months old, but research has shown that they are dependent on milk until they are least three months old. If fawns in the wild are orphaned after three months of age, they can probably survive. If orphaned before three months, they undoubtedly die.

The speed of the weaning process depends on the individual doe. I think most fawns would continue to nurse if their mothers permitted it, and some does do allow the fawns to continue long after the fawns have lost their spots. I have often seen fawns nursing at six to seven months. Some mothers are simply more indulgent than others.

When a doe is trying to wean her fawn, she allows it to nurse for a short period and then walks off. What probably helps to

Deer make use of water in summer if it is available, although they can meet their fluid requirements by eating succulent vegetation. Fawns will drink water at 2 weeks old.

prompt the weaning process is that the doe's milk production drops after the first two months. A two-month or older fawn has learned to nurse efficiently, and it takes but a minute or so to empty the doe's udder. The fawn then butts the udder, trying to get more milk. This butting, combined with the fawn's larger teeth, hurts the doe, and she just won't stand still. As the fawn's milk intake diminishes, its consumption of vegetation increases. At ten weeks, signs of wear begin to show on the premolars, indicating that the fawn is now chewing a cud and is a ruminant.

Within three to four weeks after the doe has given birth to her new fawns, her female yearlings rejoin her. If her yearling daughters have fawns of their own, they will bring their fawns along to join the family group. Deer living in woodlands tend to stay in small, blood-related family groups consisting of about three generations, led by the oldest doe. Deer living in more open areas tend to gather in much larger groups. This is probably because there is more food available in the grasslands. It is not uncommon to see five or six does or more feeding as a group while their fawns play.

Play is conditioning for later life. Most mammals engage in it, some more than others. Even if a fawn is by itself, it will suddenly run, buck, kick out, jump, and dash around in circles. Young fawns play tag, run races, and seem to have their own version of "king of the hill." Running and dodging helps to stretch and build muscles, increases lung capacity, develops reflexes, and stimulates the heart. And this is all training that will be used later for escaping from predators or other dangers. It is not done, however, in the manner of a training exercise. It is done with the exuberance and joy of living that accompany youth and well-being. Sometimes the does join in the play. Usually, however, the does watch for danger as the little ones play.

Play is most apparent in the spring, summer, and fall, while food is plentiful and easily gathered. Play has too high a cost in caloric expenditure to be done in the winter, when food is scarce. In any case deer metabolism drops in winter, so the deer just don't have the energy to play.

Working with deer—and other animals—throughout the country, I have noticed that when animals are very young, they are trusting. They have not yet learned fear. At two to three months, they become more wary, but usually depend on their mothers to warn them of danger. But at four to five months, they reach the "bashful age." Paradoxically, they become wilder than the adults. This is particularly true in national parks or preserves, where the animals are semitame but not in danger from humans. Time after time, I have noticed that the adults would stand and study me but the young would break and dash off. Whenever the young did this, it galvanized all of the animals into action. In a twinkling they were all gone. This inherent "wildness" is the quality that keeps wild animals wild.

Toward the end of August, the fawns begin to lose their spots. Blacktails and mule deer wear their spots off at about 8 to 10 weeks, while whitetails lose their spots in 12 or more weeks. All of the deer begin to shed their thin, red summer coats in August. They acquire their hollow-haired, grayish-brown winter coats shortly thereafter. The fall molting takes much less time than the shedding of the winter coat in spring so the heavier fall coat is in place before cold weather arrives. In the fall, as in the spring, the

These blacktail fawns are in the process of losing their spots; all mule deer subspecies lose their spots earlier than whitetail fawns.

difference in the number of daylight hours is the stimulus to molt—all controlled by the endocrine glandular system.

Deer don't sweat as we do, so they need another method of cooling off. To get rid of excess body heat, they breathe at a much faster rate than usual through their open mouths, in the manner of a panting dog. Heat exchange takes place in the lungs, with the incoming cool air picking up the heat from the blood and the blood vessels, and the warm air being expelled. A very hot, late summer persisting after the deer have started to acquire their heavy winter coats causes them to curtail their activities during the day.

FOODS AND MINERALS

A deer's home range is no larger than it has to be to meet the animal's requirements for food, water, and shelter. This range is smallest in the summer in most sections of the country because the requirements are more easily met then. A deer, at any time of the year, is seldom more than 400 yards (365.8 m) from cover, and in the summer cover is usually close on every side.

Food is usually extremely plentiful at this time, although the quality of some of the vegetation lessens as the plants mature. Weather is another important factor in the nutrition of plants. If the summer is cloudy or very rainy for long periods, the protein value of the growing plants lessens. Plants growing in woodlands also lose protein value as they become shaded after the leaves come out on the trees. This affects the whitetail deer in the forests but not the mule deer, because the muley spends more time in open areas. The food of the blacktail deer of the Northwest is subject to more fog, clouds, and shade than the food of any other deer. A list of common, or preferred, summer foods for the three types of deer appears in Appendix I.

All of the deer, in all sections of the country, avidly eat nonpoisonous mushrooms. It has never been understood how they differentiate between the poisonous and nonpoisonous types. Many other animals, such as horses, also eat many mushrooms but avoid the poisonous ones. Red squirrels pick poisonous mushrooms but hang them up in tree branches until the poison leaches out of them. When mushrooms are available, they often form a high percentage of the deer's diet.

Deer evidently need nitrogen, since they have a decided preference for nitrogen-fixing legumes such as alfalfa, clover, peas, soybeans, and lupines. I once saw a huge tree that had been struck by lightning; even the radial roots had been exploded and laid bare. Within a couple of days, the deer had eaten all the loosened soil from around the roots, evidently to get the nitrogen that had been concentrated in the roots by the lightning.

Every year we are made aware of the dangers of lightning and told to avoid exposing ourselves to the possibility of being struck. Every year we also hear of farm animals being killed by lightning, usually because they were standing near a barbed wire fence and the lightning traveled down the wire and jumped to the animals. What is less well known is that every year a number of deer are killed in the same manner for the same reason. In summer most deer, particularly whitetail, feed on farmland or rangeland, and most such areas are enclosed with wire. I have seen several photos of deer killed by lightning and been told of other instances. A close friend of mine had two fine bucks killed at one time. In October 1983, Sanford Olson, of Marquette, Michigan, came across three deer lying together that had been killed by lightning. The lightning had burned furrows across their bodies.

There has been much controversy, over the years, about putting out salt blocks for deer. It always smacked of illegality because shooting over salt was a traditional poacher's trick: some states still have laws on the books prohibiting the shooting of deer over salt. On two occasions, I have seen hunters putting out salt that they were going to stand over. One even had a bag of potato peelings mixed with the salt. They evidently thought the deer were all starved for salt and would smell salt and come running.

Deer do eat salt, but the latest research shows that their need for it is greatly overrated. Again, it seems to be a taste preference. Deer come to salt, but I cannot say for certain whether they locate it by odor or just come across it while feeding and then remember the place and return for more. I suspect they locate it by smell. Their need for it apparently varies with their diet. Most deer have easy access to salt put out for livestock. For example, few of the farms in northwestern New Jersey are being actively farmed today. The land lies idle or is used for horse pasture. New Jersey has more horses now than it had when horses were a major means of transportation. Almost every pasture has a horse in it—and a salt block for the horse that is available to the deer.

I have noticed that deer usually do not lick the salt block itself but eat the salt-impregnated soil beneath the block. One summer, a Scout camp where I lived used a burro for pack trips. A salt block remained in the pasture after the burro was gone. The deer ate a hole into the earth that was 21 inches (53.3 cm) deep by 24 inches (61 cm) across. They also licked clean a handful of stones that they could not remove from the hole. The stones had to be pushed aside each time the deer wanted to eat the earth beneath.

In the wild country, all of the hoofed game animals frequent seeps and springs for all sorts of trace minerals in addition to salt.

The farm right next to my home has long been locally famous for the large number of deer on it. What few people know about the farm is that it has two very large natural mineral licks that have been used time out of mind. Deer trails radiate to these springs like the spokes of a wheel pointing to the hub. The deer don't just drink the water; they eat the clays, which are impregnated with the minerals. They have eaten a crater about 12 feet wide by 15 feet long and 18 inches deep (3.7 by 4.6 by 0.5 m). This crater holds water and makes the entire area muddy, even during dry spells.

Years ago, when the D. L. & W. Railroad still had tracks crossing the river at the Delaware Water Gap, railroaders kept a barrel of salt on the trestle to be used to keep a switch from freezing. Some of the salt leaked through the base of the barrel onto the ties. The deer in the area located this salt and would walk out on the open framework of bridge ties to lick the salt. This was very

A mineral lick frequented by deer near my farm in New Jersey. The mineral-rich clay is so desirable that the deer have excavated the soil to an 18-inch depth.

unusual: most hoofed mammals will not walk on anything with openings that their legs could go through. Horses, cattle, and even antelope will not cross the cattle guards that are used out West instead of gates. Of course, the rails used in the cattle guards are narrower and more difficult to walk on than railroad ties.

A number of years ago, a writer who was not a naturalist or biologist was researching an article on deer. She came across the fact that deer have been known to eat fish. This was the kind of "revealing" data she was looking for. Evidently interpreting her little discovery as having great significance, she wrote that does take their fawns down to streams and teach them to kill suckers with their feet. One doe actually was seen catching suckers as described, and there have been additional records of deer eating fish. But fish are novelty items, not standard fare for deer.

Robert Dailey of Yorkshire, New York, reports a buck and a doe eating bluegills and perch that had been caught by ice fishermen and thrown out on the ice. C. W. Severinghaus and E. L. Cheatum, in their *Deer of North America*, give five instances of deer eating fish. Photographs are not always proof of something happening, but Tom Rogers, Jr., of Altamont, New York, has a movie of a whitetail doe eating fish that he had tossed up on the bank of Lower Sargent Pond in New York.

The oddest example of food intake by a whitetail deer involved a 2½-year-old doe that was killed in Herkimer County, New York, in August 1969. Pieces in her stomach proved that she had eaten a bird, a rufous-sided towhee. Undoubtedly the bird had just been killed when found by the deer and was eaten as carrion.

A deer is not able to convert dry food into water (as many rodents do). Deer have a daily requirement of about 1½ quarts (1.4 l) per 100 pounds (45.4 kg) of body weight in the winter, and 2 to 3 quarts (1.9 to 2.8 l) per hundredweight in the summer. Lack of water is definitely a limiting factor to the deer population in the desert areas of the Southwest. To compound the problem, the feral burros and wild horses are constantly increasing in those areas and often drive deer and wild sheep away from the few open water-holes.

Except in winter, much of a deer's water requirement is met by moisture in the vegetation it eats. Additional moisture is obtained from dew or rainwater on the vegetation eaten. In winter, deer will eat snow or lick ice, particularly if all the free water is frozen over.

In summer, particularly in the north, deer often feed in water. They eat a great many types of water plants, such as eel grass, but they particularly like the algae—the pond scum. I have often

In summer deer sometimes feed in water, and seek relief from biting insects, although they are very wary in water. This doe is feeding on algae.

watched deer eating long strings of this green ooze. The algae is very nutritious, for it is high in protein.

Research in Ohio indicated that some deer may have died of algae poisoning in 1933. I do not know what algae the Ohio deer were eating or if the pond water was contaminated. Deer do not eat the flat, thick sheets of algae floating on the surface. They prefer the gauzy, filamentous type that floats suspended in the water. It is the same algae that beaver and moose eat so readily.

Deer also feed in the water in an attempt to get relief from hordes of stinging, biting, bloodsucking insects that can make much of the forest a nightmare. While the deer may get some relief, they may also encounter serious trouble because they inadvertently pick up snails in the water. Among the parasites that infect deer, some of the most troublesome are transmitted by snails.

PARASITES

Deer are infested, and infected, by a great number of parasites. The larger the deer population and the more deer are crowded onto a range, the more parasites are transmitted from one animal to another. On overcrowded ranges, where there is malnutrition, the impact of the parasites on deer herds is greater. Research with domestic livestock as well as deer has proved that when animals receive sufficient nutrition, parasites are a nuisance but not a decimating factor. Thus, again, proper nutrition is a key to survival.

External Parasites. Deer are subject to all the discomforts imposed upon all wild creatures by deerflies, blackflies, midges, and mosquitoes. From personal experience, I know the extreme discomfort and even agony that these insects can inflict. In the rain forests of southeastern Alaska, or during the fly season in New York's Adirondack Mountains, repellents, headnets, and gloves are survival equipment. The deer have no recourse but to seek relief in water or escape to the highest windswept elevations. George Wright found a buck mule deer on Big Kaweah in Sequoia National Park, California, at 12,750 feet (3,886.2 m). That is the highest recorded elevation for deer that I can find. I have seen mule deer above 8,000 feet (2,436.4 m) in the Waterton Mountains of Alberta, Canada.

I will never forget a photograph I saw in *National Geographic* magazine many years ago, in which blackflies had eaten away the eyelids of a moose. I have seen large scabs where blackflies had

Deer flies biting the nose of a buck in velvet.

eaten large sores on the hind legs of both moose and deer. The deerflies, too, are very annoying. In the summer almost every whitetail deer that I see has a bloody spot on the bridge of its nose where these flies seem to concentrate their biting. A deer's tail is useless as a fly whisk. The deer have to rub the flies off by scraping against vegetation or by using their feet.

The nose botfly (*Cephenemyia phobifer*) is found in about 10 to 15 percent of all whitetail deer. It is also a parasite of both blacktail and mule deer, but the percentage of infection is not as high. The botfly is a beelike insect. Female flies lay eggs around a deer's nostrils and even crawl into the nasal passages. I have seen deer try to thwart these flies by snorting wildly, closing the nostrils, or brushing them away with a front foot. When the flies are attacking, the deer will stand with their heads almost touching the ground, snorting and pawing. Then they often dash off at top speed for a hundred feet (30.5 m) or so.

When the fly's eggs hatch, the larvae crawl up the nasal passages and into the pouch—the nasopharynx—above the soft palate on the rear of the roof of the deer's mouth. Feeding on the deer's mucus, the larvae develop into yellowish grubs over an inch (25 mm) in length. In the spring, these worms crawl back out of the nose or are snorted out in irritation by the deer. After about two months, the worms develop into adult flies that mate and go forth to infest other deer. The irritation of dozens upon dozens of these worms crawling around inside a deer's head, where they cannot be reached or scratched, is a torment we can only imagine.

Prior to 1958, the screwworm (*Callitroga americana*) took a devastating toll on deer across the southern half of the United States. The screwworm fly lays its eggs in cuts or wounds in warm-blooded creatures. Newborn livestock and deer were particularly susceptible, for these flies would lay their eggs in the animal's navels. They also infested sores made by ticks.

It was estimated that in Texas the screwworms wiped out 80 percent of the annual fawn crop. In Florida, the deer herd increased 60 percent after a screwworm eradication program went into effect.

The U.S. Department of Agriculture and state agricultural and game departments collaborated in a program that raised millions upon millions of infertile screwworm flies. After the captive screwworm eggs hatched, the larvae were exposed to a cobalt treatment that sterilized them. Sterilized adult flies were then released by airplane all over the infested areas. These flies mated with the wild flies and, of course, no fertile larvae were produced in the wild. Thankfully, the screwworm has now been brought under control in the United States.

The deer-louse flies, *Lipoptena depressa* and *Neolipoptena ferrisi*, are almost universally found on blacktail deer, and often on

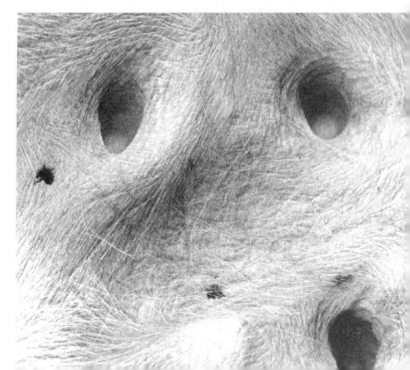

Deer louse flies on a does' udder; although they are a very serious nuisance to deer, they are not harmful to humans.

mule deer and whitetails. These pests seem to cause no real harm but are another bloodsucking annoyance. The adults have wings, which they lose when they are established on a host. From that time on they get around by crawling, and all too frequently a hunter becomes the temporary host when he is dressing or handling a deer. Louse flies do not become established on humans but they can give one a "crawly" feeling. Some blacktail deer have been known to have as many as 2,000 louse flies on them during the heavy, early-summer infestation.

Lice of various kinds are also found on deer but, again, they seem to be a nuisance rather than a threat.

Many kinds of ticks are found throughout the United States and Canada, and they can make life miserable for both man and beast. At times they can kill deer, especially fawns. In 1968, researchers found that the Lone Star tick was killing an estimated 25 to 50 percent of all whitetail fawns in eastern Oklahoma alone. They found as many as 1,300 ticks on one three-week-old fawn. The ticks usually concentrate around the eyes, ears, and mouth, though they can be found on all parts of the body; 20 percent of the fawns go blind. The fawns' flesh literally dies from a toxin injected by the ticks in their bloodsucking, and this opens the way to more infection.

Cooperation between cattlemen and the game departments has led to better control of ticks. Cattle are dipped in vats of insecticide. Killing the ticks that the cattle pick up reduces the number of ticks available to infest deer. The cattle can be dipped every month or so and, in effect, they serve as sponges to "soak up" the ticks. In addition, treated food is put out. Both cattle and wild deer get food pellets to which a systemic insecticide has been added. This gets into the animals' bloodstream and renders their blood toxic to the ticks. Much work remains to be done on this problem, but great strides have been made.

Groups of does are almost always larger than groups of bucks, yet I find less social grooming among the does than among the bucks. Although does are always cleaning their fawns, adult does seldom tolerate one another's ministrations. I have, on a number of occasions, watched two adult bucks groom each other, concentrating on biting ticks off. It is almost impossible for deer to get ticks off their own necks and heads, generally the areas most heavily infested. The deer often try to dislodge the ticks by scratching them, usually with a hind foot, but the most efficient method is mutual grooming.

Deer Ticks and Lyme Disease. In recent years it has been discovered that the deer is one of the major hosts of a disease that now plagues humans: Lyme disease.

The deer tick, *Ixodes dammini,* has a life cycle consisting of four stages: egg, larva, nymph, and adult. The tick is infectious in all stages, but it is thought that it is in the nymphal stage that it is the most virulent. In the nymphal stage, the tick is about the size of the period at the end of this sentence. In the adult stage, but before it swells up with blood, the tick is about 2.5 mm in length, about one-quarter the size of the common wood tick. It does not have white markings in the center of its back, but the female has a light edging to its body.

A similar disease called *Erythema Chronicum Migrans* was known in Europe from about 1910. It caused a red rash that lingered and grew larger. The first case of Lyme disease was discovered in this country in 1969, when a grouse hunter came down with it. The disease is named after Lyme, Connecticut, because it was there that researchers at Yale University first discovered the link between the deer, the tick, and humans.

In 1981, Dr. Willy Burgdorfer, an expert on tick-borne diseases, was working in Montana on Rocky Mountain spotted fever, a disease also carried by ticks and now found nationwide. Dr. Burgdorfer was looking for the organisms that cause the spotted fever and was checking lab specimens from Long Island, where Lyme disease is known. He saw long, threadlike spirochetes that moved with a corkscrewlike motion. He suddenly realized that he had found the organism that caused Lyme disease. The spirochetes have been named for him, *Borrelia burgdorfer.*

In each of the last three stages of its life, the tick must have a meal of blood in order to progress to the next stage, so it needs three meals in the two years that it lives. The adult female, after getting a blood meal, drops off its host and deposits thousands of eggs in the soil. They may hatch in about two weeks or overwinter in the soil. Upon emerging from the soil, the nymph crawls among vegetation until it is guided to some warm-blooded creature, such as a bird or animal. The larvae concentrate on mice, particularly the white-footed mouse and the meadow vole. It is from the tick in this stage that many humans are infected. After another blood meal, the tick again drops from its host and goes through the third change, to adult form. In adult form the tick attaches itself to any kind of warm-blooded creature. Ticks attached to birds are carried south during migration, allowing the tick to leapfrog thousands of miles across the continent. Dogs and humans become sick with the infections. None of the wild creatures do.

The spirochetes cannot, of themselves, get into a human's bloodstream; they are injected there by the ticks. If, in any of the three stages, a tick draws blood from an infected host, it passes the infection on to its next host. The reason a mosquito bite itches is that the mosquito, in order to draw blood, pumps a blood anticoagulant beneath a human's skin. It is the anticoagulant that causes the itching. It is also in the anticoagulant that the malaria organisms are passed on. If the tick has picked up the spirochetes from a previous host, it passes them on through its own mouth parts as it burrows beneath the human's skin.

A single deer may be host to hundreds of ticks through the tick season, May through August. In some areas, 100 percent of the deer tested have been found to be infected. The deer is known to be the primary host in the second, critical stage of the tick's development. In some places, such as in Westchester County, New York, a hot spot, authorities have tried to alleviate the condition by reducing the deer herd. There is even some talk of deer eradication programs. I can understand the fear that folks have of catching the disease. However, it has now been proven that dogs, cats, cows, foxes, raccoons, opossums, skunks, squirrels, rabbits and chipmunks, as well as over 40 species of birds, are also host to this tick. The eradication of deer will not alleviate, let alone eliminate, this problem.

According to the National Centers for Disease Control in Atlanta, Georgia, Lyme disease has been reported in 24 states so far, but it has undoubtedly gone unreported in many others only because most doctors are not familiar with the tremendous variety of symptoms. Most of the reported cases so far have occurred in Massachusetts, Connecticut, Rhode Island, New York, New Jersey, Wisconsin, and Minnesota. Localized outbreaks have occurred in Georgia, Texas, Oregon, and California. The disease is spreading rapidly, jumping from 226 confirmed cases in 1980 to 5,000 in 1988. Researchers are sure that the case numbers will skyrocket when proper diagnosis of the disease is established.

What makes the identification of Lyme disease so difficult is that some of the people who have it exhibit none of the symptoms. Some people seem to build up an immunity to the disease. Although seldom fatal (there has only been one reported death so far), the disease can be extremely debilitating and painful.

Many people are unaware that the tick is on them in the first place. They may not become aware of anything until the tick has become engorged and dropped off, and the itching begins. One of the first symptoms of Lyme disease is a large, spreading red rash with a light-colored center that itches. Another early sympton is lethargy: it becomes an effort to do anything or even stay awake. Lyme disease causes headaches, abdominal pain, fever and chills, and flu-like symptoms. Unless the disease is diagnosed correctly, it may be treated as the flu. This is most unfortunate because, if the disease is treated promptly, during this first stage, with tetracycline, penicillin, or other antibiotics, it can often be controlled. Second stage symptoms are heart arrhythmias (some victims have to get pacemakers), neurological disorders, meningitis, paralysis, chronic skin disease, arthritis, and pain.

Many people are panicked by the prospect of catching this disease. The surest way to avoid contracting the disease is to stay indoors. Are you going to do that? No, and neither am I. In my photography I concentrate on deer, and I am in the best deer areas most of the time. I believe in prevention, and I have taken to using Permanone tick repellent on my pant legs, cuffs, and socks. One outdoor writer living near an infected area announced that he would never use a repellent because it would scare the deer. His attitude is foolhardy; getting the disease is too high a price to pay for a shot at a deer. The deer won't smell you or the repellent either if you watch your wind direction.

In a known tick hot spot, use Permanone. Check yourself carefully when you get home and have someone check you in places you can't see, such as your hair. Remove any ticks by putting a drop of alcohol, gasoline, or kerosene on the tick itself. The ticks can be worked out with tick tweezers, but take care not to crush the tick or leave any of its mouth parts embedded in your skin. Remember that not all ticks have the disease, but at the first sign of a rash, go to your doctor for prompt medical treatment. The possibility of getting Lyme disease is just another price we have to pay for enjoying the outdoors. But you shouldn't get it if you use the proper precautions.

Internal Parasites. Internally, deer are parasitized by a vast array of worms. There are lungworms, brain worms, throat worms, eye worms, foot worms, bladder worms, colon worms, muscle worms. There are specialized parasitic worms that show up in almost every part of the body. Most of these parasites do not greatly influence overall deer populations, today, though some are deadly.

The lungworms, *Leptostrongylus alpanae,* are probably the most common internal parasites of deer. They are very similar to the cattle lungworm, *Dictyocaulus viviparus,* which the deer may also have. As many as 40 percent of deer may be infested. The female lungworms lay hundreds of eggs in the tissues of the lungs. Heavy infestations create damage to the deer's lungs, causing bronchial infections, often resulting in pneumonia. After hatching, the larvae crawl up the bronchial tubes to the trachea, where they are swallowed by the deer and pass out of the body with excrement. The larvae infest snails, which are afterward eaten by deer, completing the cycle.

Liver flukes, *Fascioloides magna,* are another parasite dependent on snails as an intermediary host. This is a parasite very often picked up by deer when they are eating water plants. Needless to say, it is not a parasite of any importance in desert regions, but it is very prevalent in the mountainous or wooded areas of the North. About 65 percent of the northern whitetails are infested.

At one time I thought it odd that deer ate so many snails, but that was before I myself ate a great deal of watercress. In some of the springs where I pick watercress, I have found hundreds of snails fastened on a couple of dozen plants. Although I now carefully clean the watercress, I've probably also swallowed a few snails now and then. I can see how easily the parasites of deer are transmitted via snails.

Liver flukes look very much like freshwater leeches. Flat and brown, they may be 3½ inches (8.9 cm) long and 1½ inches (3.8 cm) wide. They form pockets or holes in the liver. At times the liver is so badly infested that it is mushy. This can cause a deer's death.

The eggs of the liver fluke are passed out of the deer's body through the intestinal tract. The eggs hatch in ponds and lakes, and the larvae then infest the water snails. After further development, these larvae pass from the snail's body and attach themselves to the stems of water plants. The larvae develop a protective coating while they wait for the plant to be eaten. When a deer eats the water plants, it ingests the parasites. The protective coating of the larvae dissolves in the deer's stomach, and the larvae make their way to the liver to complete their adult development. Liver flukes are also found in sheep and cattle. Fortunately, they do not infest humans.

The deer brain worm, *Parelaphostrongylus tenuis,* too, uses snails or slugs as intermediary hosts. The whitetail deer picks up the worm by ingesting an infected snail or slug along with grasses. Once inside the deer's stomach, the worm's larvae leave the snail, go through the stomach wall, and get into the abdominal membranes. Traveling through the tissue, the larvae move up and get into the deer's spinal cord. The worms work their way up the spinal column and continue their development. In about forty days they emerge, as adult worms, in the spaces around the brain. There they breed and lay eggs in the tissue. Each female worm produces several thousand first-stage larvae. The eggs and larvae pass into the bloodstream and the lungs. The larvae break through the tissue into the air sacs of the lungs and are coughed up into the deer's mouth. They are then swallowed and pass out of the deer's body with the feces onto vegetation, where they again infest snails to complete the cycle.

Brain worms infest 41 to 81 percent of all whitetail deer. They are not fatal to whitetail deer because whitetails have been exposed to them for eons and have built up a resistance to them. That brain worms were fatal to moose was discovered in 1963 by a Canadian biologist, Dr. Roy C. Anderson. Elk, caribou, and mule deer also succumbed, but less often than moose. The moose ate infested snails, or vegetation on which snails had deposited the larvae. When the larvae attacked the moose's brain, the moose became disoriented, lost its fear of humans, aimlessly wandered about, went blind, lost muscle control, became paralyzed, and finally died. In area after area, as the whitetail deer increased its range, the moose population crashed. For example, years ago the moose population in Maine plummeted because of overhunting and an invasion of whitetails carrying the *tenuis* parasite. Then, as the forest habitat matured, the deer herd declined. The moose population in Maine began to explode. Over 20,000 animals were located just in the northern half of the state in 1988. Moose also began coming back in New Hampshire, Vermont, and New York State. Evidently the moose acquired an immunity to *tenuis*. Maine's deer herd is also recovering nicely.

The mule deer herd is declining over most of its range in North America, particularly in the Canadian prairie provinces. Much of this decline is attributed directly to the modification of the mule deer's habitat by man, making it more favorable to the whitetail. Part of the decline also stems from the increased exposure to *tenuis* following on invasion by the whitetail. Roundworms, flatworms, and pinworms are found in the deer's stomach.

Most of these do little actual harm except in times of stress.

Many species have evolved in such a way that they now draw sustenance directly from the deer's food supply rather than from the deer's tissues or blood. These worms are present in about 81 percent of all deer.

DISEASES

In 1924, infected livestock transmitted hoof-and-mouth disease to the blacktail deer in Tuolumne County, California. Both the cattle and the deer were shot in an effort to curb this very infectious disease. Of the 22,214 deer that were killed, 10 percent were found to be infected. Thankfully, this disease has not been a problem to deer since then, mainly because of improved control of the disease among cattle.

In August, 1955, a highly fatal disease swept through the deer herds in Morris, Essex, and Somerset counties in New Jersey. The deer that had this disease appeared to be in a state of shock. All of the deer that exhibited symptoms died. They completely lost their fear of man and usually lay on the ground with their necks outstretched and their ears drooping. Their summer coats were very rough. When they got up, they staggered around. They were all running a high fever, and most of the carcasses were found in or near water where the deer had gone go alleviate their heat. So far as was known, this disease had not been encountered before.

Dr. Richard Shope of the Rockefeller Institute for Medical Research led an investigation, aided by Les MacNamara and Bob Mangold. The autopsied deer all exhibited identical symptoms. All of the organs showed evidence of ruptured blood vessels and hemorrhaging. Most of the organs were heavily discolored—purplish-black with accumulated blood. There were also accumulations of blood in the groin area and in the intestines. A clear, straw-colored liquid was present in the abdominal cavity and the heart sac.

An extensive research program combined the forces of the New Jersey Game Department, Department of Health, Department of Agriculture, and State Police. It was thought at first that the deaths were being caused by poisonous agricultural sprays.

When several sick deer were kept in a pen with healthy deer, the healthy ones did not contract the disease. Everyone breathed a little easier because this proved that the disease was not infectious. Injections of ground tissue from deer that had died of the disease produced illness and death in the experimental deer. It then became known that the disease was transmitted via some biting vector, such as flies or mosquitoes. The vector is the biting midges of the genus *Culicoides*. With the first frost, all evidence of the disease vanished.

Continued investigation showed that the illness was caused by a filterable virus. Research also showed that it could not be transmitted to other animals or birds, nor could it be transmitted to man. As this was the first time the disease had been encountered, the research team simply named it epizootic hemorrhagic disease of deer.

Hundreds of deer died from the disease that year, and many of the carcasses were not found at the time because of rapid deterioration caused by the summer heat. The disease has since been found in many other states. It is always connected with overpopulation of deer. In James City County in Virginia, a couple of hundred deer died in one season. Parts of North Carolina have annual outbreaks in areas restricted to "bucks-only" seasons. States that have regular doe days, or where a good part of the entire deer population is harvested, are relatively free of the disease. Dr. Frank Hayes of the College of Veterinary Medicine, University of Georgia, encourages the hunting of deer during outbreaks of the disease to lower the deer population to healthy levels. Since there is no danger to man, Georgia, North Carolina, Arkansas, and South Carolina regularly conduct normal hunting operations during outbreaks.

The hemorrhagic disease hit New Jersey hard again in August 1975. A total of 358 deer are known to have been killed by this disease, and, again, many carcasses must have gone undetected. The area in the state affected most was the township where I live. It had not been uncommon to see up to a hundred deer at a time in some of the fields on neighboring farms. The deer did not suffer much from starvation because the biggest gun club in the area had an extensive winter feeding program. So lack of natural food was not a population check, but overcrowding, produced by the feeding, was checked by the disease. To compound the problem, the club did not allow hunting in the fall of 1976, which would have brought the herd to a more healthy level. The best and most efficient check on this disease is a greater harvest of deer in areas threatened with overpopulation.

In 1960 another new disease—one transmissible to man—was discovered in New York State. The disease produced boils and lesions on deer. Four people handling infected deer also developed small boils on their skin. The boils were not painful, and most of them and the resulting scabs disappeared in a week or two. There were no other ill effects to the persons who contracted this disease. Research showed that the causative agent was a *Dermatophilus*, meaning "lover of skin." This disease has been found in both wild and domestic animals in South America, Europe, Africa, Asia, and Australia. It had been encountered only once before in the United States, on domestic animals in Texas. The disease, known as *Streptothricosis*, is not considered a threat.

Anthrax, a disease caused by a bacteria called *Bacillus anthracis*, is highly contagious to deer, cattle, and humans. Outbreaks have occurred all over the world since records have been kept. Thanks to improved veterinary medicine, anthrax is not a problem in the United States, although it remains a constant threat. It is usually fatal to deer, and can be to cattle and man if untreated. Localized hot spots in this country are in areas having lowlands flooded periodically by rivers such as the Ohio and Mississippi and in limestone outcropping areas, such as parts of Texas, Georgia, and Florida. The bacterium favor a soil pH higher than 6.0. Anthrax is usually characterized by dark, bloody discharges from various body orifices. Where anthrax is suspected, animals should not be handled. On confirmation, the animals should be destroyed and their bodies burned or deeply buried.

Deer sometimes exhibit benign tumors known as *papillomas, fibromas,* and *lipomas.* These are caused by viral infection, cannot

This is a benign fibroma tumor on a whitetail doe.

be transmitted to man, and are seldom very injurious to the deer. Papillomas are the form most frequently seen; even so, fewer than 1 percent of the deer have these tumors. They are hard, warty growths on the skin, ordinarily ranging from one to a dozen or more and from the size of a marble to the size of a golf ball or even a bit larger. Almost every year, newspapers and sporting and conservation magazines carry photographs of deer with encrusted areas of papillomas, but unless the growths become massive, they do not incapacitate a deer. When a deer is skinned the growths come off with the hide, and the meat underneath is both safe to eat and as tasty as the rest. Cattle, too, have papillomas, but of a different type, and the growths are not transmitted from livestock to deer or vice versa.

The fibromas are less often seen. They develop lesions that look as if a scab had been knocked off. Fibromas do not affect a deer's meat. The lipomas are very rare. They are tumors found in the deer's fatty tissues.

Deer, like most animals, are afflicted with many other viral, bacterial, and parasitic ailments, but those I have described are the most important, and they are the ones most likely to be encountered by hunters.

Research has shown that during the period when bucks are becoming sexually mature, at the end of summer, they are more susceptible to disease. The physiological changes taking place within their bodies are somehow connected with a lowering of their resistance.

ACTIVITIES

The habits of all deer change because of the heat. In summer they feed primarily at night. The does and fawns may come out just before sunset, their shadows covering long distances on the ground. The bucks are seldom seen while the sun is still visible. The deer feed heavily for a couple of hours at dusk. They seldom have to feed long, because this is the season of abundance. They will then probably bed down in the fields or clearings where they have been feeding because the vegetation is high and affords excellent protection. The deer will feed around 11:00 P.M. and again after 1:00 A.M. A major feeding period begins just before dawn. The deer will feed until the sun is up, but when the temperature gets above 60°F (15.6°C), they head for shady cover. All head for brush, but mule deer do not seek out the heavy stands of brush that the blacktail and whitetail favor. Over much of the mule deer's range, such heavy cover is not to be found. I have often watched mule deer lying under pine trees or in the shadow of upturned rocks get up and move several times during the day as the sun's progress pulled the shade away.

In Yosemite National Park, I once had the good fortune to watch some blacktail bucks bed down for the day. There had been some mist in the morning, and, because it was cool, the deer were out later than usual. I had been out since the first crack of light, following the deer, taking notes and photographing their activities.

It was a Sunday and, with the light, the tourist traffic increased. Within an hour it had gotten almost bumper to bumper. Although I was only 200 feet (61 m) from the road, I don't believe

All I could see of a grown blacktail bedded down in some bracken fern in Yosemite. During the summer, bucks are very retiring, and often come out to feed only in the cover of darkness and early morning.

any of the tourists saw me, or the deer; I don't know if they even saw Yosemite Falls behind me.

Several bucks had been feeding together when I first found them, but an hour after dawn, they separated by about 100 feet (30.5 m), and, one after another, they bedded down. The largest buck disappeared into the forest. The other two bucks bedded down in the open area that edged the forest, in a sea of bracken ferns standing about 2 feet (61 cm) high. If I had not seen the bucks actually bed there, I could never have found them. Each buck very carefully put his head beneath the fern canopy and then sort of slid in under and around the stalks of the ferns without breaking any of them. When each was bedded, the ferns moved back into place, completely hiding the deer. It was the most amazing use of cover by deer that I have ever seen.

Comparative inactivity during periods of high temperature lessens the stress on the deer's body. The does are nursing their fawns, the fawns are converting their food to body growth, and the adult bucks have about completed their antler development.

In midsummer, the buck's antlers have reached their maximum growth. The long daylight hours have spurred the pituitary gland's stimulation of the testicles. As the testicles begin to increase in size and drop down into the scrotum, the male hormone, testosterone, is produced, and the first live sperm is formed in the semen. It is testosterone that causes a buck's antlers to harden and the velvet to dry up.

The middle of summer causes a tremendous upheaval in the life of the deer in California and Alaska, because the hunting season opens. I have spent four summers in Alaska, where the hunting seasons open early because the summers are so short. I have encountered lots of ice and snow there in August. I was surprised to find that California's season opened in the summer, however, especially in view of the fact that the weather is so hot that keeping the meat from spoiling can be a problem.

The biologists I questioned said that tradition was the main reason for the early California season; the deer have always been hunted there at that time. But there are better reasons. The blacktail deer and many of the mule deer in California are at their greatest body weight in midsummer. The vegetation dries very rapidly in late summer due to the lack of rain. Both the protein content and the digestibility of the vegetation drop, and so does the weight of the deer. And bucks taken prior to the rut provide better meat. During the rut, the bucks lose a lot of weight and their muscles are tougher because of the constant running after does.

Evidently, California hunters are used to hunting while it is hot, though hunters in most sections of the country associate deer hunting with cold weather. The early season has some disadvantages. If a deer is shot but lives long enough to move a considerable distance, particularly in heavy cover, it may have been dead for hours by the time the hunter finds it. On a very hot day, the meat may be getting "high" by then. Furthermore, I think most hunters would agree that hunting is harder before the rut because the bucks are more cautious, more wary than when they are preoccupied with the urge to mate. The major disadvantage, it seems to me, is that most hunters prefer trophy bucks, the bucks that do most of the breeding. It has to be detrimental to the deer herd to remove the genes of the finest bucks from the gene pool prior to the breeding season year after year.

Arizona, Colorado, Nevada, South Carolina, and Utah follow California, opening their hunting seasons during the latter part of the summer.

With his antlers hardened and the testosterone coursing through his body, a buck's maleness asserts itself. The most apparent change is in his increased activity and aggressiveness. All spring and for most of the summer, bucks are shy and retiring because, in effect, they have been neutered by nature. But now, instead of seeking to avoid hitting the brush with their antlers, the bucks seem bent on subduing all the bushes and saplings in their area.

The velvet is usually (though not always) stripped from the antlers within a 24-hour period, but I have noticed that most of the stripping is not done in a single night. Usually, some of the velvet is peeled off the first night so that the tips of the tines can be seen. Most of this stripping activity must be nocturnal, because I have only been able to get a couple of photos of deer with partially peeled antlers. After the second night, the brown, bloodied antlers are usually fully exposed, although some strips of velvet may still hang from parts of the antlers, usually at the base.

A buck is very thorough in removing the velvet. Picking a resilient sapling, he rubs his antlers lengthwise along the trunk. He does this in between the tines, on the inside of the antler's curve, and on the outside. He even turns his head sideways, parallel to the ground, to try to get the velvet completely off the antler bases. Frequently, bucks become very excited during this procedure. The more they rub, the more it stimulates them. The rubbing becomes faster, and the pushing harder, and the bucks really start to "work out" on the brush.

Joe Taylor told me of a buck on his preserve that completely removed the velvet from his antlers in about 10 minutes. The buck's actions were frenzied. He started to peel the velvet by rubbing against a sapling, then whirled and rubbed against the fence, then back to the sapling again. All of this was done with considerable force and great speed. This animal was the dominant buck on Joe's preserve and in excellent condition, and he peeled early. Joe thought the buck was actually peeling a little *too* early because more blood was visible on the velvet and the antlers than he had ever seen on any other buck. Joe told me he has also seen a couple of the bucks take three days to get the velvet off.

Although it is usual for the biggest, most dominant buck to peel the velvet from his antlers first, this is not always the case. Among deer on good range, where all of the bucks have had access to equally good food, some of the smaller antlered deer may actually peel their velvet first.

Like a boxer in training, the bucks begin to spar with "punching bags"— saplings and small trees. Rubbing starts with the shedding of antler velvet, but continues as a form of rutting behavior.

On command from the deer's endocrine system, the blood supply for the antler velvet is cut off and the velvet "cracks," and is then rubbed out.

For most of the deer, over most of North America, north of the 31st parallel, the buck's antlers are full grown by August 1. Then they begin to harden from the base out, so that by August 15, all but the tips are solidified. By August 21 the tips have hardened, and by the end of the month, the velvet often "cracks." Most bucks peel the velvet from their antlers between September 7 and September 15, give or take a few days, according to weather, food, and the physical condition of the deer.

Dominance among any of the wild creatures is not a static condition; it is very fluid, with all of the subordinant animals constantly pushing, testing, trying for upward mobility to increase their own status. Often this status changes overnight; often the change is of short duration; but the pressure to change is always there.

The large-antlered buck will be dominant while the antlers of all of the bucks are developing. Any upstart that cannot be put in his proper place by a "hard look" will be pounded into place by the hoofs of the boss buck.

The largest-antlered buck will again be dominant when all of the bucks have peeled the velvet from their antlers. Any subordinant is free to challenge the system, but the slashing antlers of the dominant buck quickly dispel any notion of self-aggrandizement a lesser buck may have.

There is just one time when these basic rules do not hold true. On a number of occasions, among both wild and captive bucks, I have seen bucks with long spikes or four points peel the velvet from their antlers before the big bucks. The early peeling brings with it a sense of power, a sense of opportunity. For the several days or a week until the big buck peels, the lesser buck dominates everything. It's first-peeled, first to boss.

The dominant buck does not want to accept this status change, and he usually gets raked a few times by the lesser buck's antlers. The lesser buck is quick to push his advantage and take the opportunity to push all of the other bucks around. After being pronged a couple of times, all of the other bucks are quick to give way when the lesser buck approaches.

It's a heady feeling for the lesser buck, but it is short-lived. As soon as the rest of the bucks shed their velvet, the entire social system returns to normal. The dominant buck is quick to pound the upstart into submission, if he can catch him.

Some bucks seem to have a favorite rubbing tree, and most seem to have a favorite area. This rubbing and fighting with the bushes is done far more often than most people suspect. In a short time, many saplings and bushes in many areas will bear the bright white scars of buck rubs. People who think that bucks rub trees only to get the velvet off may get the impression that there are far more bucks in the area than there really are, not realizing the role of the rubs in the depositing of scent from the forehead glands. Most bucks will scar many saplings in the process of getting ready for the rutting season.

16

Life in Autumn

T he full moon never seems as big nor as full as it does in autumn. Perhaps this is due to the clarity of the night air now
that the chill has erased the haze of summer. Perhaps it is
because this is the season of fullness. This is the time when most
living creatures enjoy the fullness of life, the season when the
growth of most living things in the northern hemisphere reaches
a climax. The fruit trees have yielded their fruit, although some
apples may remain to be picked. If it has been a good mast year,
the nut-producing trees are raining down their seeds. Most of the
wild vegetation is slowing its pace, its growth being consolidated.

For the broadleaf trees, one job remains. They must drain
their plumbing systems before the winter freeze. Moisture is re-
called from each leaf, each twig, each branch, and from the trunk
for storage underground. If this moisture were not removed, the
trees would freeze and split during cold weather. Residual sugars,
or the lack of them, will determine the color of the leaves. The
brilliant leaves of the sweet-sap trees set the forest afire in a blaze
of color. Even those trees that had bitter or tannic acid in their
veins produce a variety of browns to delight the eye with their
subtle somber hues.

While things of the plant world have passed their peak of activity, autumn is a frenetic time for the things of the animal world.
Many of the birds have gone south, not in quest of summer warmth
but in response to diminishing light, the prod that sends them to
southern climes where their stomachs can be filled. Most of those
that feed on insects have already left the North. All summer the
woodchucks have been feasting, and now, with their days of activity numbered, they compulsively gorge themselves. The woodchucks are not alone. The raccoons and bears cannot escape winter
by hibernating, as the woodchuck will, and they frantically consume everything edible in an attempt to layer the fat upon their
bodies to sustain them during their periods of inactivity. The rac-

coon will actually consume a third of its body weight in food each day if it can get it.

Deer are said to be *crepuscular* creatures, because their periods of greatest activity take place just before and after sunrise and sunset. More deer are active before dawn than after and more are active after sunset than before—particularly the bucks. Their daily activities are greatly influenced, however, by the weather and the phase of the moon. A heavy shower or thunderstorm will cause them to seek temporary shelter, though a protracted rain will not. If you see deer out feeding during a hard shower, you can count on the storm lasting several days. If they are seen feeding much earlier in the afternoon than normal, it is because the night will be dark and stormy. Deer will also feed earlier if the moon is in the new phase and the light at night will be minimal. Deer can see well in the dark, but they curtail their activities during the darkest nights. Conversely, they may not come out until after dark on the nights of the full moon; they can then feed freely all night. A full moon also makes deer more nervous, because it deprives them of their protective cover of darkness.

FOODS

I am not sure whether deer feed more compulsively in spring or in autumn. The early spring feeding is an attempt to rebuild the bodies that, in some cases, did little more than harbor life. It is certain that they consume much more food in autumn simply because so much food is available. It is now known that deer are subject to what is called *obligatory lipogenesis;* they must build up fat reserves. In late November, this accumulation of fat will come at the expense of body growth. Everything is subordinated to lipogenesis. But for now, the bounty of the year is theirs. They eat compulsively because they know instinctively that this period of plenty is brief, and if they do not take advantage of it, they may forfeit their chance of survival. It takes so little to blow out the light of life.

I can state unequivocally that, in the northeastern, central, southern and west coast regions of the United States, acorns, when available, are the favorite whitetail food. There are times when acorns comprise 80 percent of the diet. When the acorns drop, deer forsake their regular haunts and most of the other available foods to feast on these succulent brown nuts. Although they will eat all kinds of acorns, the deer of my area have a decided preference for white oak acorns, which are sweeter than others. White oak acorns

The acorns and leaves of the oaks favored by deer on the West Coast, from top left: the blue oak, the valley oak, coast black oak, and the interior oak.

Acorns and the leaves of the mast oaks favored by eastern whitetails: the red oak at left, a white oak at right, and the chestnut oak. The white oak is identified by the rounded, blunt lobes of the leaves.

contain the least amount of tannic acid. Time after time, people who are used to seeing deer feeding in old orchards on the dropped apples, or in corn fields, ask me where the deer disappear to. Bowhunters, particularly, ask, because in many states the early bow season opens in mid- to late September, which coincides with the dropping of the acorns. Bowhunters watch the deer all summer; they often place their tree stands based on the movements of the deer in summer, and suddenly the deer disappear.

Of course, the deer have not disappeared; they are just concentrating on acorns. When deer feed in orchards or fields, they are easily seen. But when the acorns drop, deer stay in the woods and up on the hilltops. They need not travel because the acorns are in their bedding areas.

The whitetail is not alone in this taste preference. Acorns are very important to both the mule deer and blacktail, as well. These nuts are more prominent in the blacktail's diet than in the mule deer's, however, because the muley spends much of its time in areas too high for oaks to grow.

Not all oaks produce a good crop every year, and not all of the oaks drop all of their acorns every year. Species belonging to the white oak group will produce and drop mature nuts each year. Those of the red oak group take two years to produce mature nuts. Quite often some of the "baby acorns" fall off the tree in their first year, but the difference between these and the mature nuts is very apparent. An individual whiteoak tree seems to produce a heavy mast crop every third year, while some other trees, such as the rock oak, have a more sustained yield. Some trees just seem to be cyclic. Rainfall, heat, and the length of the growing season also affect mast production. Sometimes the quantity of acorns varies greatly in adjoining localities.

I have my own rough way of estimating the relative production of the mast crop. I wear size 11½ boots. If I put my foot down when the nuts have fallen and cover nine acorns with one shoe, that's a good crop. When I cover a dozen, that's a fantastic crop.

I have found that the average oak tree must be 15 to 20 years old before it begins to produce nuts. A tree of that age will be 6 inches to 10 inches (15.2 to 25.4 cm) in diameter. According to William Dasmann "a stand of 27 oaks per acre, averaging 10 inches in diameter, will produce an average yield of nearly 150 pounds (68.02 kg) of acorns. This can double or triple in good production years." The best mast production comes from oaks that are over 40 years old, because those trees will have matured and will have the widest crowns. The more sunlight each tree gets, the greater its mast potential. Also, after an oak matures, more of its absorbed food can go to acorn production.

In one area, the Pennsylvania Game Commission found that in 1975, the acorn production was 8 pounds per acre (8.8 kg per hectare), and in 1976 it reached 528 pounds per acre (591.8 kg per hectare). Wild turkeys have been seen following after the deer to feed on the acorns that are missed when the deer paw down through the snow.

According to old-timers, a heavy nut crop foretells a hard winter. The old-timers may be right. The extremely good crop in 1976 was followed by an extremely cold winter, and I have seen the same thing happen before.

In my area, the deer eat all the acorns they can get, but in a preferred order. White oak acorns are eaten first, then the acorns of the pin, the red and the black, the scrub oak, and finally the large rock oak. Although the rock oak belongs to the white-oak group—and produces the largest acorns—deer simply do not have as great a liking for this species as for the others.

Acorn Usage by 400 Deer in Missouri, 1948–1953

Month	Number of Stomachs	Percentage of Total Diet
January	8	32.4
February	23	32.0
March	23	16.4
April	17	16.0
May	6	14.3
June	4	0.0
July	8	7.5
August	10	0.1
September	11	31.4
October	16	62.4
November	131	43.8
December	183	45.7

(Reprinted from "Acorn Yields And Wildlife Usage In Missouri," by Donald M. Christisen and Leroy J. Korschgen, in **Twentieth North American Wildlife Conference,** p. 345.)

H. R. Gilbert and G. H. Hart tested the chemistry of the acorns eaten by Columbian blacktail deer and found that the blacktail's preferred acorn, that of the California black oak, has only 1.9 percent tannin; it has 4.3 percent protein and 14.7 percent fat. The

Life in Autumn

interior live-oak acorn has 5 percent tannin, 3.5 percent protein, and 17.8 percent fat. In autumn, acorns make up 53 percent of this deer's diet. A surprising discovery was that in the month of May acorns form 83 percent of the blacktail's diet.

Comparison of Chemical Values of Different Kinds of Western Acorns

Species	Crude Protein (percent)	Fat (percent)	Tannin (percent)
black oak	4.3	14.7	1.9
scrub oak	2.6	4.3	4.1
blue oak	3.5	5.8	2.6
water oak	3.1	5.5	3.2
interior live oak	3.5	17.8	5.0

(Reprinted from William Dasman, **If Deer Are To Survive,** p. 64.)

The California chaparral has countless thousands of oak trees, but they do not compare in number, size, and productivity with the oaks of the eastern two-thirds of the continent. Competition for acorns is always keen, but more so in California; the dropping of an acorn sets off a race between the deer, the squirrels, and the acorn woodpeckers.

All acorns have a low protein content, but they are all high in fats and starches. They also have the advantage of being easily digestible, and their nutrients are readily absorbed. Without loosening a deer's bowels, acorns are processed and passed through the body in a very short time. This allows the deer to consume greater quantities of acorns per day than it would of other foods, which raises the daily protein intake. Conversion of the fats and starches to the deer's body fat is very rapid. You can almost see the deer putting on weight. In adults, substantial supplies of body fat are built up in less than two weeks. The first fat is stored beneath the skin, over the back and hams. Then the fat is stored under the belly skin and in the abdominal cavity around the organs. Then skeins of fat, looking like a fish net, envelop the intestines. The deer literally become "hog-fat."

Adult deer, having attained their full body growth, are able to convert to fat all of this high caloric intake that is above the needs of their basic metabolism. Young deer are still producing muscle, bone, and tissue. Their growth rate does not slow down

Both deer and turkeys feast on acorns, but overall, the turkeys are not competitors for deer forage.

until late in autumn, so they fatten at a much slower rate.

The feeding competition among wildlife is keen. Biologists have recorded 186 different kinds of birds and animals that feed upon acorns, oak leaves, oak browse, and so on. With the possible exception of the raspberry-blackberry family of plants, no other type of vegetation is so heavily utilized by as many species. In good mast years, the competition really doesn't matter, but in poor mast years every acorn counts. Poor years may be the limiting factor in the establishment of some wildlife populations, particularly the wild turkey. The turkey population is slowly expanding over much of the United States, but poor mast years and deep snows combined have, on several occasions, wiped out nucleus flocks.

Domestic pigs are allowed to roam free in most of the southern states, and in many of these states there are now many herds of feral pigs. North Carolina and Tennessee also have herds of European wild pigs. Pigs like acorns and are among the deer's main competitors for them.

Squirrels of all kinds, bears, and raccoons all fatten on mast. The death of the American chestnut some 60 years ago robbed all wildlife, as well as humans, of untold thousands of bushels of nuts each fall. Hickory trees are steady producers of mast, but beechnut trees are not. The beechnut is a very sporadic producer, although the nuts are avidly sought by all wildlife because they are so sweet. Pecans are important mast producers in the South. Walnut trees produce abundantly, but the deer do not eat the hard-shelled nuts; they are left for the squirrels. Even the squirrels gather and store acorns before they turn to walnuts. It is as if they know the walnuts will still be around when the acorns are gone. Or perhaps squirrels dislike working that hard to get a little nut meat.

Autumn's bounty includes a great many other foods that are avidly sought by one or more of the three types of deer. These preferred foods are listed in Appendix I.

In addition to the wild foods, the deer in some areas feed heavily on such cultivated crops as alfalfa, clover, soybeans, corn, and apples. Most of the corn and apples are the gleanings—dropped apples and the corn missed by the mechanical pickers. Orchards are good places to look for deer at all times of year. Except when the trees are young, deer do not cause much damage by browsing; but the bucks cause great damage by rubbing the 1-inch (2.5 cm) saplings.

Just as humans get indigestion from overeating, so do deer. The proper name for it is *rumen overload,* or *rumenitis,* or the old farmers' term "bloat." Deer get indigestion when they eat too much high-carbohydrate food, such as corn, sugar beets, grapes, pears, or wheat, too quickly. Ordinarily, apples drop from the trees a few at a time and, with a number of deer eating them, not too many are consumed by each deer. Then along comes a high wind and the apples are dashed to the ground by the bushel. The deer then eat far more than is good for them and get indigestion.

When apples and pears drop to the ground, they usually are badly bruised and the skin splits open, allowing air to get inside

This doe and her six month-old fawn are gathering another favorite fall food, dropped apples.

the fruit. When such fruit lies in the sun, it begins to ferment. Some people claim deer become intoxicated on the fermented fruit. I have never seen it happen, nor have I read of it; but it is possible. Birds are known to get drunk on fermented berries. Bees, bears, and cows get drunk on fermented apples. My family once had a Jersey cow that got so drunk she could not stand up.

A very important food, particularly for whitetails in the Northeast, is provided by autumn's falling leaves. The preferred dry leaves are the same kinds that are preferred when green. Maples are among the most important, but dogwood leaves seem to have top priority. Deer will eat the leaves when they are quite dry but prefer those with some moisture still in them and, for that reason, they will eat those leaves that are falling rather than those that dropped off a few days previously. Times without number, I have watched deer walking through the woods in autumn leisurely feeding on the just-dropped leaves, turning aside to eat one that just fell, in preference to one that came down a day before. Red leaves have a high content of residual sugars, and they are most avidly sought. People ask me how deer can tell the red and yellow leaves of the dogwood and maples if they can't see color. I believe the deer are finding the freshest leaves of their choice more by odor than by sight.

Watching deer eat maple leaves, I discovered they eat about 38 leaves per minute. By coincidence, 37–38 maple leaves weigh one ounce. There are about 608 leaves to the pound and a 125-pound (56.7-kg) deer eating 10 pounds of green vegetation per day would need over 6,000 leaves per day. A 200-pound (90.8-kg) deer would eat about 10,000 leaves per day.

I used to believe that the dry leaves of the oaks, elms, and hickories were seldom eaten because of their high tannic-acid content. Additional years of observation have forced me to change my mind. I have since seen deer feeding heavily upon the oak leaves, but only after the bulk of preferred leaves had already been eaten. Although the oak leaves are as nutritious as the maple leaves, deer do not get as much nutrition from them, because they are eaten much later in the season when they no longer have any moisture. The dried oak leaves are almost strictly a "stuffer" food—roughage; the center of each leaf cell has totally dried, leaving only the indigestible cellulose and the lignin of the cell walls. The deer instinctively know that roughage is needed for their digestive well-being. Even captive deer, offered all the grain they can eat, will feed on dropped leaves for their roughage.

Life in Autumn

Most of the fawns have been weaned in late summer, at two to three months of age, although some are still nursing late in autumn, at five months old. Most does' udders will have dried up prior to the breeding season, although W. T. McKean of South Dakota found that 83 percent of the whitetail does were still lactating in December, and up to 50 percent of them were doing so in January. As a doe's milk volume lessens, the milk's protein content continues to creep up. Deer instinctively seek out the high protein they need, and perhaps this is why so many fawns try to nurse for as long as they can.

In late autumn, the average fawns are about two-thirds the size of their mothers, and they are perfect little carbon copies. When the fawns have reached this size, they stop growing until the vegetation starts to grow again the following spring. Once their growth stops, their bodies can convert all nutrients above the needs of their basic metabolism to fat. And this fat is essential to survival.

Bucks have accumulated maximum fat by the third week in October, just before the onset of the rutting season. Does have accumulated theirs in the first two weeks of November, just before they are usually bred. Usually by the last of November the metabolism of all of the adult deer begins to decline. The deer no longer have the ability to accumulate fat.

MIGRATION AND DISPERSAL

The habits of all deer change drastically in autumn, the period of greatest daily activity. They feed longer and more often, they move about more within their ranges. And the mule deer and blacktails shift their ranges by migration.

Snow comes early in the mountainous country of the West, and it is the snow that sends the elk and deer down from the high mountain meadows, where they spent the summer, to the lower elevations. Deer and elk don't like the cold or the wind, but it is not until the snow begins to pile up that they start their migrations down. A small snow storm will not start the migration. It takes 8 to 12 inches (20.3 to 30.5 cm) of snow, or the threat of a big storm, to start the animals moving.

Once started, they waste little time. They follow the natural drainages down the mountains, funneling through draws, gaps, passes, and valleys. They may not follow the most direct route but usually choose a route of easy traveling. These migration routes

Mule deer migrations are prompted by the first heavy accumulation of snow—usually 18 inches is enough to send the does and fawns on the way to lower altitudes, though some bucks may linger in the high country.

are traditional. The fawns follow the doe as she leads them down trails where she was led by her mother, and where her mother was led before that. Much of what we think wildlife does instinctively is actually learned. So long as some members of the herd have been shown the route by the older animals, that route continues to be used. If ever a break in the chain of learning occurs, that particular migration route may never be used again.

Because migration often takes place under cover of darkness, there may be no deer in a particular area one day and dozens of deer there the next. During their fall migration, the deer do not linger to eat, as they did when they moved upward in the spring; they grab a few mouthfuls of food as they travel, rest a while during the daytime, then continue their journey.

Some mule deer migrations are 90 to 100 miles (144.8 to 160.9 km) long. The herds that winter together may not be composed of the same deer that spent the summer together. There is much crisscrossing of migration routes, and many deer actually travel much farther than they have to. Males tend to travel slightly greater distances than females because bucks usually spend their summers at higher elevations than does.

There is no migration by those blacktail deer that live in regions where the seasons do not change much—for example, the coastal areas of Oregon and Washington. Most whitetail deer do not migrate today. There is a shifting of the northern whitetails from summer areas to winter yarding areas, but this is usually less than 10 miles (16.1 km). Some whitetails in Michigan travel

Life in Autumn

about 30 miles (48.3 km) to get to their yarding areas. Years ago, some of the whitetails in Michigan had much longer migration routes.

George Shiras III recorded the longest regular migration of whitetails as being about 75 miles (120.7 km). In the mid-1880s, whitetail deer that spent the summer on the western end of Michigan's Upper Peninsula would migrate each winter down into Wisconsin. The migration was due to a lack of winter food. As with most migrating animals, travel in autumn would be direct and fast, while the return in the spring would be more leisurely and would be led by the does.

The whitetail migration stopped after 1870 because of lumbering operations. As the virgin timber was cut off, the second-growth sprouts provided unlimited year-round deer food. After 20 years, much of this food had grown beyond the deer's reach. But since the migration pattern had been broken, from that time on the deer yarded locally each winter.

Every autumn about 25 percent of yearling whitetail bucks and about 5 percent of the does disperse from the area they were born in. There is some dispersal among mule deer and blacktails, but it is not as noticeable because of their migrations.

Dispersal is always heaviest from refuge areas where there is no hunting. There, social pressure from the adult bucks forces the yearling bucks from the area. Areas that are heavily hunted have minimal dispersal because many of the adult, dominant bucks have been removed.

At the Radford Arsenal in Virginia, 40 to 50 percent of the 1½- and 2½-year-old bucks dispersed. The Crab Orchard National Wildlife Refuge in Illinois had some of the heaviest dispersal recorded. The average annual rate of dispersal for all age classes was as follows: 4 percent of fawns of both sexes, 13 percent of yearling does, 7 percent of adult does, 10 percent of adult bucks, and more than 80 percent of yearling bucks.

Sixty percent of the adult females separated from their mothers through dispersal when they were two years old; 55 percent of those that were left separated when they were three years old. The breakup was prompted by these does becoming the leaders of their own family groups. The does did not move as far from their original home range as did the yearling bucks.

Yearling males, when they disperse, establish larger home ranges than they had while with their mothers. Dispersal is usually caused by social pressure on the males, but one of the major benefits is that it prevents a lot of inbreeding or incestuous mating. The species, any species, is strengthened by having the largest gene pool possible. Dispersal by the females does not carry similar benefits, because does moving to new ranges usually suffer a higher fawn mortality, a result of being on unfamiliar ground or of being forced to inhabit marginal range because the best range is already inhabited by the older, dominant does.

A recent research project on the home ranges and dispersal of whitetail deer was conducted by K.E. Kammermeyer and Larry Marchinton in Floyd County, Georgia, on the Berry College Refuge. The deer population was very high in the summer, with about 12 deer per square kilometer, or 30 per square mile.

Thirteen deer were fitted with radio transmitters and 40 with bright plastic markers. The average summer range of the bucks was 28 acres (11.3 ha) while that of the does was 18 acres (7.3 ha). Prior to the rutting season, the bucks began to increase the size of their ranges, to 83.77 acres (33.9 ha).

All of the three types of deer have overlapping home ranges; they are not territorial—that is, they do not have inviolate territories. Each buck has a large enough home range so that it overlaps the ranges of a number of does and some other bucks.

Six out of 19 tagged bucks dispersed, but only one of the 21 tagged does dispersed. All of the dispersed deer went from areas of high deer concentrations, ample food, and protection from hunters, to areas of lower deer populations, less food, and intensive hunting pressure. The researchers concluded that the dispersal resulted from the breeding competition among the bucks.

I do not doubt the researchers' conclusions, but another aspect of such dispersals should be mentioned. Nature abhors a vacuum. There is always a dispersal from areas of high population concentration to areas of lower population, even if conditions in the low-population areas are less favorable. This is the basic reason for dispersal; it is nature's method of filling every niche that will support a particular species. A low concentration has tremendous survival value because it lessens the chance for disease to spread through the population. It also insures that a localized calamity will not affect the species as a whole.

An intensive five-year study in Texas on whitetail range and dispersal provides some very interesting data. During the five-year period, 204 deer were marked and tagged. Four hundred sight records were made on 68 of the marked deer. It was found that the average doe had a home range of about 93 acres (37.6 ha). Three

Life in Autumn

does evidently didn't want to be cramped, as two of them had home ranges of 502 acres (203.1 ha) and the third had a home range of 690 acres (279.2 ha). The bucks had an average home range of 1,079 acres (436.6 ha)—more than ten times that used by the does. The average distance moved by does was only about 400 yards (365.8 m). The maximum distance moved by a single doe was $2\frac{9}{10}$ miles (4.7 km). The maximum distance moved by any buck was $4\frac{1}{2}$ miles (7.2 km), and movements of 2 miles (3.2 km) were common among bucks during the rutting season.

Food was abundant in the spring and summer, when the bucks were growing their antlers and the does were tending their fawns. These factors reduced the deer's range at that time to about 24 acres (9.7 ha). The bucks were actually utilizing an area of about 13 acres (5.5 ha). With the advent of cool weather and the breeding season, all of the deer again became very active, and that was when the bucks moved the greatest distances. The Texas research, however, agreed with findings in other whitetail states that only a little over 20 percent of the deer population moves more than $1\frac{1}{2}$ miles (2.4 km) in dispersal.

Although most whitetails are not inclined to travel far, wanderlust seems to claim a few individuals—for instance, two does tagged by researchers in Tennessee. One doe, tagged near the Clarksville military base in 1963, was killed by a car near Kenton when she was eight years old. She had traveled 90 miles (144.8 km). The other one, killed near Dresden when she was $12\frac{1}{2}$ years old, had traveled over 100 miles (160.9 km) from her birthplace. In South Dakota, an archer named Everett Gothier shot a whitetail that had been tagged as a fawn 140 miles (225.3 km) from where he killed it. Bill Hlavachick, a biologist in Kansas, recorded the longest distance traveled by a whitetail. It was a doe tagged in Sheridan County and shot in Kingman County, Kansas, a distance of 170 miles (273.6 km).

Bill also tells of a buck mule deer that was tagged on June 8, 1970, in northwestern Kansas and shot on November 13, 1971, near the Platte River in Nebraska, a distance of 65 miles (104.6 km). If this was a dispersal record, it was a good one for mule deer. If, on the other hand, the deer was migrating, it was not exceptional because many mule deer migrate 100 miles (160.9 km) each spring and autumn.

Not much research has been done with regard to the homing ability of deer. I know of one buck, trapped by New Jersey wildlife officers, that was removed from land at Belvidere, New Jersey, to reduce a captive herd. The buck was released at Mountain Lake, a distance of 6 miles (9.7 km). The buck returned home, clearing the 9-foot (2.7 m) fence to get back in. Within a week, the buck was retrapped. This time he was released at Dunfield Creek, a distance of 11 miles (17.7 km). Again he returned home. Finally, he was transported 100 miles (160.9 km.), and this time he did not return.

THE RUTTING SEASON

The rutting season is the period of greatest activity for all three types of deer, particularly for the bucks. It is their reason for being.

At this time, the bucks are exactly what we picture a buck to be: strong, sleek, virile, in the best possible condition. They are truly magnificent animals at the time of the rut.

In November and December, a buck's thyroid, adrenal, and testicular glands all reach a peak of activity, weight, and hormone production. (I am speaking here of the northern states; the timing differs somewhat in other latitudes, as mentioned in earlier chapters.) The sperm-laden testicles of a 150-pound (68 kg) buck will measure 3¾ inches (9.5 cm) in length by 2¼ inches (5.7 cm) in diameter. The scrotum does not have the teardrop appearance that it has in cattle and sheep. Connective skin and muscle at the rear of the scrotum facilitate a raising and lowering of the testicles in accordance with the temperature.

Live (motile, or active) sperm develop within the testicles in late summer. The testicles must be kept cooler than body tempera-

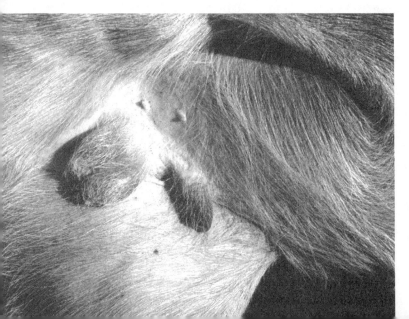

When the bucks testes decend he is ready to mate, usually before most does come into estrus.

ture if the active sperm count is to be high, however. Heat destroys sperm. As the testicles enlarge, they fill the scrotum; there is no loose, excess skin as in humans. The scrotum has a thin skin, short, thin hair, and some sweat glands to help in cooling. The scrotum drops down, away from the body. The testicles must be kept cool but not cold. Warmth relaxes the muscles, and the scrotum descends to move the testicles farther from the body and allow the circulation of air for cooling. Cold weather reverses the procedure, retracting the scrotum. Cold, dry weather definitely stimulates sexual activity in deer. It is as if they instinctively feel that winter will soon be upon them, that the season of the rut is brief, that they must mate now if the fawns are to be born at the best time.

The bucks are ready sexually long before the does are willing or able to accept them. As their frustration mounts, the bucks become much more aggressive, and they begin to travel almost constantly in search of receptive does. What seems like poor timing—this "ripening" of the bucks before the does are receptive—probably is most efficient. For one thing, mates are available for any does that come into estrus earlier than usual. For another, by the time most of the does are ready, the bucks have become so enthusiastic, so determined, that a large number of breedings will be assured.

But before that, the bucks take out much of their frustration on saplings and brush, which serve as surrogate opponents. Captive bucks hook the fences viciously. Saplings are attacked with all the strength the bucks can muster, and with movements almost too fast to follow. The fury of attack can be seen where the earth has been torn up by their hoofs.

Now the big bucks begin to make scrapes. These are roughly circular patches of earth that have been torn up by bucks pawing with their front feet. Most scrapes are about three feet (0.9 m) across and several inches deep. They are conspicuous, particularly in the forest, where the bare, churned earth shows among the leaves. Some of the scrapes are done very deliberately; others must be done in a frenzy, judging by the distance that the dirt has been thrown as the scrape was made. The tracks show that the buck pivots around while making a scrape.

Almost every scrape I have seen was made under a bush or near a tree with branches hanging low enough—about 4½ to 5 feet (1.4 to 1.5 m)—that the buck could reach up and hook them with his antlers, as described in the chapter on communication, instinct

In late summer bucks start to visit their scrapes; a primary scrape will very often be marked by an overhanging limb used as a licking branch, shown here in the circle.

and intelligence. The branches above the scrape are always broken and scarred, showing that the buck put a lot of effort into hooking, chewing and rubbing them with his preorbital and forehead scent glands.

Usually a buck will urinate, and sometimes he will defecate, in the scrapes. These scrapes are not to be compared with the wallows of elk. The elk's wallow is a churned-up mud bath, reeking of urine and semen, which the bull elk rolls in, plastering goo all over his body. Only once have I found a deer scrape impacted by the buck's lying in it.

Unlike most buck rubs, which are usually used only once, a scrape is used every time the buck is in the area. Bucks don't have a territory in the usual sense, but use the scrapes to advertise their presence.

The scrape is also made to declare their social status; most of the scrapes are made by the biggest, dominant bucks. Lesser bucks will use a dominant buck's scrape if he is not in the immediate area, but they seldom make their own. The right to make a scrape must be earned.

Numerous sporting magazines have featured articles about the making and use of "mock scrapes." I hereby verify that they work, and will describe how I have done it.

It is important to start with a well-used deer trail. Although the buck may wander anywhere, he is much more likely to use a trail in his travels than to strike out in a random direction. It is imperative to select a bush or tree near the trail with an overhanging branch 5 to 6 feet (1.5 to 1.8 m) off the ground. Some researchers, not finding such a branch where they needed it, have actually brought in a branch from another tree and nailed it in place. One even cut off the branch from another scrape and nailed it to the tree he wanted to use. This latter branch would already have been impregnated with natural deer scent. Great care must be taken not to handle these branches with your bare hands. Use a new pair of cotton gloves, rubbed with dirt and some deer scent.

Take an ordinary stick and scrape a depression directly beneath the overhanging branch, precisely as a buck would do. I make mine about 30 inches (76.2 cm) across and about 3 inches (7.6 cm) deep. I then pour about a quarter of a bottle of commercial sex scent on the overhead branch, allowing it to drip down into the scrape below. (This is the only time I use more than a few drops of scent.) Deer urine, if available, may also be put in scrapes.

If all is done correctly, chances are excellent that the mock scrape will be used by a buck, and, from that time on, it's his scrape.

Anyone observing wild or domestic animals, including deer, soon comes to the conclusion that, like humans, they are sexually motivated creatures. And as with humans, masturbation is common in the animal world. Denied sexual release because the does are not yet physiologically capable of copulation, the bucks seek relief by masturbating.

A buck's penis sheath, in the relaxed state, usually hangs straight down and is about 3¾ inches (9.5 cm) long. When the penis is not erect, a retractor muscle pulls the forward part of it back into the sheath, causing the penis to fold into an S-curve known as the *sigmoid flexure*. When the buck is stimulated, vesicles in his penis fill with blood under great pressure, and, although the penis does not increase very much in thickness or length, the S-curve is straightened out and the penis becomes rigid. There is no bone in the deer's penis, as there is in some other mammals.

The erected penis points forward and protrudes from its sheath by 8 to 10 inches (20.3 to 25.4 cm). When masturbating, a buck stands with his back humped up and his tail held stiffly erect. With a rhythmic forward thrusting of the pelvic region, he rubs his extended penis against his abdomen. Just prior to ejaculation, he lowers his back, and his final thrustings rub the penis beneath

This whitetail buck is masturbating; self-gratification is common among the highly-charged rutting bucks, as it seems to relieve some tension when the does are not receptive.

Life in Autumn

Bob McDowell, deer reseracher for the New Jersey Division of Fish and Game recorded this case of true homosexual behavior among whitetail deer. In this case the smaller deer is attempting to mount the larger buck, not merely dominate him. photo courtesy Bob McDowell.

his rib cage. Often the buck will lick the penis sheath prior to masturbation and the penis after ejaculation. The sex drive varies with the individual; some bucks masturbate occasionally, others more frequently.

Incestuous behavior among wildlife is so common that it is normal. A dominant buck, in an area where he is not hunted, may be dominant for a period of three to four years. In that length of time, he will most certainly breed some of his own daughters, and perhaps granddaughters. If he were one of a set of twins, he may breed his sister. He may or may not breed his mother, according to the doe's age when she bears him, but he most certainly will try. If he tries as a little buck, while his antlers do not protrude through the skin, or are not peeled, he is not sexually able to breed the doe. Most young bucks I have seen mounting their dams could only go through the motions; their penises were not thrust out of their sheaths.

Such dry mountings may also be displays of dominance. Many times a button buck will attempt to mount his peers, of either sex, in an effort to establish dominance over them. I have also observed dry mountings by adult bucks, and for the same purpose. If you ever see the hair on an adult buck's hindquarters all scruffed up, it will be the result of another, larger buck mounting the lesser buck to show dominance. I have observed this behavior not only among deer, but also among wild sheep, elk, and antelope.

I was raised on a dairy farm, and, as did most farmers, we kept the bull penned separately so that the cows were bred when we wanted them to be. We always knew when a cow was coming into her estrus cycle because one or more of the other cows would mount the cow. It was always the same cow, or cows, that acted the part of the bull. These were cows that had more of the male sex hormone, testosterone, in their systems than did their counterparts. I have, on several occasions, seen this happen with deer as well.

The Dangerous Time. About three weeks before the first females come into estrus, the bucks start to "run" the does. The bucks are now ranging far and wide, constantly searching for a receptive female. Most does will not allow a mature buck to get close to them at this stage. The presence of immature bucks is accepted, because they do not posture as adult bucks do.

When a mature buck is tracking a doe, he follows along the trail with his head close to the ground and with his tail pointing almost straight up, grunting as he goes. That I, as deaf as I am, can hear the grunting at a considerable distance, proves that the grunting is loud. When the buck locates a doe, he runs toward her with his head held low but extended and pointing forward. This posture is similar to the one used in aggression, with the exception

Bucks of both species are quite solicitous to does while they are in estrus.

A whitetail buck closely following a pre-estrous doe; a buck may "tend" a doe for a day or longer. When the doe is ready to breed she will stand.

that the buck's ears are held straight out to the side instead of back along his neck. When the does run from the bucks, they usually clamp their tails down and bend their legs so that their bodies are closer to the ground. They lower their profile as if they hope the buck will ignore them. He doesn't; he will chase after them in an attempt to smell their vaginas. If a doe urinates, the buck will sniff at the urine, then raise his head, and curl his upper lip. This is called *flehmening*. The upward curling of the lip traps the molecules of the doe's scent on the epithelial lining of the nostrils and of the organ at the top of the palate. This organ, the Jacobson's organ, is what allows most mammals and reptiles to detect odors and interpret them. As the doe may be tested by a number of bucks, she urinates very sparingly, just enough to tell the buck that she is not ready. By being able to ration her urine, the doe saves herself a lot of running by different bucks. The buck will smell her urine just one time and flehmen. Wild sheep will often flehmen several times with the same urine. In attempting to check out the doe's pheromones, the buck will chase her from ⅛ to ¼ mile (0.2 to 0.4 km). The buck can tell instantly if the doe is in estrus. If she is, he is interested; if not, he dashes off in the hope of finding one that is. Ordinarily, when a group of does and young bucks see a mature buck in the outstretched-neck position, they scatter before him like leaves before a wind.

Bucks are much more dangerous just prior to the breeding season than after it starts. Until they actually experience the release that comes with copulation, they are unstable—time bombs ready

This mule deer is exhibiting lip-curling behavior common to rutting bucks called the Flehmen response, perhaps to test the doe urine he has found for readiness to breed. photo courtesy Irene Vandermolen.

to explode at any moment. Wild does are seldom injured by overly amorous bucks. They can run from the bucks, and other does are available to distract them. But does in captivity are frequently killed when a frustrated buck flies into a rage. C. W. Severinghaus reports that six does were killed by rutting bucks in the pens at the Delmar Wildlife Laboratory.

John Madson, in his book *The White-tailed Deer,* tells of a buck that had been raised as a pet on a farm in Michigan. The buck was very docile and was allowed to roam about as he liked. One fall the farmer put the buck in a pen near his rabbit dogs with the idea of saturating the dogs with the deer odor to break them from running deer. One day the buck charged the farmer and knocked him into a snowdrift. When the buck attacked again, the farmer caught him by the antlers and held him off by sheer strength. It was fortunate that the farmer was a strong man. Answering the farmer's call for help, two men armed with a pick handle and a wooden beam beat the buck back so that the farmer could escape.

In an attempt to pacify the buck, the farmer put two does in with him. The buck killed one immediately. The second one was injured but was removed in time to save her.

The farmer then put in a spike buck, figuring that the young buck's spikes would slip through the big buck's rack and give the young buck an advantage. The adult buck struck first, disemboweling the spike buck on the first thrust. Not one to give up, the farmer next put in a large buck, the equal of the rogue buck in size and rack. This last buck was also swiftly killed. The "pet" buck had attacked a man, wounded a doe, and killed a doe and two bucks in two days. That must be a record of some kind.

You may recall my saying earlier that the loss of antlers seems to be an insurmountable psychological shock to a rutting buck. In this case, the farmer finally sawed the antlers off and the rogue buck became as docile as he had been before the rutting season.

Ordinarily, the removal of a buck's antlers will make the buck docile, but I have uncovered enough evidence in the literature to assert that this is not always the case. In several records, the removal of the bucks' antlers made them docile for one to two weeks, and then they became aggressive again. In one case the buck immediately lost his dominant status. In two weeks the other captive bucks shed their antlers naturally and the buck that had been dominant became the dominant buck once again.

All of this leads me to conclude that the results of sawing off the buck's antlers depend on when they are sawed off. If the antlers are sawed off just before the rutting season, while the buck's testosterone is high, he will probably become aggressive or dominant again within a few weeks. If the antlers are sawed off after the rutting season, the buck's testosterone level will probably be too low for him to regain his belligerency against humans but not against other deer.

When the United States was primarily a rural country, and more people lived on farms than in cities, it was said that the most dangerous animal in the country was a Jersey bull. One of the most dangerous animals in the country today is a pet whitetail buck. Every year there are newspaper reports of people being injured or even killed by a buck they had raised in captivity. Joe Taylor, Fred Space, and I were all put over a fence once when one of Fred's bucks became extremely aggressive. I've gone over more deer fences than I care to think about.

There are also a number of reports of wild bucks attacking people. Some of the more fortunate folks were able to scramble up a tree. A number of years ago, a woman in my area who was against hunting, bowhunting in particular, was attacked by a buck in her backyard. She was not as injured by the buck's attack as she was by the thorns of the rosebush into which he pushed her. The greatest injury was done to her philosophy; her concept of deer as gentle creatures was badly shredded.

Ordinarily deer are not dangerous animals, and for nine months out of the year, a buck has normal control over his actions—but not during the rut. Being under the influence of the hormone testosterone, a buck cannot be judged by human standards of good and bad. When he does injury to someone or something, he is not being vicious, he is merely being a buck. His actions are not deliberate. *No buck is to be trusted during the rutting season.*

Stories of aggression are common but there is one aspect of the danger that, for the sake of safety, should be given more publicity. This was brought home to me by a letter I received from a young woman whose parents have a farm at New Paltz, New York. Her family had raised a buck fawn as a pet. All went well until the buck was 1½ years old, sexually mature and with antlers that would definitely interest a hunter. This always presents a problem with "pet" deer, for the animals cannot be allowed to run loose during the hunting season. What compounds the problem is that such deer have lost their fear of humans. The family wanted to pen the animal until hunting season closed, but they discovered that the buck—now fearless—would try to mount any woman who was having her menstrual period. During their menses, women should be extremely cautious about large animals, keeping away from deer, elk, or bears encountered, especially in national parks and refuges, where such animals have lost their fear. At least one of the young women killed by a grizzly bear in one of our national parks a few years ago was having her menstrual period and may therefore have unwittingly attracted danger.

Fighting. Bucks seldom fight with the other bucks that belong to their own group. They have associated with one another throughout the rest of the year, and dominance has already been established among such bachelor groups: each buck already knows where he stands. Occasionally a younger buck may get ambitious and have to be reminded of his standing in the social hierarchy, but posturing and threats by the dominant buck usually suffice. If the dominant buck is an old one, then a younger but fully mature buck may prove to be a serious challenger. In the fight that determines the outcome, the younger buck may displace the old one.

As mentioned earlier, the extensive home ranges of mature bucks will frequently overlap those of other mature bucks. These are the deer most likely to fight. The fights I have seen were not over does, although that is often the cause. The fights I witnessed occurred because of the onset of the breeding season and one buck's invasion of another's turf.

When two unacquainted bucks meet, the first sign of aggression is a stooping or squatting position, assumed by one or both bucks. The hind legs are bent and brought forward so that the hindquarters are lowered and the back slopes below the horizontal.

The buck's head and neck are extended, and the head is held slightly lower than the level of the back. The ears are laid back flat against the neck. The whitetail buck's tail is clamped down at this stage, while that of the blacktail buck is held away from the body and often quivers. The bucks of all three types seem to bunch their muscles and appear to be under great tension at this point. The head's low, extended position causes the eyes to move to the front of the socket so that whites are plainly seen at the rear.

The bucks lick their noses constantly, their tongues flicking in and out. Elk and bison also do this, and a sure sign of extreme aggression in a moose is when its mouth is opened and the tongue is extended.

With a stiff, stilted walk, the bucks may circle each other, or they may approach each other more directly but usually with a sidling walk.

This action used to puzzle me. I wondered why a buck would turn his body sidewards, offering his opponent's antlers a bigger target. Having witnessed a great many fights, I have discovered that the answer is literally apparent. When the bucks use their sidling, circling walk, they erect all of the hairs on their bodies. This has the effect of making them appear much larger than they actually are. They circle each other, in a ritualistic manner, to make sure that each has a chance to see how big his rival really is. The circling is one last attempt by each to bluff his rival into submission rather than having to fight to do it.

While circling or approaching each other, the bucks tuck their chins in so that their antlers are tilted forward. With their ears flattened against their necks and pointing to the rear, and with the whites of their eyes showing, the bucks look "mean as hell," and they are. The hair on their bodies is erect, and this is particularly noticeable along their spines. The erect hair and the bunched muscles make a buck appear a picture of powerful, controlled fury.

The bucks snort or grunt as they are about to do battle. Or they may emit very high-pitched whines.

When the bucks finally lunge, one of them must make the first move, yet the opponents are so keyed to each other's actions that they appear to explode into action simultaneously. The distance between the bucks, when the rush is made, may be up to 20 feet (6.1 m), but most often it is 6 to 10 feet (1.8 to 3 m). The initial impact may be so great that pieces of antlers are broken off. When the bucks crash into one another it is with the force of tons of impact.

Joe Taylor witnessed a fight in which a buck lost an entire antler from one side. Joe said that the antler went flying up in the air about 20 feet (6.1 m) and the two bucks dropped to their knees and spun around in a circle like a top while fighting. In less than a minute, the buck with the broken antler ran off. Joe said the initial charge was so fast that if he had blinked his eyes, he would have missed it.

When you see bucks banging their antlers together, you are not necessarily witnessing a fight. Many times the bucks are just sparring, and this is particularly true of 1½- to 2½-year-old bucks. Bucks in this category engage in most of the head banging seen; they are constantly testing one another.

A number of clues will help you to determine whether you are seeing a sparring match or a real fight. First of all, a real fight has tremendous intensity; great effort is being put out. The legs of a buck in a real fight are widespread to prevent the other buck from knocking him off his feet. A downed buck is particularly vulnerable and likely to be gored. The body will usually be found with the legs pointing sharply to the rear because the buck was pushing forward, resisting pressure as his rival attempted to push him backward. Occasionally the two bucks will be forced to their knees, and they will continue to fight from that position.

The mule and blacktail deer do a lot of sideward head twisting while fighting; the whitetail does not. When you see whitetail bucks with their heads twisted sideways, parallel to the ground, you are seeing a sparring match, not a fight.

Most of the fights I have seen lasted 1 to 5 minutes. The longest I have seen lasted about 15 minutes. The two bucks had each come down from opposite hillsides. The younger one had to cross a small brook, and they met in the middle of an old logging road. It was apparent that one of the bucks was much older than the other, and, I think, he would just as soon not have fought. I say this be-

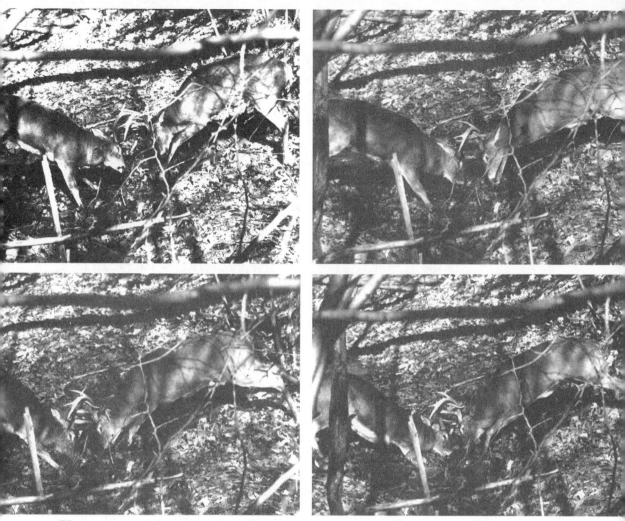

The intensity of this fight is easily seen in this sequence of photos: note the low, wide stance each buck takes to prevent being thrown, just as a wrestler takes. Look at the bunched muscles of the hind legs as they power the body thrusts in the second row of photos.

cause he averted his eyes; he did not look directly at the younger buck. The younger buck was spoiling for a fight and did not honor the older buck's signal of submission.

There was no sidling circling. The younger buck gave all the aggressive signs, including the hard look, and grunted. When the two bucks were separated by about 10 feet (3 m), they both lunged straight on, meeting with a resounding crack of antlers. From that time on it was just muscle against muscle, with each buck straining to keep his footing and to prevent being pushed backward. Initially, the older buck had a slight advantage because he had a little elevation leverage, and he finally pushed the younger buck down the sloping bank to the brook. Whenever possible deer, and particularly sheep, try to gain the uphill advantage. The dropping downhill before making contact gives added momentum to their already considerable impact.

The brook was shallow, only inches of water, but about 4 to 5 feet (1.2 to 1.5 m) wide. The battle was ferocious; the deer were up, they were down on their knees, they were up again. At times they fought to an absolute standstill, as they both pushed with all their might. Neither could gain, nor give, an inch. The old buck was bleeding from the nose, but I don't know how or when he was injured.

In the excitement I couldn't tell precisely how long it all took, but I did take three rolls of color film and one roll of black and white before it was over.

The younger buck not only had youth, and probably some muscle in his favor, he also gained the uphill advantage, forcing the older buck into the brook. The older buck realized he had lost, and his remaining efforts were directed toward getting away before he lost his life. He was able to break loose and dashed up the far bank on an angle and was gone, with the younger buck in hot pursuit. Ordinarily, the victorious buck does not chase his vanquished rival more than about 50 feet (15.2 m). This buck chased the old one until he was out of my sight.

My camera takes five frames per second, which means that whatever occurs between one frame and the next occurs in $\frac{1}{5}$ of a second. When the old buck broke free and dashed off, the young buck changed from a head-on position to a sideward position and jumped 15 feet (4.6 m) in two frames, or $\frac{2}{5}$ of a second.

As I have stated often, fights occur only between nearly equal animals; lesser animals give way before the big ones. Because of this, the bucks usually do not injure each other, although they are trying to. If one of them loses his footing and falls, the rival will try to gore him, but this seldom happens. There are eyewitness reports of rival bucks fighting for 20 to 30 minutes, until each was completely exhausted, but I have not witnessed such a fight. Usually, when one of the bucks tires, he breaks off contact and dashes away. The victor seldom chases his rival for long. He is not interested in chasing a rival he has beaten.

The literature abounds with reports of bucks locking antlers. The antlers lock when the force of the initial impact springs the antlers apart so that the tines and beams become enmeshed. This

A three-way sparring match; most likely this event started out as a match between two bucks, which attracted the third.

These mule deer are just sparring: there is no intensity in their actions and their legs are merely held under their bodies as they would be if they were walking. Also note one buck has turned his head sideways.

Eric Hall of Glidden, Wisconsin, killed this nice 10-point buck with a bow in 1984. As a bonus, he got a 19-point rack that was locked onto the 10-point rack. It is most likely that the larger deer died in a fight and the carcass was severed at the head and eaten by coyotes. photo courtesy Eric Heiting and "Deer & Deer Hunting Magazine."

locking happens with deer and moose, less frequently with elk, and I have seen photos of only one occurrence in caribou. Deer are known to lock antlers, but it is not common. Its rarity is what makes it so newsworthy, which is why it is reported in sporting and conservation magazines.

I have only actually seen evidence of locked antlers on five occasions, although, perhaps like you, I've probably seen a hundred photographs of it. I would be willing to bet that the locking of antlers occurs no more than a couple of hundred times a year among the millions of deer across the nation. All the reports of locked antlers that I have seen were of bucks with big racks having 8 and more commonly 10 or more tines.

Orrin Emerson, a veterinarian, and Marley Griffith were called to separate a pair of whitetail bucks that were locked together on Howard Harvey's farm near Humboldt, Iowa. They sawed one antler off the 8-point buck, freeing him. This buck dashed away. The other deer, a 10-point buck, was taken to Emerson's clinic but died three days later of head injuries.

Life in Autumn

The men were lucky that the 8-point buck ran off. All too often, men have tried to release locked deer only to have a buck turn on them when it was freed.

J. H. Ridley and Dan Bares, of Maida, North Dakota, were hunting pheasants when they discovered two bucks locked together in a stream covered with thin ice. The deer kept fighting as they broke through the ice, trying to reach the far shore. Both bucks were large ones with big racks, but one buck was heavier than the other. The lighter one got up on the far bank but the heavier one failed. Eventually his struggles pulled his opponent off the bank and both were drowned.

It is a shame that such magnificent animals had to drown, but at least it was a comparatively merciful death. All too often, when two bucks are locked, they suffer a lingering death from exhaustion and thirst. Sometimes one buck will die and the surviving buck may drag him around for several days before he, too, succumbs. I saw one photograph of locked bucks in which dogs had eaten a good part of the dead buck. What a harrowing period that must have been for the surviving buck!

A photo that appeared in a number of magazines showed two huge bucks that had locked antlers in Ohio. The bucks had fought so long in one spot that they had actually excavated a crater that filled with rainwater. I don't recall whether the bucks actually drowned or died of exhaustion in the mud and water.

Joseph Bruckac, a taxidermist of Greenfield Center, New York, mounted a trio of locked whitetail bucks from Jamestown, North Dakota. Joe figured that an 8-point buck and a 10-pointer were fighting when another 8-point buck joined the fray by jumping over the back of one of the fighting deer. His rack locked the three sets of antlers together, and as he twisted around, he broke his neck and died. His hanging weight kept all of the antlers locked until the other two bucks died.

A few years ago I did a program in an elementary school in Black Earth, Wisconsin. The principal there showed me the antlers of three locked whitetails that he had mounted and hung in his office. I know of one other set of three locked bucks.

Many times, with many species all over the world, I have seen two animals fighting desperately, only to have one of the two attacked by a third. I have photographs of this occurring among bighorn sheep in the Rocky Mountains; I have recorded it with Uganda kobs in Africa. The sounds of the battle attract both males and females as spectators and participants. It is because the sounds of such fights attract other bucks that rattling antlers is probably the most successful way for hunters to call in deer.

In mid-November, 1983, I photographed a beautiful big buck that had been killed in a fight. I'm sure that he was not killed by the buck he was fighting with. In the world of nature, there is no such thing as fair play; everything is done for keeps, everything possible is done just to survive.

The area around the buck I found was just so churned up that I could not decipher exactly what had actually taken place. The placement of the buck's puncture wound led me to believe that a third buck had to be involved, because it is rare for either of two fighting bucks to have the opportunity to gore the other from the rear. In this case an antler tine had entered the buck's body low and just behind his rib cage. It had just missed the paunch but had gone through the diaphragm and into the lung cavity.

This buck mostly likely was not killed by the buck he was fighting, but by a third buck that entered the fight and attacked from the rear.

Life in Autumn

Though one buck seldom gets a chance to gore another buck from the rear, I saw it happen twice in one day in the fall of 1985.

A dominant buck was tending a doe in estrus. I don't know whether the doe was just coming into estrus or whether she already was in estrus. I do know that three bucks were chasing her almost constantly. There was a young buck about 2½ to 3½ years old, the dominant buck, 4½ to 5½ years old, and an older buck, perhaps 7½ to 8½ years of age.

The dominant buck was chasing the doe constantly, grunting loudly with every jump he took trying to position himself near the doe between the old buck and the young buck. He really had his work cut out for him. I did not see any of the bucks try to mount the doe, which suggests that she was just coming into estrus.

The young buck gave the dominant buck the most trouble. The old buck mostly tagged along behind the trio. He knew he had to be there; he may have forgotten why. All three bucks had their mouths open, gulping air. Their coats were soaking wet. A couple of times, when the deer came to a downed sapling or gully, the doe and the two younger bucks would literally sail over the hurdle, while the old buck would just jump if he had to. He was always in the rear, but he proved that you can't keep a good buck down.

On two occasions I saw the old buck trail the young buck by at least 20 feet (6.1 m), a distance at which the young buck thought he was safe. Suddenly, from his inner reserve, the old buck turned on such a burst of speed that he caught up to the young buck and raked the younger buck's hams with his antlers. When the old buck started his charge, the young buck also put on a burst of speed, but he just wasn't fast enough. As in most things, the one who moves first wins. All day the old buck wore the young buck's hair on his antlers like a badge.

Breeding. During the precoital period, the bucks try to smell the hocks and genitals of every doe they encounter, and the action is not entirely one-sided. About two days before a doe comes into estrus, her vulva starts to swell and she actively seeks out a buck. She urinates on her tarsal glands frequently.

Researchers have found that 48 hours before captive does come into estrus, they increase their pacing by 28 times. Their frequency of urination is also drastically increased as they do everything possible to get a buck's attention.

One year I had my captive buck and doe in two separate pens. I had separated them several weeks before the breeding season because I wanted to be absolutely sure of when the doe was bred, as part of my record keeping. The doe was having none of that: she wasn't interested in my records; her hormones were the driving force.

On the night of the second day of her increased activity, she stretched the wire loose on the woven wire fence around her pen and made a hole big enough to jump through. Her 6-month-old button-buck fawn jumped through after her. When I went out in the morning to put her in with my buck, she was gone. Her fawn was wandering around on the outside of the pen. After we repaired the hole, we drove him back into the pen with little trouble. The doe could not prevent her young one from getting out when she did, but she had no intention of taking him along with her.

At 4:30 in the afternoon, my doe walked back out of the woods, following a wood road that came right into the yard. What thrilled me was the big 10-point buck that was standing 300 feet (91.4 m) down the road. My doe had never been out of my pen in her life. When it was time for her to be bred, she went through the fence, took up with the biggest buck in the neighborhood and, after being bred, came back home. I put my belt around her neck like a collar and returned her to the pen.

Two days before the doe comes into estrus, her vulva begins to swell.

Life in Autumn

She was evidently still in estrus, because once back in the pen, she tried to make a new hole in the fence. I confined her to a pen with a chainlink fence for the night, and by the next day she had calmed down. She was in excellent physical condition and should have been bred early in her cycle. There is a very good chance that the wild buck urged her to come through the fence by his presence. I often see tracks of wild deer, and occasionally I see the wild deer themselves, visiting with my captive deer through the fence.

On several occasions I have seen a wild doe attempt to mount another doe. Joe Taylor is sure his captive does have attempted to mount his buck. He bases this assumption on the fact that the hair on the buck's back was all roughed up over the hindquarters just before the does were bred.

Arthur Einarsen reported his observations of two mule deer does and a buck on November 21, 1939, in the Malheur National Forest in Oregon. The does were in estrus, but the buck was evidently tired. He was gaunt and his coat was rough. He showed little interest in breeding, but he was given no rest by the two does. Einarsen did not describe what the does did, but he wrote that they "continually attempted to claim the attention of the buck." After about half an hour, the buck finally mounted and bred one of the does but was evidently unable to breed the second doe. This latter doe continued to pester the buck for 20 minutes more, until they were lost from sight.

There are definite differences in the breeding activities of the three types of deer. As a rule, whitetails are not yet yarded up when the breeding season occurs, so the does are widely scattered in their little matriarchal groups. This forces the whitetail bucks to cover a large area, seeking out does coming into estrus.

Whitetail bucks in wooded areas, which would constitute about 90 percent of their range in the United States, usually seek out each doe as she comes into estrus. The doe herself actively tries to attract a buck, leaving a trail of urine and pheromones. The bucks encountering her trail track her down. Usually the dominant buck runs off any lesser bucks and lays claim to the doe. He will then "tend" her until she comes into estrus. This tending entails a lot of chasing until she is ready to accept him. Contrary to what is generally thought, the doe may express a preference for a particular buck and will not accept any other.

Research by David H. Hirth highlighted some unusual differences in the behavior of whitetails in the more open grassland areas, such as in the Texas savannas. In the more open country, the deer were inclined to be in larger groups at any time of the year, but during the rutting season the dominant whitetail buck tended small harems of three or four does. This doesn't happen in the wooded areas, where tending is a one-on-one situation.

Usually, mule deer have migrated prior to the breeding season, so they are concentrated, sometimes in large numbers. The bucks mingle with the herds and stay with them if the deer are in a national park or refuge, or if the hunting season has not started. If the buck is subjected to hunting pressure, he will seek heavy cover on the fringes of these herds and join them only under cover of darkness. Mule deer often have small harems of three or four does that they try to keep other bucks away from. But when one of the does goes into estrus, the buck will abandon the others to tend just to that particular female. After the doe goes out of estrus, the buck will again attempt to form and keep his little harem.

A mule deer tends a doe about to come into estrus; her 7 month-old fawn tags along. Muleys sometimes gather pre-estrous does into small harems.

Life in Autumn

Some blacktail bucks behave like woodland whitetails, while others attempt to gather a harem of does, as do elk. I have not seen such harems, but a number of authorities have reported their occurrence. They have also reported that while the herd buck is battling a rival, a smaller buck may slip in and service a doe that is in estrus. And occasionally a big buck will share the favors of the does in his harem with a subordinate buck.

There are records of two whitetail bucks breeding one doe without animosity between the bucks. I have never seen this, but Joe Taylor has. Another friend of mine reported that he and several others witnessed a "gang rape"—the breeding of a whitetail doe by nine bucks near Belvidere, New Jersey. The incident occurred in a fenced area of several thousand acres, adjacent to a manufacturing plant. An exceedingly large deer herd had built up there. The doe came down off a hill pursued by the bucks. Her mouth was open and her tongue hung out from her exertions. She could not outrun the bucks, and every time she lay down one of them hit her with his antlers or feet and forced her to get up. She got no respite, and in the 15 or 20 minutes she was in sight she was mounted by three different bucks. There was no fighting among the bucks.

When a doe comes in estrus, she is always pursued by many hopeful bucks. Usually hope is all they have, because they just are not large enough to compete with the dominant buck. When I was in Louisiana one January, during the peak of the rut, I saw a doe pursued by two large bucks that appeared to me to be equal in size, strength, and antler size. But one of the two bucks was the better buck, because he kept the other buck and, occasionally, all the lesser bucks away from the female. Most of his attention was centered on the other big buck. During one of his longer forays after his main rival, an exceptionally persistent 4-point buck quickly mounted the doe and bred her. When the dominant buck saw what was happening, he broke off chasing his rival and charged, full speed, at the 4-pointer. He was too late; intromission had been accomplished. With the dominant buck thundering down on him, the little buck badly bent his penis as he withdrew and took off from the doe at top speed. But a few minutes later he was back, trying to get close to the doe again.

When a buck finds a doe in estrus and smells her hocks and licks her vulva, the doe usually urinates immediately. The buck licks the urine and then raises his head and curls his lip, flehmening. Even though the doe is ready to be bred, she usually does not stand for the buck the first time or two that he attempts to mount her. His tongue constantly flicks in and out as he licks his lips. He caresses the doe with his tongue about the head and body, and she also caresses him. Sometimes she will rub her body against the buck, further stimulating him. It is well known that the tactile stimulation of pressure or the rubbing of bodies, especially if returned, increases sexual tension. At times the entire body seems to be an erogenous zone.

When the doe finally allows the buck to mount her, intromission is not always accomplished. The buck's penis has a very flexible tip and actively seeks out the doe's vagina. As Leonardo da Vinci said, "the penis seems to have a mind of its own, if not a life of its own."

There is no mistaking when copulation is completed. After entering the doe, the buck gives only one or two preparatory thrusts and then lunges, plunges deeply into her and ejaculates. The ejaculatory thrust is so violent that many times the doe is thrown forward to her knees and the buck's hind feet leave the ground. The buck immediately withdraws. The doe then either stands with her back humped up or squats as if to urinate, straining hard with her tail upraised over her body. Although I have never seen evidence of it, some researchers report that the doe discharges a yellowish secretion at this time. If the doe is a young one that has never been

After mating, the does of both species usually squat and strain as if to discharge the semen, as this California mule deer is doing.

A whitetail breeding a doe.

bred before, she may stay in this squatting position for 20 minutes or more, straining as if to eliminate that which had just been thrust into her.

Again I must stress that each animal is an individual and its actions reflect that individuality. After copulation, some bucks pay very little attention to the doe, while some I have seen are as attentive to her afterward as they were before copulation. Some bucks may breed a doe only once or twice in the 24 hours that she is receptive. I know of one buck that bred a doe eight or more times in the daylight hours. Undoubtedly he bred her many more times during the hours of darkness while she was in estrus. I also saw a buck breed a doe twice in about 15 minutes.

Most of the breeding is done at night, because that is when deer are most active, particularly if hunted. Where they are not hunted, the bucks are active both day and night during the breeding season. In heavily hunted areas, trophy bucks don't travel in the daytime at any time.

The tremendous activity of chasing after the does, fighting, and breeding leaves little time for the bucks to eat. The rutting season usually lasts about 60 days. During this time, a mature buck may breed with only four or five does or perhaps with as many as 20. In captivity, bucks have been observed to breed with two does and even three in a single day. Bucks have been known to have their sperm count drop because of overbreeding so that they become temporarily infertile. In the wild, however, bucks are seldom so fortunate as to find 2 or 3 receptive does in one day. A

A muley buck here covers a doe: the actual mating is a matter of seconds, although it may be repeated while the doe is in estrus.

buck usually stays with a doe for a day before she comes into estrus, spends a day copulating with her, and then goes off to seek another doe.

If the doe has conceived, she has no further interest in bucks. If she has not conceived, she will come back into estrus 26 to 28 days later. She is usually geared to the 28-day lunar cycle. It is most unusual for a doe not to be bred during her first or second estrus period. If she is not bred or does not conceive in either of these two periods, she will come into estrus a third, and perhaps even a fourth, time before remaining unbred for that year. By March most bucks are no longer capable of breeding any more that season, and become sexually impotent. But that's not always the case.

The length of the rutting season helps to ensure some fawn survival in the case of a catastrophic happening during the birthing period the following spring. Although the bulk of the breeding takes place during a 60-day period, the season may extend 120 days or more. A study of 894 deer in New York State by Lawrence W. Jackson and William T. Hesselton showed that the earliest breeding date was October 1 and the latest was February 9, with the peak of activity occurring between November 10 and December 15. This length of time is typical of all deer, all species, in all sections of the country, although the dates may differ in particular sections. Some blacktail deer have been seen breeding in July in

Life in Autumn

the southern latitudes, while the northern populations are capable of breeding until April. But these are extremes and are not typical of the species.

Speaking of extremes, I want to give you three examples that are really extreme for the whitetail.

Most hunters have seen spotted fawns in the November and December hunting seasons, fawns that were born in late August or September, fawns from does that were bred in their third to fourth estrus periods.

Robert Buss, from Hawley, Pennsylvania, reported in the July 1983 issue of *Pennsylvania Game News* that he, and many other witnesses, saw two fawns in spotted coats in the third week of March of that year. Bob works for the game commission and estimated that the fawns had been born in either late November or December. That means that the does were bred in, or around, May evidently because they were not bred in their first estrous periods.

Bill King, the Richland County, Ohio, game protector, reported that he and a number of folks saw a whitetail doe and two still-wet fawns that were born on January 27, 1985. The temperature was below zero, and the authorities were afraid the fawns would freeze to death. The doe was lying down with her two little ones snuggled up against her for warmth. The report was written up two weeks after their birth, and the fawns were doing just fine. That doe had to have been bred in early July.

Rocky Figgolair, of Locke, in Cayuga County, New York, witnessed a doe giving birth to a single fawn on February 17, 1979. The doe was small, and this was undoubtedly her first birth. The temperature that morning was $-20°F$ ($-29°C$). During the next two days the temperature hit a low of $-22°F$ ($-30°C$) and a high of $9°F$ ($-13°C$). This doe was one of a number of deer that had been coming into the area behind Mr. Figgolair's home to feed on the corn and apples that he put out there for them. The snow in the area was 6 to 18 inches (15.2 to 45.7 cm) deep but hard-crusted enough that the deer were able to move around freely. Because the deer were used to him, Mr. Figgolair checked the little fawn when it was two days old and found it to be a buck.

E. L. Cheatum and G. H. Morton, doing deer research in 1946 for New York State, found that the earliest breeding date ranged from September 22 to 28 and the latest from February 16 to 22. The doe at Mr. Figgolair's had to have been bred around July 20 to 22. The only month for which I haven't found a breeding date among northern hemisphere deer is August.

Cold weather stimulates the onset of the breeding season but photoperiodism, the response to light, is the primary trigger. From my personal observations, I have concluded that excessive dark, rainy, overcast autumns advance the rut by about five days. I ascribe this to the fact that the deer's eyes are getting less light, as if the hours of daylight were actually shorter. We know that photoperiodism is the basic determination of the time of the rut. Every step is orchestrated—a precise fitting together of season, conditions, responses. A poor diet may cause a late or irregular breeding season. An overabundance of food may also delay the season or delay pregnancy in an individual doe.

Researchers have found that obesity will cause a doe to come into estrus later than normal. Excessive fat retards development of the follicle in the ovary, where the egg develops. Fat in the doe's reproductive tract may prevent a fertilized egg from reaching the uterus or becoming implanted if it does get there. This may be why, after a particularly good autumn with a superabundance of food, many does may not breed during their first estrus period. The colder weather before a doe's second estrus period burns up some of the excess fat so that pregnancy then becomes possible.

The bucks, which enter the breeding season as magnificent specimens, quickly lose weight and vigor, becoming quite gaunt, listless, and rough-coated. A weight loss of 25 to 30 percent is not uncommon. As the rut wanes, the buck's swollen neck decreases to its normal size, and the buck loses interest in the does. Although the bucks are capable of breeding does both before and after this period, the bucks are really "in rut" only while their necks are swollen. A short time after the swelling of the buck's neck goes down, his antlers will drop off. His main interest now is in eating all the food possible to regain his weight and fat reserves before winter sets in. The depletion of the body reserves is one of the main reasons why the most vigorous bucks (and the largest bull elk, too) are often the ones to die off in an exceedingly hard winter. They have given too much of themselves for the perpetuation of the species. With the next generation assured, nature sacrifices those that made it possible. It was their reason for being.

A fawn's diet determines whether it will breed at six to seven months of age or have to wait until the following year. If the fawn had a 16- to 18-percent protein diet, it will be in condition to breed by the time it is six or seven months old.

Whitetail fawns have a much higher breeding potential than do mule deer and blacktail fawns, because many whitetails have access to a much better food supply. Blacktail deer have the poorest record of all. On some of their poorest ranges, they do not breed until they are 2½ years old.

On an average range, about 40 percent of whitetail doe fawns breed. On the best of ranges, 65 to 74 percent breed. If a female fawn can attain a weight of 80 to 90 pounds (36.3 to 40.8 kg) she usually comes into estrus, though she is likely to do so about a month later than is normal for the mature does. The yearlings that give birth contribute about 30 percent to the annual herd increment. In one recorded instance in British Columbia, a blacktail fawn doe was bred by a fawn buck. This is the only such record I have found for blacktails.

Ordinarily, bucks do not become physically mature until they are 4½ years old. They usually become sexually mature when they are 1½ years old, but many are sexually precocious when only a few months old.

I have observed hundreds of little bucks attempting to mount other fawn bucks, fawn does, and adult does, including their mothers. When the fawns are young, these actions are part of their play pattern. After four or five months, the advances are made in earnest. Usually the subordinate fawns just run away from the sexual advances of a dominant buck fawn. Even some adult does will run.

When a precocious buck fawn attempts to mount his mother, she usually whirls around and strikes at him with a front foot. Her rebuke does not always have much effect. Some of the little bucks just keep trying. I think that as more research is done, it will be found that more buck fawns are capable of breeding than previously thought possible.

Interbreeding. Ordinarily, in states that have populations of both mule deer and whitetails, the ranges of these two deer don't really overlap. Their food preferences and habitat requirements differ enough so they remain more or less segregated. Where they do encounter one another, the whitetails give way before mule deer, not because of any aggression but simply because the mule deer are larger.

Interbreeding between the mule deer and the blacktail deer is common and widespread, for the blacktail is really a subspecies of mule deer—at least at this stage of the blacktail's evolution. But if, as most experts agree, the blacktail is an emerging species in its own right, interbreeding with the mule deer will probably become less common in the distant future. Hybridization between

A crossbreed from the mating of a whitetail and a blacktail doe: note the whitetail antlers and the blacktail-length metatarsal.

these two deer is evidenced primarily in the size of the offspring and in the color pattern and shape of the tail and rump patch, as noted earlier.

It has long been thought that interbreeding between the mule deer and whitetail occurred infrequently and that the resultant offspring were infertile. But, increasing numbers of whitetail-mule deer crosses are fertile, as well as increasing numbers of whitetail-blacktail crosses.

Jerome J. Pratt reported a number of generations of fertile crosses between the Coues whitetail and the desert mule deer. The original interbreeding occurred in captivity, and some of the fertile offspring escaped into the wild. In 1963 the last two hybrid bucks were turned loose. These hybrids proved fertile.

R. Havel, a biologist from Nebraska, reported 10 hybrids out of 17,039 deer that were examined in that state. W. C. Peabody, a

Life in Autumn

biologist from Kansas, found 2 hybrids among 983 deer that he examined. The highest percentage of hybrids was reported by R. Webb, who found 6 hybrids out of 1,000 deer that were checked in Alberta. Unfortunately, none of the reports could say whether the hybrids were first- or second-generation hybrids or whether they were fertile.

Researchers have discovered that most of the crosses between these two species are between whitetail bucks and muley or blacktail does. When either of these does come into estrus, they will attract whatever bucks of either species are in the area. The bucks of both species chase the does as part of the courting ritual, but the whitetail always chases harder and longer; as it generally happens, the mule deer gives up the pursuit while the whitetail continues and, in the end, does the breeding.

As mentioned, the mule deer population is declining in the Canadian prairie provinces, while the whitetail population is increasing. There is a very good possibility that the whitetail will further reduce the mule deer of the area simply by breeding it out of existence.

This idea is not as farfetched as it seems. The black duck is closely related to the mallard duck and the two species interbreed very readily. The black duck population has been declining as the mallard population has been burgeoning. Many ornithologists are concerned that the mallard will breed the black duck out of existence by the year 2000. In certain areas of Alberta, this could also happen to the mule deer. Although the first offspring would possess the characteristics of both species, by the third generation, if the male is a whitetail in each case, most of the mule-deer characteristics would be eliminated.

The offspring of any of these crosses reveal the hybridization in a number of ways. Most frequently it shows up in the antlers. A hybrid may look like a whitetail but have the bifurcated rack of a mule deer, or vice versa. The size and placement of the metatarsal gland is also different than on either parent. And the tail and rump patches may be confusing.

In 1984, I gave a deer seminar in the state of Washington. While there I had the opportunity to photograph a cross between a whitetail buck and a blacktail doe. I photographed one crossbred buck and saw two others. The buck that I photographed had the perfect whitetail antlers and ears, and the blacktail's tail and rump hair. The clinching characteristic was the metatarsal glands, which were the 3-inch (76.2-mm) crescent-shape of the

blacktail. The front half of the deer had all the whitetail characteristics, the back half all those of the blacktail, as if someone had
joined the two halves together.

The preorbital glands of a mule deer are much larger than
those of a whitetail, being almost 1 inch (25.4 mm) in length. The
gland on the whitetail is about ¹¹⁄₁₆ inch (17.5 mm) in length while
those of the crossbreeds are about ¹³⁄₁₆ inch (20.6 mm).

A final note on the subject of the breeding season comes from
a Pennsylvanian named John Miller, who contributed it to the
January 1977 issue of *Pennsylvania Game News*. Some friends
were visiting Miller in late October, and en route they had seen a
number of deer crossing the roads. One of the visitors, a lady,
asked why so many more deer than usual were moving about.
Miller told her it was because of the rut. "What do you mean, *rut?*"
the lady asked. John explained that the deer were mating. After a
thoughtful pause, the lady said, "If you are going to get into a rut,
I think that's the best kind."

THE HUNTING SEASON

Perhaps the greatest disruption in the life of the deer is the opening of the hunting season. In many areas, the small-game hunting
season opens first, so the farmland deer and those of the fringe
areas are preconditioned to the influx of hunters and the noise of
shooting. But nothing quite compares with the opening of the deer
season.

People assume because I spend so much time photographing
and studying deer that I can easily go out and shoot a deer as soon
as the season opens. Not so. It would be comparatively easy to get
a deer at any other time of the year, because then the deer are
living predictable lives. All of that changes on opening day.

In my area, even though some tracts of land encompass several square miles, there are enough roads so that any area can be
approached from all sides. And they are. When the hunters get out
of their cars along the perimeter of such an area, they usually push
the deer ahead of them to the tops of the small mountains. Then,
unless the hunters are willing to climb the ridges, they wonder
where all the deer have gone that they saw so often in the lowland
fields. The habits of all of the deer are changed drastically during
the hunting season.

Bucks are almost always warier than the does and doubly so
in those states that have bucks-only hunting. Adult does are war-

ier than the yearling bucks because, even if not hunted, they have been subjected to more pressure by hunters. The more a deer has been hunted, the greater its "flight distance" becomes. I have also found that the more a whitetail buck is subjected to hunting pressure, the more it runs with its tail down. The younger bucks make up the greatest proportion of the annual hunters' harvest, not only because they are less wary but also because there are a lot more of them.

Where there is no hunting, the ratio is 100 bucks to 160 does. Where there is hunting of either sex, the ratios do not change greatly: most hunters would rather take a buck than a doe. Where only bucks are hunted, the adult bucks comprise 10 to 15 percent of the herd, adult does about 50 percent, and the fawns 35 to 40 percent.

Though there are fewer bucks than does, a hunter who sees a doe move past should assume that a buck will follow. I am not saying a buck *will* follow, but there is a good chance that one will. Where the hunting season coincides with the rut, the chance is all the better. But even when a buck is not intent on mating, he is likely to bring up the rear, while does are likely to take the lead. This "order of march" is true for many ungulates. I doubt very much that the buck is being altruistic when be brings up the rear. Among many animal species, the males bring up the rear because it is the most dangerous position. The males are there to intercept the predators that are stalking the herd from the rear, or downwind. The deer experiences very little predation of this type. The bucks follow along at the rear because the buck is not going to expose himself to danger when he can let the females discover the danger for him. On occasions I have seen the big bucks following along behind the smaller bucks. This was definitely a danger-avoidance tactic.

More deer escape hunters merely by standing still than by any other means. Deer blend into their backgrounds amazingly well, and most hunters walk right past them. Some hunters don't really see anything in the woods at any time. Others are looking for deer—picture-postcard deer. They are not looking for an ear, an antler, the movement of just a tail that may be all that is visible. At other times the deer are aware of the hunters' locations, and they circle around and come in behind the hunters.

Many states have conducted controlled hunts in fenced areas perhaps a mile square. The number of bucks in the enclosure is known and a specified number of hunters are allowed in at a given

A whitetail hunkered down in the middle of a field: bucks would much rather hide from danger than run from it.

time. Then the entire proceedings are watched by biologists from high towers placed at strategic points in the enclosure.

Six hunters at the Cusino Wildlife Experiment Station in Michigan hunted for four days before any of them even saw one of the nine bucks that were enclosed in the mile-square area. On the fifth day one of the hunters did kill a buck. These men were experienced hunters. They were hunting in good weather and had the advantage of a tracking snow. Many people thought that hunting the deer in an enclosed area would be like "shooting fish in a barrel." But deer are elusive animals, and this and other experiments conclusively proved that point (although it needs no proving to anyone who knows deer).

When anti-hunters declare that the deer never have a chance, they just don't know what they are talking about. It is true that our modern weapons give the hunter more clout than in yester-year, if the hunter sees the deer. Humans have weapons and superior brains. In all other aspects, the deer are far superior. Their sense of smell is something of conjecture. The deer's sense of hearing is far keener than a human's. Its eyesight is better than ours on moving objects, because deer are more alert. They are stronger, faster, have more stamina, know the area better. And I'm not so sure that their innate knowledge, coupled with their learned experience, isn't the equal of human intelligence in the escape and evasion situation. Deer have the ability to stand still much longer than humans do. Deer are not pressed by time. Deer don't have to be in or out of the woods at a certain hour, they don't have to meet someone at a certain time, they don't have to be in a certain place at a certain time. Today means nothing to them; they don't even have to eat. So when a deer suspects danger from a certain point, it can stand or lie all day in one spot without moving—until the danger betrays its presence or departs.

Biologists report that most of the time the bucks stand still and allow the hunters to walk past. Then they may remain in that spot or sneak off to one they think may be safer. Bucks have been seen to circle behind hunters and follow them at a safe distance. The bucks often crawl into thickets or under blowdowns, or get behind or under cover that looks too skimpy to conceal a deer. Unless startled from a hiding place, or shot at, deer almost never run. If they do run, it is only far enough to reach the next piece of cover. Most blacktail and mule deer stop just before plunging into cover. Some whitetails do, others don't.

Deer that know they are being tracked will watch their back trails, sometimes even circling back to watch the trail from the side. The only way a hunter can hope to get a shot at such a deer is to make large circles from the track, arching around where the deer might be in the hope of getting ahead of it.

Windy days are bad for hunting because the deer are as skittish as a panful of Mexican jumping beans. Conditions are ideal after a rain has softened the fallen leaves, or after a light snow. Even then, the still-hunter should do a lot more looking than walking. A light, soft snow is a tremendous advantage to the hunter because it softens the leaves, allows the deer's tracks to be seen and followed, and makes deer stand out from their surroundings. But of course, just as snow makes the deer more visible to a hunter, it also makes the hunter more visible to the deer. Wearing white camouflage clothing is of no help if the hunter moves, because a white object walking in front of the black trees is also easily seen. Many companies are coming out with snow camouflage that is basically white but has black streaks through it. This pattern has proven to be more successful than the all-white outfits. As noted earlier, a certain amount of blaze orange—"safety orange"—is required in many states. But if you can sit still, the color of your clothing doesn't matter.

A really cold snow does allow the deer's tracks to be seen but the creak of the sharp edges of snow compressing under a hunter's feet is as noisy as dry leaves rustling. Deer can stand still in cold weather a lot longer than a hunter can, so again the advantage is with the deer.

Many hunting parties conduct deer drives. Drives may be slow and silent or fast and noisy. A slow, silent drive gives the drivers as well as the standers a chance at a shot, while a noisy drive usually provides only the standers with shots. One thing is certain: You don't actually drive deer, you only stir them up; they go where they want to. The most successful drives are generally those that have been held by the same groups on the same territory for a number of years. These groups get to know where the deer prefer to go when stirred up and can place the standers on the productive spots.

In parts of the Southeast where driving deer with dogs is not only legal but a must, it is important to know the natural crossings. Another requirement is that the dogs be "open trailers" so the stander can hear them coming and be prepared. The hunter must also remember that the deer may be quite a distance in front of the dogs. Ordinarily, the cover is so dense that unless the stander knows the direction the dogs are coming from, the stander may not be alert enough to see a deer go by. Fast-running dogs should not be used because the more slowly the deer is pushed, the more likely it is to stay in the hunting area instead of hightailing to the next county. A slow-moving deer makes a much better target, too, even though the hunter will probably be using a shotgun instead of a rifle. With the deer quickly ghosting through exceptionally dense cover, a rifle is often useless for such hunting. Moreover, the standers (and occasionally the drivers) in a hunt with dogs may be too close for safety with a rifle.

The hunter who stalks mule deer in the western states will concentrate on the brushy draws and coulees because western deer favor such areas. The deer know every gap in the high country, and they know the shortest way over the ridges. The hunter had better know, too. Mule deer do not "stay put" as tightly as whitetails. There is less cover for them to hide in or behind, and they are accustomed to moving out in front of danger.

A lone hunter can often flush a mule deer out of a patch of cover by making noise while approaching the cover and then

Life in Autumn

standing motionless. Mule deer cannot take much of this kind of "war of nerves." If a buck knows he is in danger but can't smell the hunter—doesn't know where the hunter is or what the hunter is doing—it won't be long before that buck seeks shelter elsewhere.

Mule deer can usually be driven from their beds in the thickets below the rimrock if a hunter up on the ridge chucks small stones down into the brush while walking along. Without the disturbing sound of the stones coming through the brush, a buck would probably lie still and let the hunter walk by. The buck would feel safe because of the distance to the ridge. Small stones should be used, because large ones will frighten the buck badly and drive him out at full speed, whereas he will probably try to sneak out if only disturbed. From a vantage point on the ridge, the hunter will probably be able to spot the deer before he gets out of range.

Lone hunters who prefer to stand have to know the deer's trails in the particular area and the times when the trails may be used. The lone stander is also in position to take advantage of deer stirred up by other hunters and should also realize that during periods of intense hunting pressure the normal trails may not be used at all. And the direction of the wind is of vital importance to the still-hunter who attempts to stalk or follow a deer and to the lone stander who hopes to ambush one.

Most of the deer that are taken in any particular area are killed during the first two days of the hunting season. Each Saturday is also a big day, because most of the hunters don't have to work that day. The deer's habits and patterns, completely disrupted by intensive hunting pressure, revert to normal when the woods quiet down for even a couple of days. Normal areas for feeding will again be used, although most of the feeding will be done after dark. Deer may be almost anywhere and may go almost anywhere, but they use their established trails whenever possible. These deer trails are not the result of random wanderings but are the easiest way for the deer to reach a given destination. Depending on where a trail goes, or why it was used, it may not be used today or even for some time. Yet the main trails will be used by the deer that made them for as long as the deer and their offspring live, or as long as the trails continue to serve their purpose.

The noise made by two bucks fighting attracts any other buck that hears the sound. Sometimes the third buck will join in the battle; at other times he is just curious. More and more hunters are taking advantage of this by rattling deer antlers as a hunting technique. This practice has always worked well in the Southwest, where it was employed by Indians of that area to aid in their hunting. In the Southwest it works equally well on whitetails and muleys. Rattling antlers is only effective during the rut, particularly during the five-day peak, but most states' hunting season corresponds to the rutting season.

Rattling antlers has always worked better in the Southwest than anywhere else for the same reason that the bucks there make more scrapes; the buck-to-doe ratio is closer to being even, which means that there is a lot more competition. But rattling can be done successfully in all parts of the country.

A few years ago I was doing my deer seminar at the Dixie Deer Classic in Raleigh, North Carolina. That evening I was at a reception given by my host, Dr. Carroll Mann. Included among the guests were Fred Bear, Dr. Rob Wegner, Dick Idol, Tom Fleming, and many other knowledgeable deer people. The talk turned to calling deer by antler rattling, and Tom Fleming was persuaded to show us his technique. Afterward Fred asked me what I thought about its effectiveness, and I replied that I still had to be convinced. Tom then offered to prove the technique's effectiveness to me the following rutting season. I agreed to be his student on one condition. If he was successful I'd tell the world; if he was not successful, I'd still tell the world. Tom's technique proved to be as good as he claimed, and I've been telling the world ever since.

Arthritis has prevented me from drawing a bow for years, although I am again getting back into archery. I went out with Tom on land open to the general public, and I mention this to indicate that it was not exclusive club land where hunting conditions would be more favorable. The rutting season was on but had not reached the peak period. Windy days are not good antler-rattling days because the sound is blown away from all but one direction and many bucks simply won't hear it. Under ideal conditions bucks can hear rattling antlers at ½-mile (0.8 km) or more. In three days Tom called in four bucks, the largest being a 6-pointer. The size of the antlers didn't matter; what did matter was that Tom had said it could be done anywhere, by anyone, and he has proved it. Tom has called in many good-sized bucks with his rattling.

Deer occasionally respond to loud noises out of curiosity, in contradiction to the normally valid rule that a hunter must be silent. Once I got a nice buck that came up to see what was making all that noise. I was. I hadn't planned on hunting that particular day, but a friend came by before dawn so I agreed to go out with him for a couple of hours. We got into the woods while it was still

Life in Autumn

dark. When the morning mist dissipated, I did not like the spot I had chosen because there were no "alleys" where I could shoot through the dense rhododendron. So I moved to a spot that looked better. Since I had already made some noise walking on the leaves and forest duff, I figured a little more wouldn't matter. I didn't really expect much action anyway until after 9:00 A.M., when most of the hunters would get cold and tired of sitting and start to move around.

Wanting to be comfortable, I knocked down an old dead chestnut tree. I propped this up against another tree and used it for a seat. I made no attempt to be quiet. I was seated for about two minutes when, to my surprise, I saw a young buck walking through the brush toward me. Evidently he was drawn to the spot by the racket I had made. He was equally surprised to see me but I recovered and moved faster and was done hunting for that season.

This was not an isolated instance. I have talked with many hunters and have read many accounts of deer walking up to investigate loud or strange noises. Deer are very curious creatures. I don't recommend banging on the bushes to attract deer, but at times it has worked.

Many times deer have been attracted by the smoke of a campfire. Ordinarily, smoke must be an alien odor that spells danger to deer, yet at times it attracts them. There are rural areas where most of the homes (as well as deer camps and vacation cabins) have wood stoves or fireplaces, and perhaps in these regions the smell of wood smoke is so constant that deer lose any fear of it. Sometimes, there is no accounting for what wildlife is likely to do.

People often asks me if the deer scents that are sold for hunting really attract deer. As explained earlier, some of the scents are made from the food that deer eat, such as apples, corn, and acorns. Some of the cover, or masking, scents may be made from neutral-odor products such as walnut, hazelnut, or from animals such as fox. Most of the scents are made from deer glands, secretions, or excretions. Do they work? Yes, they do. All of these scents are successful when used properly. But there is no point in using scents on a deer drive. Driven deer have no time to be attracted by the scents. Scents work best for the lone stander.

There are also a number of commercial deer calls. Most of them imitate the low, raspy, reedy, blatting call of a buck. Some are higher-pitched and sound like a fawn. I have not been able to call in a buck with any of these during the hunting season, but I have called in both does and a couple of bucks when I tried the calls prior to the hunting season. I have, during the rutting season, called in bucks with the grunt calls. Care should be taken not to overdo the calling, because the sound may help the buck pinpoint your location. You want to get his attention, and stimulate his curiosity; you don't want to drown him in sound.

"SMARTNESS"

There is a great deal of discussion about the relative "smartness" of the three types of deer. The whitetail is undoubtedly the smartest, mainly because of where it lives. The proximity of man has forced the whitetail to become more wary. The fact that this deer is also found in or near heavy cover means that it has a better chance to hide and stay hidden.

This muley seems to have had the idea of camouflaging his antlers in a bush. photo courtesy Irene Vandermolen

Life in Autumn

Mule deer are usually found in more open country where they can be seen for greater distances and where there may be far less cover to screen them. The human population puts far less pressure on the mule deer, and when this deer is high in the mountains, it may not even see a human for months on end. Furthermore, the mule deer has had only about half the number of years of association with Europeans and their descendants and their firearms. Mule deer subjected to increased hunting pressure have proved that they can "smarten up."

The blacktail receives more hunting pressure than the muley and lives in closer proximity to larger human populations. The blacktail inhabits some of the densest cover of any of the deer and is as elusive as the whitetail. He has had the opportunity to become "smarter" than his cousin, the mule deer.

The Wounded Deer. Killing a deer is what makes me a hunter; the appreciation of the beauty of the outdoors is why I am hunting in the first place.

A fact of life that we all have to live with is that there are "slob" hunters in the woods as well as genuine hunters. Nonhunters and anti-hunters don't all realize that the genuine hunter—the sportsman—is as much against the slobs as they are. The slob is the one who blazes away at anything, at any time, at any distance. Slobs also have many other unsavory characteristics, such as disregard of private property, littering, and so on, but we are concerned here with another issue: decency toward wildlife. People who hunt have a moral obligation to know their quarry, their weapons, themselves, and their limitations thoroughly. Decent hunters will not try for a "pot" or "luck" shot but will be sure that their weapons can kill the game cleanly and quickly if they shoot.

Unfortunately, not all shots are killing shots. This is true for the best of hunters. Every hunter must check to make sure whether the deer fired at was missed completely or wounded. If the animal was wounded, the hunter should spare no effort to follow, locate, and dispatch it.

There are no hard rules about a deer's reaction to being hit. There are differences in bullet and arrow weights and the speed and shocking power with which they are delivered. The actions of the deer before being hit also have a tremendous bearing on its reactions. A deer that is shot while standing and not alarmed can be dropped in its tracks much more easily than a deer that is excited and has been running. An excited deer has adrenaline pump-

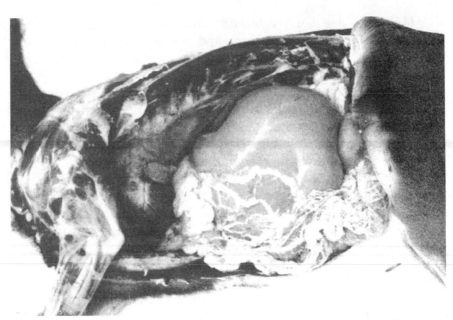

A whitetail doe with her left ribs removed showing the heart, collapsed lungs, spleen, and paunch. On an average size deer such as this, a shot placed two inches behind the foreleg elbow and two inches up into the body will be in the center of the heart.

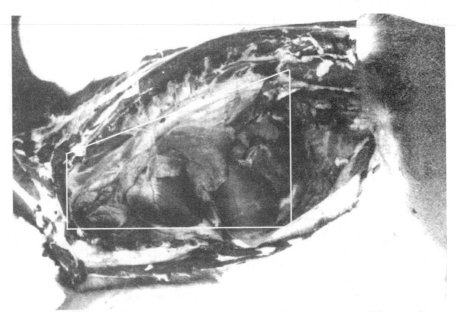

The doe with her left ribs removed to show the heart, collapsed lungs, liver, and spinal column. Any shot in the boxed area would be fatal.

ing through its system and may run hundreds of yards with its heart shot out. All the same, most deer will react in the manner that I will describe.

Brain, neck vertebrae, and spine vertebrae are the surest, fastest-killing shots when it is possible to get them. When hit in any of these three spots, a deer will drop instantly, although it may not die instantly. Unfortunately, these are also three of the most difficult shots to make. And if a hunter feels that the deer in his sights is worth mounting as a trophy, he is unlikely to shoot at its head or neck.

A shot in the lung-heart area is almost always fatal, but the deer probably will not be killed instantly. When hit in the heart, the deer may hunch up and kick out with its feet. The blood from a heart wound will be a deep red, and the hair that may be cut off by the bullet will be the regular long guard hair. Sometimes a heart or lung shot will knock a deer off its feet, and occasionally a lung shot will cause a deer to rear up. Blood from a lung shot will be bright red and often frothy. Deer shot in these areas should be followed immediately because they are not going to go far, since the effort of running itself exacerbates the loss of blood.

How far can a deer go that has been hit in the lungs or heart? There are too many conflicting factors that preclude an accurate answer. The main factor is whether the animal was frightened before it was shot so that the adrenaline was in its system. There are limits to what a deer can do, based on physiological parameters.

A deer has approximately 1 ounce (28.35 g) of blood for each pound of total body weight. A 150-pound (68-kg) buck would have 150 ounces of blood, or a little over 9 pints (4.3 liters). The average man weighing about 175 pounds (79.4 kg) has approximately 3,600 cc of blood or approximately 9 pints (4.3 liters).

A human heart will pump 60 to 80 cc per beat, but a deer's heart is almost double the size of a human's. As this is so, then the possibility exists that a deer's heart would be able to pump 120 to 160 cc of blood per beat.

Half of the deer's blood would be filled with oxygen as it passes through the deer's lungs and is delivered to the cells via the arteries. The other half would be carrying carbon dioxide being returned by the veins. A human being can live 2½ to 3½ minutes before losing consciousness due to a lack of oxygen. A deer presumably can do the same.

It has been shown that a human heart beats about 70 to 80 beats per minute when at rest, and a deer that is standing or moving about is also in the range of 60 to 80 beats per minute. A running deer has a heart rate of 180 beats per minute.

Under normal conditions an adult doe that is peacefully lying down, doing absolutely nothing, has an average heart rate of 37 beats per minute. If it is chewing its cud, the heart rate is increased to 41 beats per minute. A deer standing quietly has a heart rate of 45 beats per minute. While foraging, taking a few steps, then taking a few mouthfuls of food, its heart rate is 50 beats per minute. According to the rate that it is walking, its heart rate may range between 60 and 78 beats per minute. A doe that is not being pushed, but is just running to catch up with the other deer, for example, has a heart rate of 104 beats per minute. When that same doe is forced to run hard, its heart rate will increase to speed oxygen to the muscles at a rate of 138 to 160 beats per minute. The highest recorded heart rate for an adult deer was 210 beats per minute.

Fawns have a faster heart rate than do adult does at any time of the year. Bucks normally have a slower heart rate than do does because of their larger size. Bucks are always more active than does, and this too causes their hearts to be larger and to pump more efficiently. Athletes almost always have a slower heart beat than nonathletes because their hearts are more developed due to the demands made on them.

The lighter-weight deer, of either sex, of any age, at any given season of the year, have a slower heart rate than do their heavier contemporaries.

The deer's heart rate is slowest in January and February, when their basic metabolism is also at its lowest. The average heart rate is highest in May and June. The heart rate slows down in the warmest part of each day. After extreme exertion, it takes about 15 minutes for the deer's heart to return to a normal beat. *Bradycardia* is the word used to denote the slowed heart beat in January and February. *Tachycardia* denotes an excessively high heart rate.

A deer's heart that has been destroyed by a bullet or an arrow loses its pumping ability at once, although there may still be electrical impulses. The damaged heart would leak blood, as would the lungs, but the deer could live for approximately 1¼ to 1¾ minutes on the oxygen that is in its brain and bloodstream.

A deer shot through the heart may be knocked off its feet by the rifle's shocking power, but it is not likely to stay down. It makes little difference if the deer was standing or running; it will

Life in Autumn

try to run at top speed as soon as it is hit, at a speed of about 35 miles per hour (56.3 km per hour). The deer doesn't bother with trails when shot. It bounds off into the woods in any direction and frequently zig-zags. It may run into objects. If you are tracking a wounded deer, remember that a badly wounded deer does not follow trails, but one with superficial wounds may.

The heart will not be functioning, so the deer will run until its supply of oxygen is exhausted. Running at a speed of 35 miles per hour the deer could potentially run 3,850 feet (1,173.5 m) in one and one-quarter minutes or 5,390 feet (1,642.9 m) in one and three-quarter minutes. Although it is not likely the deer will run this far, the potential exists.

If the deer is shot in the lungs, it will not be able to run as far because the heart will be pumping the blood until traumatic shock is produced by a lack of blood and a corresponding lack of oxygen. Here is where the big difference comes in as to whether the deer was unsuspecting or agitated when shot.

The heart rate of an unsuspecting deer is 70 to 80 beats per minute. When shot, its heart rate will jump up to 180 beats per minute within one heartbeat, according to the size hole and whether the missile, either bullet or arrow, went through one lung or both lungs, and whether it exited through the far side of the deer's body.

The loss of a third of the blood, or 3 pints (1.4 l) is fatal, although the animal may die with a loss of 2 pints (0.9 l) of blood plus the shock of the wound. If the deer's heart is pumping out 120 cc (approximately 5 ounces) per beat, the loss of 1,200 cc (3 pints) could conceivably be achieved in 10 beats. At 180 beats per minute, the 10 beats would take a little over three seconds. If the unsuspecting deer was standing still when shot, it may take all of that three seconds to get up to a top speed of 35 miles per hour (56.3 km per hour), for an average speed of 12 miles per hour (19.3 km per hour). At that speed, a deer could cover 17.6 feet (4.8 m) per second for a distance of 58 feet (17.7 m) A lung-shot deer would probably go at least twice that distance. It takes approximately 22 seconds for the hormones from the deer's adrenal glands to get throughout its system, and the deer would probably be dead before that occurred.

A deer running flat out would have a heart rate of at least 180 beats per minute that should pump the blood out of its system faster. But the adrenal glands would have pumped the hormone epinephrine (adrenaline) throughout the deer's system, which would cause faster coagulation and vasoconstriction, cutting down on blood loss. The deer would already be at a top speed of perhaps 35 miles per hour (56.3 km per hour), so it would be able to travel at least three times the 58 feet (17.7 m) for 174 feet (53 m). The deer would probably go two or three times that distance, because of the action of the epinephrine making the supply of oxygen last longer and postponing lactic acid buildup, which, coupled with blood loss, causes death.

Some deer do not run off at top speed. Almost every deer will give some reaction to being hit, such as humping its back, kicking out with its hind feet, or rearing its forequarters. These signs may not be noticed as the deer tries to get away at its best speed. Some deer will run as long as they can; others will run a short distance and bed down; others may walk off from the beginning if they do not detect the hunter. The figures above are for a "worst-case" scenario to encourage hunters to look for a much greater distance than they previously may have thought it possible for a wounded deer to travel.

The most extensive simulated testing of a deer's bleeding was done by my good friend Dutch Wambold, of Pennsylvania. In his experiments, Dutch took beef blood, mixed in citrate-of-soda to prevent coagulation, and laid down a series of trails, dispensing the blood from a squirt bottle in as near a duplication of bleeding as possible. He made squirts, dribbles, drops, continuous bleeding, and pools of blood. His findings were a revelation, to say the least. A pool of blood 12 inches (30.5 cm) across, with the fluid applied liberally to counteract its being soaked up by the snow, required a mere 6 ounces (0.18 l) of "blood" to create. Remember it takes about a 42-ounce (1.8-l) loss of blood for a deer to bleed to death. With many of his trails, Dutch concluded that the deer would go over a mile (1.6 km) before dropping.

I did some further experimentation of my own to determine the rate of blood loss of wounded deer. I took an eyedropper with an internal diameter almost identical to the external diameter of an arrow shaft, $^{310}/_{1000}$ inch (measured by micrometer). I did not have access to blood at the time, and so I used water. I realize that blood is thicker than water; but different liquids forced through an opening of the same diameter form drops having approximately the same diameter.

By actual count, it took 398 drops of water to equal 1 ounce (30 cc). Let's round that off to 400 drops. At 400 drops, it takes 6,400 drops to equal 1 pint (0.47 l). It would take 19,200 drops to equal 3 pints (1.42 l), the amount of blood loss that is fatal to deer. That number of drops—as Wambold's exercise showed—will make a long trail when distributed by a running deer. Of course, there are other variables under actual hunting conditions; the style of broadhead and the amount, and kind of tissue it cuts through each affects the rate of blood loss.

That is why it is imperative that the deer be hit in a vital area; it must be made to hemorrhage. A deer hit in a nonvital area may later die from trauma, infection, or even from blood loss, but it may take several days to die and is not likely to be recovered by the hunter. And it is exceedingly important that a bowhunter be able to drive an arrow all the way through a deer so that there is an exit hole to leave a blood trail. Many deer are shot when they are almost under the hunter's tree stand. You must realize that, in shooting down on a deer, you have a very small target. An adult doe's rib cage is about 10 inches (25.4 cm) wide, while that of a buck is 12 inches (30.4 cm). From that you must deduct 2 inches (5 cm) for the ribs themselves, giving you an 8-inch (20.3-cm) and 10-inch (25.4-cm) target, respectively.

If your arrow does not exit through the bottom of the deer's rib cage, there will be no blood trail whatsoever to follow. The deer will die quickly because its lung area will quickly fill up with blood, in effect drowning the deer. If you hit the deer in the chest area, watch the deer's exit path, as far as you can see it, before getting out of your stand, and then follow immediately. Go to the spot where you last saw the deer and then you will have to follow its tracks. The deer should be within 100 to 200 feet (30.4 to 61 m). A deer hit in the spine will drop in its tracks, although another shot will be needed to kill it.

A deer shot in the paunch usually hunches its back and clamps its tail down. But the clamped-down tail is not a sure sign of a hit, because scared deer often do this, too. The blood trail will be dark in color, with a greenish or brownish tinge from the stomach contents. This is an unfortunate area in which to hit a deer, and it is one that no hunter ever tries for deliberately. A shot through the entrails creates more shock, as it severs more of the deer's tissue than does a paunch shot.

There will almost always be tufts of hair cut off by the arrow, bullet, or slug. A deer is darkest on its back with the color shading lighter to pure white on its belly. Hair from high on the deer's back is slightly longer than that on its side. The hair on the belly is the longest of all body hair, that of the tail is longer yet.

Shots that go through the shoulders usually hit either the lungs or the spine. Shots in the hams may sever the femoral artery. If so, deer will soon bleed to death. Deer that are shot in the lower legs often escape. If only one leg is hit, the deer may bleed, but it can run almost as fast on three legs as on all four. Deer that lose a leg are crippled when they walk, but the limp is seldom noticeable when they run.

Deer that are wounded will seek the nearest heavy cover they can find. If you cannot find your deer the first day, perhaps because of darkness coming on, look the next day along the nearest water. The fever that is created by the wound will cause most deer to seek water. Usually a wounded deer will go downhill, simply because it is easier for it to do so. A deer with a broken front leg will try to keep on level ground because it has trouble going downhill.

There is a lot of controversy as to whether or not a deer should be followed at once. I believe that they should be, for a number of reasons. In my area, the hunting pressure is so great, particularly in the gun season, that a deer that is not dropped in its tracks is not likely to be claimed by the one who first shot it. The progress of a deer can be followed through the area as each hunter it passes fires at it. Although traditionally the hunter drawing the first blood has the right to claim the deer, this is rarely what happens any more. Nasty confrontations may be avoided by accepting that the first hunter to reach the deer is going to claim it. It may not be right, but those are the facts of life today on public land. On private land, tradition still holds.

I think that a deer should be followed at once, to keep the animal moving, to keep its heart beating at its highest rate, to pump out the greatest amount of blood, to prevent the blood from clotting, to cause the deer to deplete its oxygen supply, and to prevent the wound from healing at all or being treated by the deer.

It is said that a hunter should wait from 20 to 30 minutes, even up to an hour, after hitting a deer to give the deer a chance to stiffen up. No deer I ever found had stiffened up until rigor mortis set in, usually about one hour after death had occurred. The deer does suffer from shock and this, coupled with the loss of blood and oxygen, causes it to go more slowly. If the deer is bedded, it may not be able to rise again for any of the above reasons, but it does not stiffen up.

I learned my lesson on this score the hard way. I lost only one of the deer that I hit while bowhunting, but it was the biggest buck I had ever shot at. I was hunting in an orchard at Camp Paha-

Life in Autumn

quarra in Warren County, New Jersey. The buck came in to feed earlier than usual because a storm was brewing; but, because of the approaching storm clouds, it was also dark earlier than usual. I got a good lung shot that buried the arrow almost up to the feathers. The deer took off as if it were turpentined, heading for a swamp. I figured the deer would bed down in the heavy cover.

I waited the prescribed 20 minutes. I waited, but the rain didn't. It poured in torrents. What should have been a fantastic blood trail was completely obliterated by the downpour. I searched for about one hour with no luck. The next morning I enlisted the aid of two scout troops that were spending the weekend at camp, and we thoroughly combed the swamp and the surrounding hills for hours. We never found the buck, or even any sign of the buck. Even later, I never found the remains, although I am sure the deer could not have survived the hit.

In trailing a wounded deer, walk to one side of the track, and try not to disturb the trail you are following. If possible, walk on the opposite side of the trail from the sun. Tracks and disturbances of the leaves show up much better against the sun. If you are following a blood trail and you suddenly lose it, tie your handkerchief to a branch as a marker. Then make ever-widening circles until you find another sign. If you fail to find a new sign, you have your marker to tell you where the last sign was. If no more blood or tracks can be found, the best bet is to go in the same general direction as the trail was going and check all possible pieces of cover.

If you have to follow your deer's trail after dark, you can do a much better job if you use a gas lantern than you will with a flashlight or a fluorescent light. I don't know just why it is, but the light from a kerosene lantern causes a blood spot, even if dry, to reflect and to be seen much more easily.

A new dimension has been added to trailing wounded deer, one that has been tried in a number of states and has been very successful, and that is in the use of special trailing dogs. Although different breeds of short-legged trailing dogs are used, the long-haired dachshund is a favorite. A hunter unable to find his wounded deer can put in a call to such organizations as Deer Search, in states where this service is legal. As soon as possible after receiving a call, a volunteer with dogs will join the hunter. The dogs are kept on a leash while trailing. Even the use of a good trailing dog does not guarantee the recovery of the deer. The success ratio runs from 75 to 90 percent. At times the deer is not wounded seriously and is able to go on, but the recovery rate is high enough so that this method is greeted with enthusiasm by hunters all over.

How many wounded deer are lost each year? The figures are appalling, ranging all the way from 33 percent up to 80 percent of the total hunters' take. It is next to impossible to get an accurate handle on the number of wasted deer because, unless the carcasses are found, and many are, there are no statistics other than those given by the hunters. Some hunters may inflate the number of deer they hit; others are most reluctant to admit hitting a deer that they did not recover.

I hope that any deer you shoot drops in its tracks. But make sure your deer is dead and not only stunned or wounded before you attempt to dress it out. Many "dead" bucks have recovered and dashed away. Many have attacked the hunter. Always approach a downed deer ready to shoot again. If the deer starts to get up, shoot. An extra shot is as nothing compared to the trouble a wounded deer can cause. To make sure a deer is dead, touch its eye, with a long branch. If there is no involuntary response, it is safe to commence dressing it out. A dead deer's eye turns greenish, but not until about half an hour after death.

17

Life in Winter

Snowstorms take a variety of forms. Some are herded in by a "Norther"—their flakes slanting in, almost horizontal. The wind presses the attack from all sides, and wind-driven snow is capricious. A field may be swept almost bare, while beyond the fencerows the drifts are piled like frozen ocean breakers. Such storms shriek as if consciously intent on locking half a continent once again in the grip of an Ice Age. The cold penetrates. It is driven through all living creatures, for the wind-chill makes a mockery of thermometer readings.

Other snowstorms look like storybook pictures, blanketing everything in deep, white silence. The fenceposts stand like soldiers at attention under their shakos of snow. The evergreens are transformed into tepees, their snowladen branches bent to the ground. The world looks like a fairyland, but it can be a deadly fairyland. Some creatures reap a benefit from it; for others it wreaks havoc.

The meadow mice love snow. Under its protective white shield, they link the area together with networks of tunnels through the grass. They are safe from the prying eyes of hawks in the daytime and owls at night. The mammalian predators take a toll, but it is only a fraction of what it was before the earth was mantled.

The smaller predators, deprived of such fare as meadow mice, intensify their pressure on rabbits, hares, and the grouse. The rabbits are eager to share a woodchuck's den, and the hibernating woodchuck will not even realize that it has a tenant. As soon as the storm ends, the hares will be as busy as a state road crew tramping out their highways and byways. A grouse may seek shelter under the tepee evergreens, or it may plunge completely beneath the insulating blanket of snow.

To the deer, snow may be many things. It may force deer to move or it may keep them from moving at all. Deep snow may be a trap or a blanket. It can provide the deer with food by bending

This buck is bedded down
on a hillside protected
from the wind.

This is a typical winter
bed. Unlike moose, which
envelop themselves in
winter snow, deer scrape
away the snow so they can
lie on leaves.

down the branches of trees and bushes, or it can deprive them of food by burying it.

It was the threat of deep snow that forced the migration of the mule deer and blacktails down from the high country in autumn. It was the cold wind before the coming of the deep snow that forced many of the whitetails to abandon their regular haunts and seek shelter in the swamps and in the gulleys and draws along watercourses.

Yarding areas are not chosen because they provide food. They are used despite the fact that often little food is available there. The main requirement of a yarding area is protection from the wind. Cover is more important than food in areas of cold weather and deep snow, and the nutrition of the food available is low.

Not all whitetails yard up every winter. It depends upon the winter. My area of New Jersey is on a dividing line. Deer in the areas south of here almost never yard up. The winter of 1960–1961 was exceptionally cold, with deep snow, and the New Jersey deer yarded up for most of the winter. We have had comparatively mild winters since that time, with only a couple of years when the deer retreated temporarily to sheltered areas. Our deer were yarded up in the winter of 1976–1977 because of the extreme cold and strong winds, although we had only a moderate amount of snow. Fortunately, our deer were able to move out to seek food as soon as the weather moderated. In New York and upper New England, extreme cold and wind, coupled with deep snow, took a terrible toll of the deer. Thousands upon thousands perished.

In most warm-blooded creatures, man included, the basic metabolism rate speeds up as the temperature drops. Basic metabolism is a measure of energy requirements since, if a creature's body temperature is to be maintained, heat production must equal heat loss. It is like running a furnace under forced draft. You can raise the temperature, but this requires extra fuel. This is why it is vital for anyone who expects to be out in cold weather to carry emergency food to prevent hypothermia. A few years ago, biologists discovered to their great surprise that a deer does not respond to cold in this manner.

It had been known that young animals of all species have a higher basal metabolic rate than adult animals, and that the larger species have a lower basal metabolic rate, in proportion to their body size, than the smaller species. This is another reason why the northern members of a single species are larger than their southern counterparts. They are simply more energy-efficient.

Males of many species—including deer—are larger than the females, and yet they have a higher basic metabolic rate. Bucks are usually larger, grow faster and are more active, curious, and independent than does. They leave their dam sooner than the females, and they get into more trouble through their curiosity and independence. Owing to all these factors, bucks burn up their body reserves faster. Most adult bucks do not go into the winter season with as good body reserves as do the females, unless they have access to almost unlimited food after the rutting season. Unfortunately, this is seldom the case, and winter mortality is therefore usually higher among adult bucks than does. Nature seems to have foreseen this situation and counteracted it to some degree by providing for the birth of more male than female fawns.

The basic metabolic rate for a deer requires about 1,140 calories per day for each 100 pounds (45.4 kg) of body weight, if the air temperature is 32°F (0°C) or higher. This metabolic rate *drops* instead of speeding up when the temperature gets below freezing. Because of their lowered metabolic rate, deer will lose 12 to 15 percent of their body weight in cold weather even if they have an abundance of food. They simply do not take in as much food, nor could they utilize more even if it were available in the winter. And an abundance of deer food is seldom, if ever, available in the winter.

The glands of the deer's endocrine system—the adrenals, pituitary, and thyroid—are at their smallest and are mostly inactive during January and February, the period of the coldest weather. When winter begins, the deer have not adjusted physiologically to withstand the severe weather, and they respond by heavy feeding. If the temperature drop is slow and steady, the deer adjust to the slowing down of their endocrine system and become adapted to the cold. If the temperature drop is very sudden, severe, and prolonged, they may go into shock and die. Once the physiological adjustments to the stress of winter's cold are made, the deer voluntarily fast, seek cover, and hold their daily activities to a minimum. After the endocrine adjustments have been made, they can then withstand a sudden and severe drop in temperature.

During a heavy snowstorm, deer will lie down and not get up, unless disturbed, until the storm passes, even if it is of several days' duration. They literally become buried in the snow. The insulating qualities of their coats are so efficient that the snow won't even melt. And the covering of snow itself also acts as insulation, giving the deer additional protection from the cold.

By winter the common family unit will consist of a doe, her yearling doe, and the fawns of the year. The yearling bucks will be in bachelor bands.

It has been found that deep, soft snow is important to the survival of moose. They lie down in it so that the snow actually envelops their bodies. Deer don't do this. Deer paw down through the snow to the leaves below, if possible, before lying down. Moose will not use their snow beds a second time, whereas deer use the same beds again and again. On numerous occasions I have seen a large buck and even an adult doe drive another deer out of its bed and then lie down in the same spot. This habit of pawing out a hole in the snow costs the deer vital calories, because the snow would be a more effective blanket if it enveloped the animal's body. Deer do not take advantage of the snow, as some other species do. This supports the theory that the deer living in the North today got there through range expansion, an expansion that is still going on. They did not evolve in the north.

Their habits change drastically in the winter; most of their activities are now during daylight hours. They remain bedded through the long, cold nights. They do not bestir themselves even at dawn. About 8:00 A.M., as the sun comes up, they begin to feed if any food is available. Lying still at night conserves energy, and feeding during the daylight hours also helps. Even on days that

A doe paws through the snow to locate some acorns she may have scented.

are clouded, the air temperature generally rises so that fewer calories are burned for heat during the periods of activity.

The deer will avoid windy areas wherever possible. They like to feed on steep mountain slopes that face south, southeast, or west. They will sometimes use almost vertical slopes. The snow is usually lighter on these slopes and usually melts off faster, and vegetation is usually heavier. Deer will also choose these areas as bedding sites to take advantage of the sun. If the temperature rises above 40°F (4.4°C), they will seek shade, even in winter, because they become uncomfortably warm in their winter coats.

WINTER FOODS

The fantastic growth of summer vegetation gives a false impression that deer food must be very plentiful. This vegetation is important, but the critical vegetation is that which is available during winter, particularly that which is available in the yarding areas for the deer that are forced to yard up.

In the winter most deer revert almost entirely to browse because they are forced to. Most of the other herbaceous food is just not available. Protein is high during the plant's growing season, highest in late spring and early summer and tapering off in the fall. The protein content drops as the plants begin to dry up, at the time when deer need it most.

Not only does the protein level of the plants drop as much as 25 to 40 percent, but their digestibility also lessens. In the springtime the deer may be able to digest about 70 percent of the plants consumed. The digestibility ratio may drop to 45 or 50 percent in August and as low as 12 percent in the winter. Low digestibility means that even less of the protein content can be utilized. If deer could eat more vegetation, they could obtain more protein, but there is even less food to eat in the winter and, with their lowered metabolism, the deer couldn't eat a lot of food even if it were available.

The protein content remains proportionately higher in the dormant browse than in the herbaceous plants and forbs. Deer are selective feeders, and research has proved that they instinctively select the food with the highest protein.

Many states have performed research operations on deer whereby tubes, bags, or openings were placed in the animals' throats or stomachs so that food could be removed for testing. In this way it was proved that the deer instinctively ate more nutritious food than what the researchers selected for them. When allowed to feed freely on research plots, they selected a 17 percent protein diet, compared to a 7 percent protein diet collected by the researchers from the same vegetation on the same plots. There is also the possibility, however, that the acids and the fermentation in a deer's digestive processes add to the protein content of the food eaten.

Deer need a mixture of forage. It has been proved that the very best of foods will sustain them for only about two weeks without ill effects if it is fed as a single-item diet. In the wild, deer always eat a mixed diet if at all possible.

There are other very important factors besides the protein content of a food. White cedar is lower in protein than balsam fir, yet deer prefer the cedar and will thrive on it, while balsam is a starvation food. White cedar has only 2.7 percent protein, while aspen has 5 percent protein, yet deer can eat three times as much white cedar as aspen and will thrive on it while starving on aspen. This is because cedar is much more digestible than the aspen.

The same is true of bitterbrush and sagebrush. Bitterbrush is lower in protein than sagebrush, but mule deer much prefer it.

Sagebrush is an important winter food, but deer that are fed sage exclusively soon stop eating, as it decreases their appetite. It is also known that the aromatic, volatile oils of the sage, which impart its distinctive odor, are antibacterial and therefore inhibit the microflora and bacteria of the deer's digestive system. The more volatile oils a plant contains, the less it is favored by deer.

I have found that 6 to 8 inches (15.2 to 20.3 cm) of snow seems to be the limiting depth through which the deer will paw to feed on grasses. I have seen them go down through 16 to 18 inches (40.6 to 45.7 cm) to feed on rape plants, but the rape plants themselves were almost that height, with their top leaves just below the surface of the snow. Rape and all other members of the cabbage family are avidly eaten by deer, and their heavy odor makes it easy for the deer to locate them, even under the snow.

I have also found that, although deer do feed on grasses, they much prefer broadleaf forbs (nongrass herbaceous plants), even after the forbs have died and dried. I have often seen deer stripping the dried leaves off the dried stalks.

Deer are forced to do more browsing in the winter. Fortunately, winter browsing, while the plant is dormant, does less damage to the plant than browsing during the growing period. Various types of plants have different degrees of tolerance to browsing. Many plants die if they are 30 percent browsed, while others are actually stimulated by the same degree of browsing. Mountain maple is an excellent deer food, and it can withstand 80 percent browsing each year. White cedar and yew, both favored deer foods in parts of the North, are severely damaged by even moderate browsing. Browsing is a form of pruning, and as gardeners and orchardists know, each variety of plant must be pruned differently. In the wild there is no such control, so the most highly favored plants will usually be overbrowsed and may even be killed off. Then the deer are forced to feed on less beneficial plants. Where the deer population is high, the best food plants are soon eaten or killed off. And as long as the population remains high, there is no way for the favored plants to come back.

Deer seem to prefer sprouts that grow directly from the stumps of trees over those that grow up from old root systems. Probably the protein content is higher or the taste is better, but the reasons are not known for certain. Sprouts of either type grow much faster than seedlings because they have the old tree's entire root system to put to use. Before the tree was cut, the roots had to gather enough nutrients to provide for an entire tree, whereas now all the nutrients go into a few sprouts. The sprouts compensate by growing fantastically large leaves—four to eight times larger than normal. Unless browsed heavily, in just a few short years the sprouts will grow beyond a deer's reach and the deer will then be forced to feed on seedlings.

Many of us are again heating our homes with wood, and many woodlots and forests are again being opened up that had reached the climax stage. An old maxim of deer management is "feed 'em with an axe." From the end of World War II until the oil crunch of the 1970s, many people switched to gas, oil, or electric heat. I heat seven rooms in my home with just one Vermont Casting stove.

If you are heating with wood, and cutting your own hardwood, you might want to do as I do. Don't cut the tree off as close to the ground as you can get it; cut it off about 4 feet (1.2 m) high. A low-cut stump will send up perhaps 2 or 3 sprouts; a high-cut stump may send up 8 to 12 sprouts. Don't worry about the firewood you are losing, because you can go back next year and cut the stumps close to the ground. In the meantime, you will have provided a lot of good deer food.

In Wisconsin it was calculated that the whitetail deer were crippling or eliminating over 600 million tree seedlings every year. Where the deer populations are high, neither natural nor artificial reforestation is possible. The deer have the capability of changing the entire composition of the forest areas they inhabit.

It is during the winter that competition between deer and domestic livestock becomes most acute. Northern whitetail deer are seldom affected, because most livestock in their range are kept in barns during the winter. The blacktail and particularly the mule deer have to compete with range cattle, sheep, and, in the Southwest, goats. Goats especially are competitors of the deer all year because browse makes up over 50 percent of their annual diet.

The competition is highest in autumn, when both sheep and deer are feeding heavily on forbs. Deer and cattle compete mainly in the spring, when both are feeding on newly sprouted grasses. In the other three seasons of the year there is minimal competition, but in the winter on overgrazed ranges, even cattle are often forced to feed on browse. At such times livestock may prevent deer from occupying the same ranges. Since more and more beef cattle are now being raised in the southern states, this problem is also found east of the Mississippi River. In some regions it is acute. The problem does not occur where the range is not overstocked, since local forage plants replenish themselves over the winter.

It is at this time of year that blacktail deer feed most heavily on the seedlings of the Douglas fir and western red cedar, much to the consternation of lumber companies that are trying to reseed huge tracts of land where they have cut the timber. In some areas, the deer actually prevent the reforestation that is essential to sustained timber yields.

When possible, whitetails will pick up whatever corn has been lost to the picker, and they also feed on rye grass, alfalfa, and winter wheat. Dry leaves, when available, are eaten by all three deer in the winter even though they are low in protein and digestibility.

For a list of the common winter foods preferred by the three types of deer, see Appendix I.

In my home area, red cedar is browsed as high as the deer can reach, yet red cedar is only a "stuffer"—starvation food. I have heard countless deer hunters refer to rhododendron and mountain laurel as good deer food. These plants, too, are "stuffer" foods. Deer are able to eat these evergreens, even though the leaves contain a poison known as *andrometatoxin,* which is deadly to livestock. But the fact that deer can eat something—and may eat great quantities of it when other forage is unavailable—does not make it good deer food. I have opened the stomachs of countless deer that had died of starvation and found them stuffed with red cedar and rhododendron. Unfortunately, these plants, as well as mountain laurel, are low in nutrients. Deer often fill their paunches with these plants for lack of better food and then slowly starve.

Harley Shaw of Idaho reported that on five different occasions during the winter of 1961–1962 he found places where whitetail deer had pawed down through 3 to 6 inches (7.6 to 15.2 cm) of snow to eat wintering colonies of ladybug beetles. There were no food plants in the deer-scraped areas, and the rumens of three out of eight deer collected had large quantities of beetles in them, proving that the deer had actively sought out the beetles as food. This is the only instance of insect-eating that has come to my attention, but it is not totally surprising since insects are virtual reservoirs of protein.

DEER YARDS, DOMINANCE, AND AGGRESSION

Deer yards are usually in dense evergreen swamps, draws, gulleys, or along brushy watercourses. The deer's main objective is to get out of the wind. Snow depth is usually less in these sheltered areas because a lot of the snow remains on the tree branches. Often, more than half the snow depth is held aloft. And when the snow starts to melt, the area under the trees will be bare long before the surrounding area is free of snow.

These spots are always warmer because they trap whatever warmth the sun gives. The dark trunks of the trees absorb and hold the warmth instead of reflecting it away as do snow areas. The branches also provide a protective umbrella that prevents a rapid heat loss at night. You may have noticed how much faster the temperature drops on a clear winter night than when there is cloud cover. Even in winter, some heat is given off by the trees in the form of thermal radiation. In effect, the trees create a microcosm that works to the deer's advantage.

Although the wind and cold cause the deer to yard up, it is snow depth that determines how long they remain yarded. They will stay in the yard area until the snow melts sufficiently, and once they leave it they will not gather there again until forced to do so the following year.

They do not deliberately tramp out a network of trails through their yarding areas; the trails are a result of their search for food. There is almost always a food shortage in the winter yard areas, and this creates belligerence as the deer respond to the most important life-governing force—the survival instinct. Not only is there a food shortage and physical stress from the cold, but also the psychological stress of crowding. Whitetails, unlike blacktails and mule deer, are not herd animals; they gather into large groups only when forced to.

Aggression is very common in the deer yards. Bucks turn on does, and does even turn on their own fawns, and fawns turn on other fawns. Dominance in a deer yard is in a constant state of flux. It is readily achieved but has to be reestablished at almost every contact because of the desperate need for food.

Bucks usually dominate the does simply because they are larger and heavier. Yet some of the large adult does are so used to being leaders that they will not give way before the bucks. The matriarchal does will drive subordinate does away from food. Fawns of the dominant does are dominant over fawns of subordinate does. It is much like human society. The children of the rich and powerful tend to be dominant over the other children even though they have done nothing on their own to warrant this rating. An often overlooked factor in this social standing is that dominant parents, in both human and animal populations, can more easily provide what the young need, so these young have an early

An older doe exercises dominance over a buck; the maintenance of the social order is an almost year-round activity, and this foreleg flailing is a common means of its expression, even among antlered bucks.

advantage that they tend to maintain all of their lives. The condition in which a fawn enters the winter is very important. Its condition and size will play an important role in establishing its rank on the dominance scale.

Another interesting fact about dominance is that when a subordinate is not a threat it will be tolerated and granted more privileges than a more aggressive deer that is higher up in the social standing. I have seen a big buck tolerate a yearling buck and allow it to eat, though the big buck would drive off all other large or older bucks. New deer moving into an area are usually dominated regardless of their sex, age, or size. After they become established as residents, they may fight their way up the social ladder. But once dominance has been established in a yarding area, the status quo is usually maintained for that winter. Dominance is not a leadership trait but merely "might making right" and "survival of the fittest."

The dominance of adult bucks sometimes shows up in strange ways. The New Jersey Game Department found that in any livetrapping program, bucks constituted 80 percent of the catch for the first week or so. Livetrapping is most successful when natural food is scarce. The deer can then be lured into the traps with food. The adult bucks showed their dominance by driving other deer away from the traps. Their dominance gave them the first chance at food and the chance to be trapped first.

The fact that the bucks lose their antlers prior to or during the winter months of hardship proves that the antlers are sexual weapons. If they were to be used for protection, they would be retained until spring at least. With his antlers gone, a buck fights or achieves dominance by striking out with his front feet, just as the does will at all times of the year.

In achieving dominance, all deer of either sex usually follow the same patterns. In threatening another deer, the aggressor usually drops its ears down and back until they lie against its neck. Then the aggressor stares at the other deer, giving it the "hard look." If the aggressor decides to carry the action further, it sidles

When food becomes scarce, deer will drive their own young from it, as this doe is doing.

up toward the lesser deer. This is usually followed by rushing and snorting. Then the front foot is raised and used to strike out and down. If the lesser deer does not give way, the aggressor will stand on its hind feet and flail out, alternately striking with both front feet. Most often the objective is achieved by just a single strike, and sometimes the aggressor will skip the preliminary steps and just rush at its opponent and strike out. At times the attack will come from the side or from the rear with no warning at all, and often patches of hair are knocked out in these encounters. Sometimes both deer rise up and flail away. The larger, heavier deer wins out.

It is hard for the fawns to understand why the mother that has taken such good care of them now drives them away from what little food is available. The doe cannot understand it, either; she is just instinctively responding to the necessity of obtaining food for herself. Intolerance increases as the winter goes on and the food supply decreases. It is from the fawns that nature extracts the greatest toll.

The legs of an average deer are 18 to 22 inches (45.7 to 55.9 cm) long. The legs of a seven- or eight-month-old fawn are only 16 to 18 inches (40.6 to 45.7 cm) long. If the snow is 18 inches deep, the adults can walk through it even though it is hard going. The fawn's chest drags, and walking through snow of that depth is almost impossible for it. Its only means of moving about is to jump over the snow, and this is exhausting. Even the adult deer can move only with great difficulty when the snow reaches 24 inches (60.9 cm). Snow above that depth imprisons them.

Henry Leabo, of Lancaster, California, witnessed mule deer bucks using teamwork to negotiate deep snow. Henry was doing some winter camping in the mountains of Utah. He saw three deer making their way through the deep snow. One deer would lunge forward for about 50 feet (15.2 m), then lie on his side while the other two deer passed him. A second buck would then take the lead for about 50 feet. The first buck now brought up the rear. Then the third buck broke the trail, followed by the first and second. The bucks kept up this alternate system for as far as Henry, watching through binoculars, could see them.

Some researchers have concocted a descriptive formulation of winter conditions. They use the term "severe winter" to mean 60 days with a 15-inch (38.1 cm) depth of accumulated snow or 50 days with a 24-inch (60.9 cm) depth of snow. A "moderate winter" is 60 days with 12 inches (30.5 cm) of snow. I can understand try-

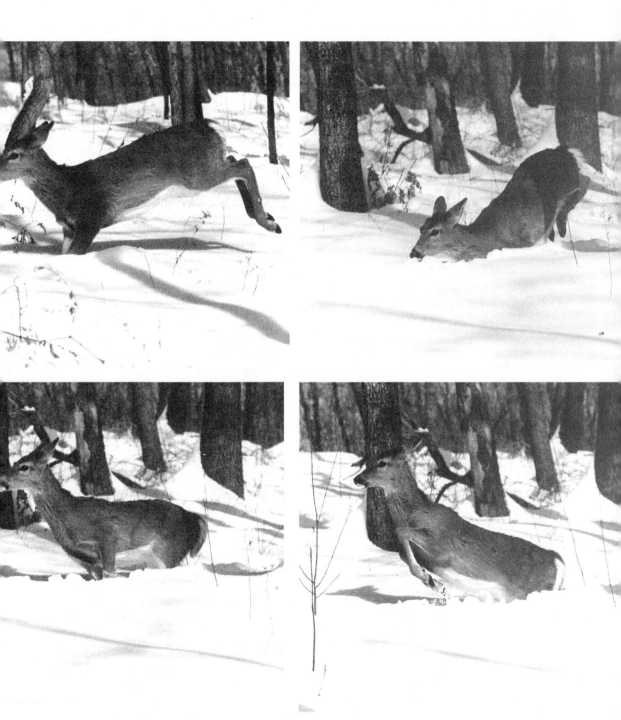

This sequence shows the extra effort a deer must exert to move in deep snow; this kind of exertion can cost up to 600 extra calories per hour at a time when deer are on a starvation diet.

ing to work out a common denominator, but the two categories for a "severe winter" are not the same thing. A deer can walk in 15 inches of snow but cannot in 24 inches. That's a tremendous difference.

Wisconsin deer researchers have worked out a Winter Severity Index based on the depth of snow plus the air temperature. The biologists add the number of days between December 1 and April 1 that have 18 inches (45.7 cm) of snow coupled with the number of days with a low temperature reading of 0° F (−17.8°C) or lower. This data is collected at 12 weather stations throughout the state. The totals are added together, then divided by 12 to give an average Winter Severity Index. Mild winters have less than 50 such days, moderate winters have between 50 and 80 such days, and severe winters have over 80 such days.

Louis Verme, of the Cusino Research Station on the upper peninsula of Michigan, has come up with a more complex Winter Severity Index that has gained widespread usage by several other states and Canadian provinces. His system is determined by measurements of heat loss through air chill, the depth of accumulated snow, and the compaction of the snow on a weekly basis.

Pressure cookers are fitted with an electrical heating element and a thermoregulator to maintain distilled water in the container at a constant temperature. These "chillometers" are placed just 5 feet (1.5 m) above the ground on a shelf that is exposed on all sides, but is roofed over. Official temperature readings are taken under winter conditions. The electric meter records how many kilowatt-hours are used to maintain the desired water temperature, the loss of heat being referred to as the *air chill*.

To test the compacting of the snow, a 1-inch (25.4-mm) pipe is weighted with 3 pounds (1.4 kg) of lead. The pipe is dropped on the snow from a specified height and the depth to which it sinks is recorded. This shows the impact force of 3 pounds to the square inch. If you go back to my formulation of the heavy surface of a deer's hoof in the "In the Tracks of Our Deer" chapter, you will see that 3 pounds (1.4 kg) to the square inch is about half the bearing surface of a 140-pound (63.5-kg) deer. The impaction depth and the snow depth are combined to give a weekly snow hazard. When the index exceeds 100, widespread mortality can be expected.

All of these indexes are used only as a base for comparison, because there are so many variables that can affect the deer in any particular area. Though the indexes vary, they do provide a good data base from which to start.

Verme and a colleague, John Ozoga, also found that the effects of these indexes were most severe at the beginning and the end of the winter: in the beginning of winter because the thyroxine levels had not dropped enough to have lowered the deer's basal metabolism rate; and at the end of winter because increased thyroxine had raised the basal metabolism rate. During the middle of the winter, the deer were conditioned to withstand the extreme cold, reduced food, reduced activity; they were, in effect, in a state of semihibernation and increments of winter chill had less effect than when their basal metabolism was running higher.

In their Michigan Department of Natural Resources Research and Development Report #237, Verme and Ozoga stated:

> Based on our data, the relative severity of a winter can best be gauged by analyzing the seasonal weather pattern. Extreme winter severity during the midwinter period, in itself, is not especially deadly for deer. At this time they are geared-down physiologically to withstand great adversity. Conversely, a yarding season that begins unusually early and ends late can cause havoc because deer simply "run out of gas" over this interval. A severe winter, therefore is one which subjects deer to prolonged, unrelenting attrition; consequently, it is physically deteriorating.

The deer are usually forced into their yarding areas before the snow becomes too deep for them to get there. Most whitetails travel no more than 2 or 3 miles (3.2 to 4.8 km) to yard up. The maximum distance is about 15 miles (24.1 km). In their quest for food, a network of trails is soon established, but only within the sheltered area. Food can be available ¼-mile (0.4 km) away, but once the deer yard up, that food might just as well be on the moon. They will not go out into the exposed areas to eat. Depending on the size of the yard and the number of deer in it, all food that is easily reached may soon be consumed.

STARVATION

When the easily reached food is gone, deer stand on their hind legs to reach overhead branches. A seven-month-old fawn, standing on its hind legs, can reach up about 5 feet (1.5 m). An adult doe can reach up about 6 feet (1.8 m), an adult buck about 7 feet (2.1 m). Naturally, the fawns get the least food. When all the available food

This doe appears to be in poor condition as she reaches for the remaining browse. In extreme cases, starving does may actually reabsorb the fetuses they carry, but it's more likely malnourished fawns will be stillborn.

is eaten up to the 7-foot height, a "browse line" is apparent. This is also called "high lining," but it really means starvation.

It has always been a puzzle to me why deer have not developed the habit of "riding down" browse as moose do. When a moose wants to reach the uppermost branches of a sapling, it will stand on its hind feet and put its chest against the sapling. With its body weight, the moose then pushes the sapling over so that it can easily feed on the upper branches. This often kills the sapling by breaking it, but it helps the moose now and in the future because it creates holes in otherwise dense stands of same-age saplings that would otherwise grow too large for the moose to push over or to reach the lower branches. New saplings soon grow in the holes thus created.

I have often watched deer balance on their hind legs to get a few mouthfuls of food, but I have never seen them even attempt to put their feet against the trunks of saplings as domesticated goats and some African antelopes do. Many of the saplings a deer browses could be pushed over, yet deer have not developed this behavior.

Deer prefer browse that is no thicker than a wooden matchstick. A study showed that they prefer red maple twigs up to 4.17 mm in diameter, quaking aspen up to 3.84 mm, northern red oak up to 3.47 mm, black cherry up to 2.95 mm, and gray birch up to 1.26 mm. When hunger is severe, they will eat browse up to the size of a wooden pencil (7.5–8 mm).

But eating browse of this larger diameter is a losing battle for several reasons. First there is the fact that the smaller twigs are the newest growth and their bark has the highest protein content. The older bark of the larger twigs not only has less protein but is less digestible.

A deer gets its nutrition out of the bark and not out of the cellulose of the wood. Larger twigs have less bark in proportion to their volume. The amount of bark on a twig can be calculated by multiplying the diameter of the twig by 3.14 to find its circumference, and then multiplying the circumference by the length of the twig. A matchstick-sized twig of 3 mm diameter and 100 mm length has 942 square mm of bark. A pencil-sized twig of 8 mm diameter and 100 mm length has about 2,512 square mm of bark, or a little less than three times as much bark.

To find the volume of the twig, you multiply 3.14 by the radius of the twig, then multiply the resulting product by the radius again, then multiply that product by the length. Thus a twig of 3 mm diameter and 100 mm length has a volume of about 707 cubic mm, while a twig of 8 mm diameter and 100 mm length has a volume of 5,024 cubic mm, or about seven times as much bulk.

In other words, the larger twig has seven times as much volume for less than three times as much nutritious bark. What it means is that when a deer is forced to eat the large twigs, it is actually getting less than half of the nutrients out of each paunchful of browse because it has to consume so much more bulk for the amount of life-giving bark. And this is *if* the deer can find enough of even the larger twigs to fill its paunch.

There is one last problem, and this is often the one that pushes deer over the brink of starvation. When a deer eats any browse during bitterly cold weather, it must produce extra body heat to thaw out the frozen twigs in its paunch before they can be utilized. When the deer is eating the large twigs, the body heat and energy

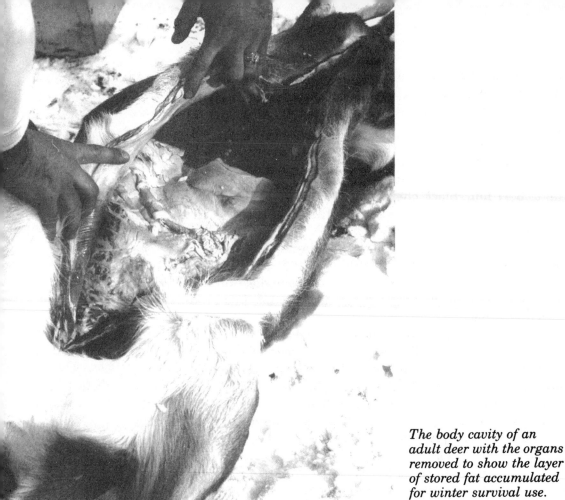

The body cavity of an adult deer with the organs removed to show the layer of stored fat accumulated for winter survival use.

lost in converting this food to assimilable nutrients may be greater than the benefits gained.

The same thing happens when the deer eats snow while in a weakened condition. Water is imperative, because much of starvation weight loss is actually caused by dehydration. The drain on a creature's body to thaw the snow and convert it into water is tremendous. It is for this reason that a person who is tired and cold should not eat snow to alleviate thirst. Doing so can bring on hypothermia and under some conditions cause death. Young mammals require a higher percentage of water per body weight than do adults. Because of all of these factors, fawns comprise about 60 to 80 percent of the deer that die of starvation.

It is normal and expected that deer, like range cattle, will lose 12 to 15 percent of their body weight due to the cold winter weather even if they have all the nutritious food they can eat. A deer can lose up to 30 percent of its body weight and survive. The critical point lies between 30 and 33 percent. A loss of a full third of its total body weight is almost always fatal. Yet a deer can survive a higher overall weight loss if the habitat is such that the deer is exposed to less body-heat loss.

Most deer, in most areas, go into the winter with some fat on their bodies. The amount of fat is determined by the quantity and quality of the food available before the onset of winter. The amount of fat, the amount of winter food available, and the severity of the winter are the three crucial factors determining survival. A deer with poor fat deposits may survive a mild winter because it will be able to move about more freely and have access to more food. A deer with poor fat deposits cannot survive a protracted and severe winter.

As the temperature drops and food intake is lessened, the fat deposits are utilized. The fat surrounding the back and hams is the first to go, and then that of the abdominal cavity. Essential fatty acids (aceto-acetate and hydroxybutyrate) are derived by the oxidation of these body fats.

A winter group of 34 deer pawing through snowy fields for unharvested cabbages at a produce farm; note the dominance contest going on at the right edge. Winter is a good time to observe deer since they are active in daylight and cover is less dense.

A deer with any fat on its body will have a very high fat content in the marrow of its bones. The marrow of a deer in good condition is white and contains about 95 percent fat. This fat looks like suet. After the body fat has been utilized, this marrow fat will be withdrawn. The color of the marrow changes in accordance with the percentage of fat that is left—from white to yellow to pink, and finally to bright red. The bright red color has the look and the consistency of cranberry sauce and denotes a fat content of about 1½ percent. A deer with marrow like this is dying or already dead. To find out for certain whether a deer has died of starvation, one has only to check the marrow of the upper leg bones. Break the leg bone and if the marrow is red, no further research is needed. The carcass need not be fresh to prove this point. Even in late spring or summer, when nothing is left but scattered bones, starvation can be proved by this method. The marrow will have dehydrated but the remaining dried shreds will still show the red color.

At the same time that the fat is being withdrawn from the marrow, the body is also forced to "cannibalize" its own muscle protein. As starvation continues, the deer's liver is also affected. Ordinarily the liver produces glycogen from glucose and proteins and stores it to be released to the muscles as energy when needed. Without glucose or glycogen, the deer develops hypoglycemia—an abnormally low level of blood sugar. A deer with severe hypoglycemia is subject to trembling and may become too weak to stand. This condition almost always foretells death. Deer in such a weakened state become increasingly susceptible to pneumonia and other diseases, and to predation. They also lose all fear of man, mainly because they just don't have the strength to escape. This loss of fear may also be attributable in part to brain dysfunction, since the brain is starved along with the rest of the body.

Starvation of deer is extremely common, and in many areas it is growing worse each year. Those states that allow does to be hunted and are harvesting a greater percentage of their deer herds are reducing the losses due to starvation. In states that still have a bucks-only law, the situation is becoming increasingly bad. A. Starker Leopold, a zoologist at the University of California, has estimated that in a severe winter as many as 2 million deer die in the United States from starvation.

We are learning that one of the most important things we can do for deer faced with starvation is to leave them alone and make sure everything else leaves them alone, if possible. It has been proved that, except where highly nutritious natural food is abundant, deer usually lose more by active foraging than they gain, because of the energy expended. We also have learned that, where the deer's endocrine system has had a chance to adapt to the cold, the animals are geared to be sedentary. The key to their survival is their inactivity. Any activity burns up calories, which when not replaced by food, have to be taken from the body tissue. Strenuous activity burns up much larger numbers of calories and also produces stress.

For a variety of reasons—which will be discussed in the section on management—even artificial feeding can lead to undue activity and stress. Moreover, it can maintain an unhealthy overpopulation which will cause greater starvation problems later on. Such feeding should be restricted to real emergency situations. For a fuller discussion of this topic, see the chapter on artificial feeding.

Some fawns have been known to go more than a month without food and still survive. Some adult does have been known to go over two months without a mouthful of food and still live. So long as the weather stays extremely cold, a deer will not move any more than it is forced to. It is, in effect, idling its engine so that fewer calories are consumed. Any kind of disturbance that forces the deer to move out of their beds during severe weather lessens their chance of survival. For this reason, the disturbances sometimes caused by snowmobiling—even indirectly—can be extremely detrimental.

There is nothing wrong with using snowmobiles, provided they aren't misused. Snowmobiles provide great sport when used in open fields or on woodland trails designated for them. They should not be run all through the old woods roads that lead back into the swamps and other deer yarding areas.

Thankfully, there are not many irresponsible snowmobilers who actually run deer with their machines. But, unfortunately, there are some. Many snowmobilers have harmed deer inadvertently just by disturbing them, causing the animals to leave their beds and try to move out of the area. Such people would not dream of harming the deer; they merely want to see them. Many snowmobilers have only realized the harm done after I have explained to them how important it is for the deer to remain as inactive as possible. There are times when just the noise of the snowmobiles is enough to disturb the deer. Research has shown that deer forced out of an area on weekends by snowmobiles will usually return after a day of quiet. It is unfortunate that the deer have to expend the energy to leave any area and then return to it at a time when they are fighting for survival.

Helenette Silver, a New Hampshire Game Department biologist, determined in 1959 that a 112-pound (50.8-kg) deer at rest needs 1,323 calories in a 24-hour period. Arthur C. Guyton calculated that a 154-pound (69.9-kg) human at rest needs 1,850 calories. On this basis, C. W. Severinghaus and Benjamin F. Tullar, of New York State, figured that a man and a deer of equal weight would need about the same number of calories if both were at rest. Carrying this supposition further, they calculated that a deer and a man engaged in the same type of motion would expend similar amounts of energy and require the same number of calories.

Charts prepared by Guyton indicate that a man walking very fast requires over 6 times more calories than when resting. If he walks up steps he will burn 11 times more calories than at rest. The researchers figured if a deer was panicked and ran from a snowmobile, it would burn up 6 to 11 times more energy than while at rest. The running deer would burn up about 900 calories per hour.

Snowmobiles are used most extensively on weekends. On weekdays the deer may spend 12 hours bedded down, for a cost of 1,200 calories; 6 hours of standing relaxed, for a cost of 630 calories; 4 hours of light exercise, such as feeding, for a cost of 680 calories; two hours of running, for a cost of 1,140 calories—a total of 3,650 calories. Severinghaus and Tullar calculated that if snowmobiles went into the deer yard for four hours on any day, the deer's energy expenditure would skyrocket to 6,830 calories per day. This is almost double the normal daily expenditure.

If deer were disturbed by snowmobiles on the weekends during December, January, February, and March, those weekends would cost each deer as many extra calories as would ordinarily be used in one month of normal living. In effect, it would subject each deer to one additional month of winter. We are losing countless thousands of deer to starvation because, with the food available, they cannot produce enough calories to survive a normal winter. A winter of weekend snowmobile harassment, which burns up an additional 100,000 calories, means that more deer are going to die needlessly.

Dogs are another menace and particularly so during the period of deep snow and cold. Although there are wild dogs that live on deer year-round, in the winter most of the dogs chasing deer are somebody's well-fed pets. I've seen two collies and an Irish setter,

An indulgent doe allows her 7-month old fawn to nurse in December. By the onset of winter, the fawns have reached two-thirds of their adult weight, and all body growth stops until March. Note that the snow is accumulating on their hair without melting, so efficient is the winter coat's insulation quality.

wearing collars and tags, come running through my land searching the bottomlands for deer. Such dogs are just out for a romp. They may not even be interested in killing the deer; they just enjoy the chase. And it is the chasing of weakened deer that hastens their death.

The stress of fear alone would surely be detrimental, even if the deer were not run hard. When they run, their mouths gape, their tongues often protrude, and they breathe heavily. Sometimes these same symptoms are visible when they are frightened but not running. The surge of adrenaline entering the nervous system during stress may produce the symptoms of a hard run.

When the snow is deep and fluffy, hungry wild dogs are at as much of a disadvantage as the deer, but well-fed dogs have a distinct advantage. As explained in an earlier chapter, all dogs have the deer at a tremendous disadvantage when the snow forms a crust hard enough to support a dog's weight but not a deer's. Dogs,

The trail into an over-browsed yarding area of cedars. Once deer bunch up in yards they are subject to many dangers and diseases. Trails like this also lead predators to the weakened deer.

for their body weight, have a much larger foot-surface than deer, and the deer's sharp, pointed hoofs break through all but the strongest snow crust. Under these conditions, the dogs or other predators wreak havoc on the deer.

Yet another major fault of snowmobiles is that the dogs follow the packed-down tracks the machines make in soft snow, and are thus enabled to get back into areas they otherwise could not penetrate. Dogs, snow, and deer are always a bad combination; when these components are coupled with severe cold, they are most lethal. Deer movements are not as devastating if the weather is moderate.

The remarks above regarding snowmobilers are equally true of cross-country skiers, and, in some instances, cross-country skiers are even more disruptive to the deer. Research has shown that, although snowmobiles disturb the deer, the deer actually move sooner and farther when the person is on foot, such as on skis or

384   *Life in Winter*

snowshoes. Most deer have become so accustomed to the sounds of motors in the distance, and even up close, that they pay little attention to them. A person on foot cannot be located or kept track of as easily as can a machine, because the person moves much more silently. Most deer associate a person on foot with hunting and hunters—predators to be avoided.

The weather may give the deer a respite. Even the most severe winter usually includes one major period and two lesser ones when the weather moderates. The midwinter thaw usually occurs around the third week of January. This break may be no more than a day or two or possibly three days long, but it usually sends the thermometer up above the freezing mark.

As soon as this happens the deer start to move about, and if there is any food available, too, they will feed all day long. Usually, February has a couple of short periods of moderating weather. Every moderate spell, no matter how brief, gives the deer a new lease on life.

The Indians called February the "Starvation Moon"; it was thought to be the month of the greatest winter hardship. There is

This deer was found dead of starvation, even though food was brought to the vicinity of its winter yard.

no doubt about it, February can be brutal. It is usually the month of greatest snow accumulation. January is usually the coldest month, and the snows that have fallen have little chance of melting. February is a snow month, and its snows pile up on top of the base laid down in January. But most of the deer that die of starvation do so in the last part of February and the first half of March. March is a month of promises seldom delivered.

In southeastern Alaska, the Sitka blacktail deer are forced from the steep mountainsides by the deep snows of a hard winter down to the sea beaches—and to their death. The beach areas are the only space where the snow is shallow enough to allow the deer to move. The deer feed on the remnants of the western red cedar and the hemlock and then are forced to feed on the kelp and seaweed washed up on the beaches. Except for the latter stuffer foods, the beaches are bare, and, soon, so are the bones of the deer, bleaching whiter than the driftwood they are scattered among. Along the Nakwasina Passage and Deadman's Beach, the remains

Many victims of winter stress show up on the fencelines that trapped them; this is the skeleton of a muley fawn.

of as many as 10 to 12 deer per mile are commonly found each spring following a hard winter.

The Gunnison mule deer herd in south-central Colorado has probably been studied as extensively as any herd in the country. All kinds of management practices have been tried there. Overpopulation and range destruction were major causes for the decline of deer population. In one canyon, in one spring after a severe winter, biologists found 253 carcasses of mule deer that had died of starvation.

The literature on the whitetail deer abounds with records of death by starvation. The spring of 1977 added pages of statistics. Although the winter of 1976–1977 was the coldest on record, in my area of northwestern New Jersey, we did not have a great snow depth. Our deer were fortunate. They did seek out the sheltered draws during the prolonged periods of cold, but they were able to move out as soon as the weather moderated. Our deer also had the advantage of going into the winter hog-fat from a fantastic acorn crop. They were still feeding on acorns in mid-February.

The winter of 1960–1961 caused the greatest die-off from starvation I have ever personally seen. The 63 inches (160 cm) of snowfall compacted to about 4 feet (1.2 m), and that depth remained for most of the winter and late into the spring because of the cold weather. Many of our deer wintered in the red-cedar-covered lowlands, which became death traps for many of them but allowed a viable nucleus to survive. When spring finally melted the snow, I found 27 dead deer in one square mile.

When you hear preservationists talk about "letting nature take care of its own," you know they have never witnessed deer dying of starvation by the dozens, by the hundreds, by the thousands. They have not seen a beautiful deer turn into a dull-eyed, listless rack of bones, wrapped in a rough coat, stoically awaiting death. The deer don't comprehend what is happening to them. There is nothing they can do about what is happening, so they only attempt to endure.

These preservationists have not seen the accidents that happen to starvation-weakened deer. Fences that the deer once cleared with ease become insurmountable obstacles. Many deer catch their feet in the top strands of fence wire, and they hang and struggle until death releases them from shock and cold. In mule-deer country, hundreds of skeletons are hung up on the fences each winter, like laundry bleaching in the sun. In the early spring, trout fishermen find the rotting carcasses of deer that tried to get

Even fences like this one are not usually an obstacle to deer, except when they are startled or weakened.

some of the new vegetation growing along the streamsides. The deer that slipped into the water left their carcasses there.

The preservationists have not seen the deer being eaten by dogs while still alive but too weak to offer any resistance. They have not seen the deer try to rise, only to fall at once, too weak really to struggle. They have not seen these things, or they would not be preservationists but conservationists. They would realize that deer are a crop that must be managed, harvested, and made

Life in Winter

use of. The deer are better served by biologists with facts than by people who act only from emotion and sentiment.

The number of fetuses a doe is carrying depends on her age and her general health, as previously discussed. When a doe is subjected to starvation, several things may happen. During the first four months of a fetus's life, nature is interested in saving the life of the doe if a choice has to be made. If she is carrying twins, one of the fetuses will die and the nutrients that were the fetus will be absorbed into her body and utilized. Occasionally the dead fetus will be aborted, but this is a waste of nutrients and seldom happens. The remaining fetus will continue to live, growing very slowly. Before the critical point in an extended period of starvation is reached, the remaining fetus will also be sacrificed and absorbed. The doe will be barren for just that year. Barren does are a rarity, and are usually the result of starvation and not of old age or lack of impregnation.

A fetus that is carried four months usually survives because by that time the new sprouts of spring are available. In March the deer's endocrine system reverses itself and the basic metabolism speeds up, forcing the doe to be more active and to feed more to secure the nourishment that is now needed by the fetus as its growth speeds up. After four months of gestation, the demands of the fetuses take priority over the bodily needs of the doe.

The skull of a mule deer with 20 pounds of fence wiring entangled around one antler.

A young victim of starvation stands in the characteristic humped-back position that indicates it is malnourished.

Deer that are in the last stages of starvation stand with their backs humped up against the cold. Their hair stands on end in a futile effort to increase the depth of insulation. The fawns are short-faced, because they have not reached their full growth, and with their hair standing on end this shortness is emphasized. The eyes of a starved fawn appear to bulge out of their sockets. The deer stand until they are too weak, and then they collapse.

One fawn that I found during a severe winter was so weak that he staggered as he walked. His chest was skinned and cut from falling on ice and rocks. He had been trying to get some food from the few blades of grass that grew in the warmth of a spring hole. When the fawn staggered off into the deep snow, he fell and could not get up. As I walked toward him, he struggled to get to his feet and bawled loudly. I picked him up and he lay in my arms without a struggle.

Life in Winter

The little buck was between eight and nine months old. On good food, he would have weighed about 80 to 90 pounds (36.2 to 40.8 kg). Good food was something he did not have, and he weighed 34 pounds (15.4 kg). His skin hung in loose folds; his backbone protruded so sharply that every segment could be felt. He had the typical winter-starved look, the hair of his head on end giving him the typical fuzzy appearance.

I took the little buck home and put him in a sheltered pen. He could not or would not eat any of the good maple sprouts and browse that I brought him. I tried soft commercial feeds and even tried to feed him warm milk concentrates, but to no avail. Too much had been taken out of his body; he had been pushed beyond the point of no return. He died five days later.

I thought the reason was that, through starvation, the microflora and bacteria in his paunch had died, and without them he could no longer digest food. This was the common belief held by the biologists until 1971. Reports and scientific papers in the most prestigious wildlife journals lent credence to this theory. It was believed that a starved deer could reintroduce these symbiotic organisms by direct mouth-to-mouth contact or mouth-to-anus contract with a healthy deer—by eating browse that another deer had chewed or by eating the feces of a healthy deer. Deer do exchange their stomach organisms in these ways, but we now know that this is not needed.

In 1971 David DeCalesta, Julius Nagy, and James Bailey, all from the Department of Fishery and Wildlife Biology at Colorado State University, shattered that myth for all time. These researchers starved nine penned mule deer under controlled conditions. After the deer died, fluid from their rumen was compared with the fluid taken from nonstarved deer. Naturally, the microorganisms in the rumen fluid of the starved deer were lower than in the nonstarved deer. But there were still more than a billion viable organisms in the rumen of the starved deer. The starved deer had maintained viable bacteria in their digestive systems without having to reintroduce any from outside sources. This proved that starved deer would be capable of digesting food if they had access to it.

In another experiment, the researchers starved 26 mule deer fawns and does. Four fawns died after having no food for 33 and 34 days, but the other nine lived after being starved for 36 days. One doe died after being starved for 54 days, but all of the other does lived, including one starved for 64 days. The main purpose was to observe and record the effect of the reintroduction of food.

The researchers had already proved that the deer had the organisms in their stomachs to digest food, but tests were needed to prove that the starved deer could actually handle it. Equally important, the tests proved the deer did not suffer from digestive problems such as bloat after they again had access to food. The crucial factor was that deer that had lost less than 30 percent of their total body weight could survive, but a loss of more than 30 percent doomed them. They were not doomed because they could not handle the food but because with a 30 percent weight loss, too much had been taken out of their systems to allow survival.

These two experiments have begun to change many concepts of deer management because, for the first time, biologists can calculate the deer's chances of winter survival. Given the known factors of the condition of the deer going into winter, the amount of winter food available, the weather and snow conditions, and the estimated days of starvation, biologists can more readily estimate deer survival and herd population. This information is vital in setting long-range plans for the harvesting and the management of deer.

The effects of a severe winter are insidious. The obvious loss is the carcasses of the starved deer that are found in the spring, their bleaching bones scattered about. Less obvious is the much-lowered fawn production rate the following spring. Most of the poorly fed does will give birth to a single fawn, instead of the normal twins; that may halve the annual fawn recruitment rate. Some does may have lost both fetuses and are barren that year.

A well-fed doe, after a moderate winter, will produce fawns that weigh 6 to 8 pounds (2.7 to 3.6 kg), and most often they will survive. A doe that has been starved over the winter will produce fawns, or a fawn, in the 4 to 5 pounds (1.8 to 2.3 kg) weight class. Most fawns weighing less than 5 pounds (2.3 kg) at birth die within a week. In fact as many as 68 percent of all fawns die after an extremely severe winter in our northern states. Some of the adult does will die; many of their last year's fawns will die; and some of the biggest bucks may die because they did not regain their body weight after the rut.

Even if the following winter is as hard as the first, there will be a higher fawn survival rate the second spring because the does, not having had the strain of lactation, will survive the winter better. The deer herd will nevertheless be greatly reduced in overall numbers. The size of the antlers of the bucks will also be much smaller during the fall following the first severe winter, because the bucks that survived had to replenish their body weight at the expense of their antler growth.

The overall take by hunters will also be much lower the second fall because of the number of seven- to nine-month-old fawns that have died, some of which would have been legal yearling bucks. And the yearling bucks always make up the bulk of any state's take.

Even if the third winter is moderate, the fall following will still see an exceedingly low harvest of deer. Only by the fourth fall, after the third and moderate winter, will the deer herd begin to recover. Even a single severe winter will have long-lasting effects.

Those surviving deer that have lost the greatest amount of body weight make the greatest weight gains in the spring when food is again available. Deer have a tremendous tenacity; they are tough. Although many will die, more will live, and it has always been so. Individual areas may be hit very hard, but this has little effect on the total population of deer. They have lived with adversity for many millions of years, and have evolved because of it.

III
TOWARD SOUND
DEER MANAGEMENT

18

A Historical View
of Management Problems

T he easiest part of wildlife management is managing the
wildlife; the hardest part is educating the public so that the
wildlife can be managed.

Before World War I, most Americans lived in rural areas; they
had their roots in the soil and they knew about the relationships
that living things have with the soil. Today most Americans live
in urban areas. When we were a farm-based people, we understood
that all living things were products of the land or the water. We
knew that the number of living things such areas could support
was directly related to the food available and to the protective
shelter provided for each species. We understood the meaning of
the word *habitat* in an immediate way, without its present vague,
fashionable associations.

We find it humorous that some city children think that milk
comes from cartons and can't imagine taking it from a cow. We
don't find it humorous when these same city children become vot-
ing adults who make decisions concerning wildlife and our natural
resources, for they make those decisions with little more information than they had about the source of milk.

Some of Walt Disney's nature movies are excellent educational films. But Disney also created the "Bambi syndrome" that
pervades the thinking of millions of our citizens who grew up with
the utopian view of deer promulgated by the movie *Bambi*. That
classic children's movie, which endowed wildlife with human emotions and intelligence, was intended to be a beautiful, sentimental
fairy tale involving cartoon characters no more true to life than
the "Big Bad Wolf" that Disney created to go with his "Three Little
Pigs" or the three bears encountered by Goldilocks. Unfortunately
(and incredibly to those of us who have spent any time observing
wildlife) a large segment of the public actually sees wild creatures

through the sentimental haze of a cinematic fairy tale. This works to the great detriment of wildlife rather than to its benefit.

The most important reason for deer management is that "the deer is its own worst enemy." The deer is an exceedingly adaptable creature, one that can thrive almost anywhere. Deer populations are limited only by the availability of food, water, and cover. If these three requirements can be met, deer have the reproductive capability to increase their numbers to the point that they destroy their range and devastate their own herds. Given good habitat and a lack of predators, a deer herd will almost double its numbers every year.

PREDATORS

The coyote probably kills more deer than any other wild predator—simply because there are more coyotes than any other predators. Today coyotes are found almost as commonly across the continent as the deer are.

For years I thought that the wolf was the smartest wild animal in North America, and they are smart. But the coyote is not only smarter, it is also adaptable, while the wolf is not. The coyote has survived in spite of, or perhaps because of, humans. We have eliminated the coyote's major predator, the wolf, from most of the coyote's range. We have provided it with countless easily gotten meals in the form of domestic creatures. The bumper sticker seen on western pickup trucks reading, "Eat lamb; 10,000 coyotes can't be wrong" is apt. We have also provided the coyote with a wide selection of cultivated fruits, berries, melons, grains, and vegetables. And we have altered the coyote's social structure.

In areas where the wolf and the coyote are not molested, dominant males prevent the lesser males from breeding. Coyotes do not form as large a pack as do wolves, but the dominant "alpha" adult males have control over the males of their own family group. Where coyotes are constantly harassed, this family social structure is broken down. The family members scatter; all of the males are free to breed. Consequently, persecution of the coyote has actually spurred their population growth. More coyotes breed, and their litters are larger.

For years, the old Biological Survey, now called the Fish and Wildlife Service, has fielded hunters and trappers and government personnel in their efforts to control the coyote. In the 1930s as many as 80,000 coyotes were taken a year. Today, as many as

The eastern coyote is larger and heavier than its western counterpart. All coyotes are efficient predators of fawns and account for the more deer kills than any other predator except feral dogs.

80,000 coyotes are still being taken each year, and we have more coyotes than we ever had.

I earnestly support the control of individual animals that are preying on livestock; but I can never condone attempts to eliminate a species. Nor should the control be done with poison, particularly the compound 1080, which is a chain-reaction poison, killing everything that eats the bait or the victims that have eaten the bait.

Coyotes are now found in every one of the continental states, Alaska, and most of the Canadian provinces. For years we have been talking about endangered creatures and how we are losing many, yet the eastern coyote is a brand-new animal, put together by God within the last 35–40 years.

The western coyote weighs an average of 35 pounds (15.9 kg). This animal moved north and then eastward. Some of the female coyotes bred with the small eastern timber wolf in Ontario. Wolves will kill the male coyote; they usually will not kill the females in estrus. The resulting cross is almost identical to the red wolf found in Texas and Louisiana, which sprang from similar parents.

This crossbreed moved into the Adirondack Mountains and into the New England states. Again the females bred with large dogs when male crossbreeds were not available, producing coy-dogs. The coy-dog is gradually being bred out of existence because the female offspring of the mating of a male dog and the wolf-coyote cross come into estrus around the first of November. When the female breeds with a dog, she whelps her pups at the beginning of January. Wolf, coyote, and fox males always help to feed both the female and the pups; dogs don't. The coy-dog female, being unable to get enough food for herself and the pups, usually loses her pups in a hard winter. In most areas, all traces of the dog have been bred back out of the bloodline. What we have today, according to the scientists at Harvard and Yale, is a brand new subspecies of the coyote that weighs between 45 and 55 pounds (20.4 and 24.9 kg), and it is classified as the Eastern Coyote.

When these new coyotes first came into New Jersey, they were not protected simply because there were no laws to protect them. None were needed; we didn't have coyotes. In 1975 protectionists pressured the state into passing a law protecting these new coyotes. The law was repealed several years ago when it became apparent that the new coyote was here to stay, and that it was an efficient predator of deer. Most of the eastern states are facing similar dilemmas. Maine, where there are thousands of coyotes, has

This winter-killed whitetail was probably scavenged by foxes.

probably encountered the most problems, exacerbated by the fact that the Maine deer herd is only now recovering from a sharp decline.

There are voluminous records of predation by coyotes in research literature, and the predation is by no means just of fawns. Coyotes are smart enough to team up and sometimes hunt in packs. Usually all it takes is a pair of coyotes. A single coyote can kill an adult deer, but coyotes like to gang up on such large prey.

Joseph Dixon reported that a pair of coyotes hunted a mule deer that had returned to the mountains in the early spring. There was bare ground except over the mountain crests, where the snow had drifted to considerable depths. The coyotes did not try to catch the doe on the bare ground but teamed up to drive her up over the crests, where she floundered in the soft snow and was killed.

Dixon also reported a muley doe that drove a single coyote out of the area where she had a newborn fawn. When first sighted, the doe was chasing the coyote and lashing out at it with her front feet each time she got close. When the coyote crossed the road, the doe had to stop to avoid hitting Dixon's car. The doe ran around the car and continued her pursuit of the coyote, which sought shelter under a pile of brush. Undaunted, the doe jumped high into the air and came down on the brush heap, holding all four feet close together. She did this repeatedly until the coyote finally dashed away again and this time made its escape.

Jean M. Linsdale and P. Quentin Tomich report a number of instances in which adult blacktail deer were chased by coyotes and several occasions when the deer chased the coyotes. They did not report seeing any deer killed by coyotes. Deer hair was found in 10 percent of all coyote scats checked on the Hasting Reservation in California, but the coyote is a noted scavenger, so hair in the scats is no proof of a kill.

In December 1976 I received a telephone call from my friend Charlie Summers, who lives in Denver, Colorado. Charlie and his photography partner, Joe Branney, had been in Yellowstone National Park right after Thanksgiving. They had gone there to photograph wild sheep and mule deer, and one incident had so impressed them that Charlie wanted to tell me about it.

They had been watching a small group of mule deer feeding on a hill when they noticed two groups of coyotes that were also interested in the deer. Unfortunately, the distance was too great for them to take any pictures, so all Charlie and Joe could do was sit and watch the tableau unfold.

The coyotes had their strategy worked out in advance. The group that was below the deer carefully stalked within striking distance, keeping concealed from the deer by high grasses, sagebrush, and other vegetation. The coyotes that were to be the drivers made no attempt to conceal themselves but trotted openly along the skyline, diverting the deer's attention from their cohorts.

When the coyotes at the bottom were in position, those on the ridge charged down at the deer. The deer bounded to the bottom of the ravine—right into the coyotes waiting there. When the melee was over, a buck was thrashing around on the ground. All of the coyotes shared in the bounty of the hunt.

In a study of mule deer wintering along the Green River in Utah, researchers found concentrations of 81 to 135 deer per square mile (2.59 sq km). In the course of the study, 89 dead deer were found, of which 19, or 21 percent, had been killed by coyotes.

The study showed that in this area the cougar was the chief predator of deer. Twenty-six of the 89 dead deer, or 29 percent, had been killed by cougars. Cougar kills are usually easy to identify because in most cases the back has been broken or the deer has been killed by being bitten through the throat. The cougar, which takes its prey by stalking, will eat its fill of a deer and then cover

A cougar with its kill; cougars take deer routinely. The cougar will now cache the carcass so that it can return and feed later.

the carcass by scraping dirt, grass, leaves, sticks, or snow over it. This caching practice is puzzling because the job is never completed; at least a part of the deer remains visible. Usually, the mound covering the carcass is so conspicuous that it attracts attention to the kill instead of concealing it. The cougar usually returns to the kill until it has consumed all of the meat. Stanley P. Young and Edward A. Goldman, in a famous study of the cougar, found that one of these big cats will typically kill 50 or more deer per year. Frank C. Hibben, in a study of the predation of mule deer in New Mexico, agreed with these figures but pointed out that most of the deer killed were old, sick, or infirm. The theory that predators kill defective prey exclusively or chiefly is widely accepted by the public but is being closely questioned today by biologists. J. Burton Lauckhart, in a study of blacktail deer in northwestern Washington, found that most of those killed by cougars were adult bucks in their prime. Ian McTaggart Cowan suggested that the reason for this might be that the adult bucks range higher in the mountains than the does range and so are more likely to be encountered by the cougar. Lauckhart found that the cougar in that region killed about 35 deer per year. Cowan's research on blacktail deer on Vancouver Island in British Columbia substantiates that figure as well as the fact that adult bucks are the deer most frequently taken.

The cougar has been nearly extirpated over most of the range of whitetail deer, so cougar predation on this species is almost nil. In the past couple of years the cougar has reestablished itself in some of the eastern states. The province of New Brunswick has always had a small population of cougars, and the big cats have now been reported in Maine, New Hampshire, and Vermont. If these sightings are reliable, there is a good chance that the cougar could reestablish itself in the Adirondack Mountains. Alex McKay photographed what looks to me like a cougar in New York State in the summer of 1972. Because of the high grass, the body and long tail do not show in the photograph, and I can understand why authorities cannot accept the photo as proof of the cougar's return. According to New York authorities, there are no cougars in that state today, although there have been "sightings." A cougar killed in Pennsylvania in 1967 had undoubtedly escaped from captivity. It was a young cat and very much underweight. Evidently it was having a hard time living in the wild on its own. In Florida, the cougar is on the endangered list, and the population is steadily declining. But the cougar has reestablished itself in the Great

Smoky Mountains of North Carolina and probably Tennessee and West Virginia. A number of authenticated sightings have been made in North Carolina.

Dr. Maurice Hornocker of Idaho has conducted what is probably the most thorough recent study of the cougar's predation on deer. His findings are that a healthy cougar kills a deer every 7 to 10 days—35 to 50 deer a year, substantiating the work of earlier researchers. Hornocker found that the cougar was taking only 4 percent of the deer available to it. He noted that the cougar was successful in 37 of 45 attacks on mule deer and elk, an 82.2 percent success ratio. This proves that the cougar is a very efficient predator. The African lion, by contrast, has only a 10 percent success rate.

A famous study of the Jawbone herd of mule deer, on the western slope of California's Sierra Nevada Mountains, showed the annual increase in the herd population to be 32 percent. But because of tremendous overbrowsing by the deer, the habitat could not stand any increase in deer numbers. The study showed that 23 percent of the herd died of starvation each year, 7 percent were harvested by hunters, and 2 percent were taken by predators.

This is where the cougar plays an important role. In many areas that are too rugged to be hunted, the cougar effectively fills the predator's role for which it was designed, helping to keep the deer population in proper relation to the food supply.

Wolves at one time played an important role in the control of the deer population in the United States. This is no longer true except in parts of Minnesota. In a recent study of wolves in Ontario, G. B. Kolenosky showed that the wolf needs about 6 pounds (2.7 kg) of meat per day. Eight wolves hunting as a pack were successful in 46 percent of 36 chases. The wolves killed 29 deer in 63 days, a ratio of one deer for eight wolves every 2.2 days. In the time that the wolves hunted in this area, they removed 10 percent of the deer herd. That figure should not be taken to mean that the wolves were reducing the total deer population by 10 percent, because they were hunting in the winter, at a time when many deer die of starvation. A number of the deer that the wolves killed would have been part of the annual herd surplus that would have died of starvation had they not been killed.

I personally can attest to the fact that wolves can limit the expansion of deer. Years ago, when I started guiding wilderness canoe trips in Quebec, there were no deer to be found north of the Barrier Dam area of Verendrye Provincial Park. As the lumber companies opened up the country, the wolves' howling disappeared from the night and their tracks from the beaches. In a short time, with wolves gone, deer tracks were seen on the same beaches that formerly showed only wolf tracks.

Research done in Alaska shows that where there are no wolves the blacktail deer are subject to heavy winter mortality because the habitat is usually heavily overbrowsed. Where wolves are present, their predation keeps the herd in better balance with their habitat, so there is more winter food and less winter mortality. The deer therefore have a much higher reproduction rate.

In November 1982, Michael Nelson and David Mech witnessed unusual predation of a deer by wolves in the Superior National Forest in Minnesota. They were following, by airplane, a radio-collared wolf that was one of a pack of seven walking on the ice of Thomas Lake. Although the temperature was just below freezing, most of the lake was free of ice.

An adult doe was swimming in the lake, where she had gone to escape pursuit by the wolves. The wolves, however, were traveling around the small lake in an attempt to intercept her when she got to the other side. When the deer became aware of what the wolves were doing, she turned back toward the center of the lake.

When the researchers returned 3½ hours later, the deer was still swimming in the same area while the wolves rested on the nearest shore. When the doe swam within 245 feet (74.7 m), one wolf jumped in and swam toward her. As the doe turned away, the wolf returned to the shore. The doe then headed toward an island that the wolves were able to reach by crossing the ice. Another wolf jumped in and swam toward her, only to return.

An hour later, when the researchers again returned, the doe was again swimming toward the island, with the wolves running to intercept her. The tired doe rested in shallow water at the foot of a small bluff, which allowed the wolves to approach undetected. When she bounded into deep water, one wolf followed and caught her, attacking her head and neck. A second wolf swam out and also attacked the doe, but it returned to shore in about one minute. Within 10 minutes of the time the first wolf had attacked the deer, it had killed her. It then towed her to shore, where the entire pack fed on the carcass.

There are numerous accounts of wolves chasing deer into the water, but this is the first record of a swimming wolf killing a swimming deer. In all of the other instances, the deer easily outswam the wolves. This deer was in frigid water, swimming almost

Management Problems

constantly for at least 4.7 hours. She may have been swimming much longer than that.

In another incident, one wolf entered the water 27 times over a 3½-hour period, but could never catch up to the deer. The researchers also noted that the wolves were perfectly capable of estimating just where the deer would land after swimming a pond or a lake, and would be there to intercept her. They concluded that open water may not always be as much of a refuge for deer as was previously thought.

Lest you think that the wolves always get off scot-free, Nelson and Mech reported a wolf that had been killed by a nine-point buck, although the buck itself was killed by the other three members of the pack. In December 1983, Tom Pearson found the carcasses of both animals on a freshly plowed road just south of Silver Island Lake in the Superior National Forest. The snow in the woods was 17½ inches (44.5 cm) deep, and the temperature was around 0°F (−17.7°C). The deer had run down the road, pursued by two wolves. The buck might have outrun these wolves but was intercepted by two more wolves coming out of the forest. After a brief fight, the buck was killed, but not before it had gored a female wolf that weighed 75 pounds (34 kg). There was blood on the buck's antlers and perforations in the wolf's rib cage that had filled the cavity with blood. The buck was 90 percent consumed when discovered, but all indications were that it had been a prime animal with no previous injury to incapacitate it. Its bone marrow had a high fat content, one sign of a well-nourished deer.

This is not the only instance of a wolf being killed by a deer. J. H. Frijlink, working in the Algonquin Provincial Park in Ontario, Canada, reported in 1977 having found a wolf that had been killed by a deer striking it on the head with a hoof. The wolf was diagnosed as being rabid, but it was killed by the deer, not by rabies. It is not known if the deer had suffered any injuries, but it had not been killed.

The bobcat and lynx have been known to kill both adults and fawns of all three deer varieties but most of their predation is on fawns. Minnesota has reported 12 authenticated records of adult deer being killed by bobcats. New Hampshire has reported 40 deer killed by bobcats in seven years, but most of the kills were fawns. During periods of deep snow, when the deer are yarded or have bogged down, they are much more vulnerable. No one should be surprised that the bobcat and lynx can kill a deer. These cats weigh up to 40 pounds (18.1 kg), as much as the largest coyotes.

Under these same snow conditions, the fisher and wolverine have also been known to kill deer. Sometimes, however, a deer is more than a match for a predator. Jean M. Linsdale and P. Quentin Tomich tell of several instances in which mule deer does chased and treed bobcats that happened to be in the area where the does had their fawns hidden.

Deer hair is frequently found in fox scats because the fox is one of the chief scavengers of dead deer. In winter, every fox trail will sooner or later lead to the carcass of a deer that has died of starvation, or been wounded during the hunting season but died later, or been killed by an automobile.

When foxes are caring for their kits in the spring, they carry their kill or their scavengings back to their dens. In a short time, the area around the den is littered with scraps of fur, feathers, and bones. When Joe Taylor was one of New Jersey's predator-control men, he found a den that had the feet and hoofs of whitetail fawns. This might lead one to think that the fox had killed the fawns, but in this instance Joe knew the fawns had died of starvation because he had removed a nursing doe from a nearby highway after it had been killed by an automobile. The orphaned fawns had died, and the fox retrieved the carcasses as carrion.

Both red and gray foxes probably do kill fawns on rare occasions. In the early summer of 1975, John Powless of Carlisle, Pennsylvania, was at his cabin in the Seven Mountains in Mifflin County, Pennsylvania. As he sat drinking a morning cup of coffee on the cabin porch, he heard a loud bleating. He was amazed to see a whitetail fawn running toward the cabin, pursued by a gray fox. He ran out and chased the fox away. The fawn was exhausted and stood trembling with its mouth wide open and its tongue hanging out. After the fawn got its breath back, it ran back into the woods. The doe had evidently been away feeding at the time, or the fox never would have dared chase the fawn.

Black bears will kill an adult deer if they can get one at a disadvantage. Under ordinary circumstances, a bear is too slow to catch a deer. Black bears methodically search western mountain meadows for the young of both deer and elk.

There are several records of golden and bald eagles attacking both fawns and adult deer. These instances are so rare that I mention them only because of their oddity. Bald eagles often feed on deer that have been killed, particularly on the ice of lakes and rivers. I have seen this on a number of occasions. Michigan has a sizable wintering population of bald eagles, and the birds make

Management Problems

good use of some of the deer that are killed on highways but are unfit for human consumption. These deer are placed out on the ice as feed for the eagles at the Seney National Wildlife Refuge.

In my experience, the oddest case of predation on deer was a case of mistaken identity, I am sure. In June of 1975, one of my neighbors phoned and asked me to rush over with my camera. This neighbor and his son had been driving a tractor and a wagon in one of their fields about 10 A.M. on a clear, sunny day. Suddenly they noticed a great horned owl dive out of a tree and plunge onto something hidden in the high grass at the edge of the field. As the owl dived into the grass, a whitetail doe dashed out of the woods and attacked it, striking out with her forefeet. The bird had dived at the doe's fawn. The owl turned from its attack on the fawn to fight the doe, but it stood no chance against the slashing hoofs of the mother. The owl was knocked into the grass away from the fawn. The fawn, uninjured, scrambled to its feet, and both deer retreated into the woods. At this point, my neighbor called me.

I caught the owl without effort; it was in no condition to put up any fight. The left wing, though not broken, must have been dislocated because it hung down and could not be used. Or the breast muscles may have been badly bruised. The doe's hoofs had knocked clumps of feathers from the breast. There is no doubt that if the doe had continued her attack, the owl would have been killed. My theory is that the owl mistook the back of the fawn for a cottontail rabbit.

Dogs probably take a greater toll of deer than all of the other predators combined. Many of the culprits are wild dogs that have been abandoned or have strayed and reverted to living in the wild. Some, however, are well-fed pets that chase and kill deer, following an innate desire for the hunt. Healthy adult deer can usually outrun healthy adult dogs when there is no snow or ice on the ground. It is during the time of deep snow or during the birthing period that the deer are most vulnerable.

I remember my first experience with wild dogs. It occurred in 1939 or 1940. The pack consisted of a huge Airedale, a German shepherd, a big black mongrel, and two smaller, nondescript mongrels. The deer herd was just reviving after a decline in my area of New Jersey, and the dogs wiped out the nucleus. They then turned their attention to sheep and cattle. The situation reached the point where the farmers feared for their families, and everyone kept a gun handy. After the dogs were finally killed, it took the deer two more years to repopulate the area.

In parts of the Deep South and parts of Canada, swamps are so extensive and the vegetation so thick that dogs are used for hunting deer. There is no other way to get the deer out of those areas, and the use of dogs does not guarantee anyone a deer. Hunters from other sections of the country should not condemn the use of dogs in the areas where they are being used legally, because terrain and vegetation have to be considered.

John Sweeney, Larry Marchinton, and James Sweeney of the School of Forest Resources, University of Georgia, conducted extensive tests on radio-monitored deer to learn the deer's reactions to being chased by dogs. Their work gives important insight into the behavior of deer. Three things must be stressed: The dogs used were the type most commonly used by hunters in the South. They were dogs trained for tracking deer, not dogs trying to kill deer. They were well-fed dogs that did not depend on killing deer for food. Moreover, the conditions were favorable to the deer; they were not on a starvation diet, and there was no snow or ice to give pursuing dogs an advantage.

These deer had all been trapped earlier and had radios attached to their collars. The researchers then radio-monitored each of the deer till they knew the animal's range, habits, and constant whereabouts. Then the dogs were turned loose on the deer repeatedly, and none of the deer was ever caught or injured by the dogs. I found the report fascinating because these researchers determined actions and reactions that had only been conjectural before radio transmitters were employed.

The escape behavior of the deer was catalogued in five divisions: holding, long-distance running, circuitous zigzag running, separating from the group, and using escape habitat.

The dogs chased the deer for an average run of 33 minutes and covered 2.4 miles (3.9 km). The longest run recorded was 13.4 miles (21.6 km) for a time of 155 minutes. Where the deer population was high, the runs were short because the dogs kept switching to new deer as they encountered them. Where water was available the deer used it, and this almost always stopped the chase. Although the deer could be chased out of their home ranges, they came back in a day or less.

In another study, conducted in California, Raymond Dasmann and Richard D. Taber found that the holding pattern of blacktails differed. Mature blacktail bucks in heavy cover would stand and fight off the dogs; whitetails did not. They would lie still until they were sure the dogs were on their trail, then they ran.

The deer that ran in circles or mazes were usually does and fawns that were reluctant to leave their home areas. Bucks were more likely to take off at high speed and run in a more-or-less straight line. During the rutting season, bucks have a much larger range, so although the bucks may have been leaving the area they usually frequented, they were not really venturing into new territories.

In running in circles, deer often crossed their old tracks or encountered other deer that laid down new tracks also. This pattern was very confusing to the dogs and threw them off the trail. When the pressure was taken off the deer for just a few minutes, they would stop and rest. They would run again only when they had to.

Studies of this type were also conducted in Virginia and in Arkansas. The report from Arkansas had this opening statement. "The domestic dog, *Canis familiaris,* is the most abundant and widespread predator, or potential predator, of the whitetail deer, *Odocoileus virginianus,* in North America." Yet the Arkansas studies, like those in Virginia, confirmed the finding of the earlier study done in Georgia that—in terms of actual effect on the deer population—the dog is *not* a major predator of deer. This is not as contradictory as it seems. The apparent acquittal of dogs as serious predators of deer applies only to certain regions and under special conditions.

Using radio-tracking and harassing the deer with packs of trained dogs, the Arkansas researchers found that the dogs were able to catch only 1.1 percent of the deer chased. They found that pregnant deer does did not abort their fawns after being chased. They also found that deer did not suffer ill effects from taking to cold water to escape the dogs after being heated up by running. It had often been stated that such deer were likely to get pneumonia. Not so, claimed the researchers, and they have proved it, though only for a region with a warm climate. (Researchers in Vermont have verified that deer chased into the water in winter sometimes contract pneumonia.) The Arkansas biologists concluded, as many others have, that predation by dogs is of a relatively "sanitary" type, weeding out the sick and unfit, because dogs are not as efficient predators as coyotes, wolves, and cougars.

The dogs used in these experiments were deer hounds that had been trained to follow deer and to bark while tracking so that the hunters would know where the deer were. This meant the deer also knew where the dogs were. The researchers were quick to point out that wild dogs may not bark or betray their presence to the deer. Moreover, I must stress again that dogs used in the experiment were trained *not* to kill deer; they were well-fed; and they were running the deer under conditions not adverse to the deer.

That wild dogs hunt differently, particularly in the North, shows up in the tallying of the kill. In 1969, the state of Minnesota found that dogs killed more than four times as many deer as wolves and coyotes combined. These kills were made in January, February, and March, when snow, ice, and lack of food all work against the deer. Conservation officers found that 519 kills were made by dogs, compared to 126 for wolves and coyotes.

In just one month, January 1962, dogs killed 77 deer in Pennsylvania. Bear in mind that the figure given represents only *located* kills. At least that many must go unreported.

A "pet" dog attacking a deer: even otherwise tame dogs can become a threat to deer, by harrassing them in winter. Feral dogs probably kill a larger number of deer than all other wild predators combined.

This young deer was hamstrung by pet dogs, and it had to be killed.

In the September 1984 issue of *Pennsylvania Game News,* Game Protector Gary W. Sinuth of Lebanon Township reported that two dogs were chasing three deer that ran into a small farm pond to escape. The dogs would not let the deer out of the pond but kept harassing them whenever they tried to do so. Eventually all three deer drowned.

A report from Nova Scotia stated: "An alarming increase this year was noted in the number of dog kills. The known dog kills increased from 15 in 1974 to 53 in 1975."

In one three-year period, Vermont had 4,267 known dog kills. In 1969 there were 1,406; in 1970 there were 1,498; and in 1971 there were 1,363. The dog-kill figures dropped to 631 in 1972, to 541 in 1973, and down to 340 in 1974. Intrigued by the drop in the dog-kill rate, I phoned the Fish and Game Department in Vermont and talked to my friend Bob Candy, head of the Education and Information Division. Bob confirmed what I thought. The years 1971 and 1972 had had exceedingly hard winters with up to 8 feet (2.4 m) of snow. The years 1973, 1974, 1975 had had mild winters, easy on the deer, and without any advantage to running dogs.

In 1971 and 1972, the Vermont deer herd plummeted through starvation. The weakened survivors were easy prey for the dogs. Bob told me that the state game wardens claimed the kill figures represented only about a tenth of the damage actually done by dogs. Many deer had been chased and had died of exhaustion, as was proved by the frothy blood filling the lungs. And many deer that were chased into the icy water developed bronchial illnesses such as pneumonia. They were in such poor condition that when they got soaked thoroughly, they died. Often, in both cases, the dogs did not actually attack the deer but they had effectively killed them anyway, as a result of the chase.

The laws were changed in Vermont in 1970 to allow game wardens and their deputies to shoot dogs seen chasing after deer for the period of December 1 through May 31. Prior to 1970, this had been permitted only from February 1 through April 30, so the new law extended control for an additional three months. Dog owners have no recourse to law if their dogs are shot while chasing deer. Many townships are now passing "leash" laws similar to those in New Jersey which prohibit owners from allowing their dogs to run loose. This has also helped to curb dogs from running wild.

As chief gamekeeper of Coventry Hunt Club, I personally have investigated dog problems and have witnessed dogs pull down deer. In the 1950s and early 1960s, New Jersey had a lot of snow each winter. In the mid-1960s, the jet stream changed our weather pattern. Prior to this, most of our storms came from the west. Now most of the storms swing south through Tennessee and up into Virginia, sweep up across south Jersey, and miss our area. Years ago, 24-inch (61 cm) snows were common. An adult deer's chest is only 20 inches (50.8 cm) from the ground, and six-month-old fawns are even shorter. The deer could move through this snow depth only by leaping, which was exhausting. While the snow remained powdery, dogs could not move any better, but the weather in New Jersey seldom stayed cold enough to keep the snow powdery. The surface would soften on warm days and then freeze to a crust at night. The deer's sharp hoofs would break through a crust that could support a dog, and the slaughter was on.

One family in our club owned big black mongrels that were a cross between a shepherd and black Labrador retriever. These dogs were seen pulling deer down. One day I found them eating the hind leg off a young deer. The deer was still alive, but too weak to get up. It lay there and watched the dogs eat its legs. The dogs ran at my approach but the deer could not be saved. Its anus and the inside part of the ham had been eaten. When I confronted the man who owned the dogs, I told him his dogs would have to be penned up or I would get authority from the state to shoot them. He never knew how revealing his answer was: "I can't pen them up, they don't even come in to eat anymore." The dogs were finally "controlled," not because of the deer damage but because they hit the owner where it hurts, in his pocketbook. The dogs were seen killing sheep and a young heifer belonging to a neighboring farmer.

The laws are anachronistic. If a man's prize bull, worth thousands of dollars, gets out and is killed by a car on a highway, the farmer must pay the car's owner for the damages. If a man's dog destroys that same farmer's bull in a pasture, or the public's deer, the dog's owner is not liable for the damages, legally, in most states. In most states, anyone shooting a dog for the destruction of personal property can be sued. These laws need changing.

In 1969 the State Board of Health in Georgia figured there were over 300,000 stray dogs in that state. There were several attacks by these wild dogs on people. Robert Nash, vice president of the Georgia Cattlemen's Association, stated that of 11,610 head of cattle destroyed by dogs and other predators in 1967, $888,058 worth of damage was done by dogs. The damage to the swine industry was $238,014 that year, a total loss of $1,126,072 worth of cattle and hogs. The loss of deer could not be calculated, but Hubert Handy, chief of Georgia's Game Management Division, has said that the state's deer population of 120,000 would be 60 percent larger if it were not for the casualties inflicted on the deer herds by wild dogs.

In the August 1984 issue of *Deer & Deer Hunting Magazine*, fellow biologist Larry Marchinton points out that Handy had evidently overestimated the predation by feral dogs and underestimated the breeding potential of the whitetails. Marchinton allowed that there were still 300,000 feral dogs in the state, but pointed out that the annual harvest of deer by hunters is now 144,000 animals of an estimated population of over 800,000. Yet it has to be reiterated that wild dogs in the South hunt deer under different conditions from dogs in the north, that it, without the advantage of snow on the ground. None of this debate can negate the fact that feral dogs affect deer herds. All across the continent, the reports implicating dogs in the deaths of thousands of deer each year continue to come in.

I am not a hater of dogs. I love dogs. I raised springer spaniels for years. But I also love wildlife. Throughout most of New Jersey, the dog problem has been alleviated, although not eliminated, by licensing and by prohibiting free-ranging dogs. Almost all townships also have dog wardens to take care of wild dogs. The most effective law is the one prohibiting dogs from running loose, though if dog owners all used good sense, there would be no need for such a law. People who profess to love their dogs should love them enough to want to know that their dogs are safe at home. Since dogs on private land can run loose when accompanied by the owner, there is no very severe restriction placed on either the dogs or their owners. Laws merely help ensure that the dogs are not free to kill deer during the winter or to kill fawns in the spring. Even the researchers in Georgia, Virginia, and Arkansas had reservations about what the trained dogs, let alone wild dogs, would do to newborn fawns or to the adult does just prior to, during, or immediately after they gave birth.

Much of New Jersey is rapidly becoming urbanized. More and more of our farmlands are disappearing to development. Each new home usually has at least a dog or several cats. The Humane Society estimates that there are 80–100 million pet dogs and cats in the United States, producing 2,000–3,500 young per hour. They make no estimates of the feral animals, but the totals would be staggering. Although New Jersey has tried to legislate control of dogs in the state, many of the city folks do not keep their dogs under control when they move to the rural areas. In 1973, in his *The Ecology of Stray Dogs* A. M. Beck concluded that ⅓ to ½ of all dog owners allow their pets to run free at all times.

In 1974, R. N. Denny studied the impact of uncontrolled dogs on wildlife and livestock. He estimated the population of feral and free-roaming dogs in the United States at between 12 million and 30 million. And most dogs, despite thousands of years of domestication, still like to chase deer and other game animals.

A good friend of mine, Lou Stout, lives just a few miles from where I was raised on a farm, north of Belvidere, New Jersey. Beaver Brook flows right past his house. This stream is large enough that Lou was using his canoe to ferry some firewood home. Because of the beautiful October foliage, Lou had his camera with him while he gathered wood, and he photographed the events that followed.

While going upstream, Lou heard a dog barking up on the bank. He tied his canoe and climbed up over the bank to discover a large yellow dog, similar to a big golden retriever, barking at an eight-point buck it had chased into a farm pond. When the buck saw Lou, it dashed from the pond, only to be pursued and brought down by the dog. The buck was able to regain its feet, but the dog felled it again, biting it around the neck and face. The buck turned back to the pond and was dragged down again. By the time the buck had reached the water it had been bitten repeatedly and had a broken right front leg. The dog would not enter the water, nor would it stop harassing the buck. The dog's owner was called, and he removed the animal. A game warden killed the buck, judging that he could not survive his injuries. Although this scene occurs thousands of times each year, it is seldom documented pictorially.

Not all dogs are killers of deer; indeed most are not. Many dogs will have absolutely nothing to do with deer, and some actually develop affection for certain deer. I have often been told of fawns raised by people that became good friends with the family pooch. Back in the late 1950s, my good friends Doris and Art Dickinson raised a fawn that became very attached to their big collie. The fawn and the dog would eat dog pellets out of the same bowl.

The oddest incident that I have proof of, although I have heard of others, was related by my good friend Charlie Alsheimer of Bath, New York. Charlie's neighbor Cynthia Palmer has a wildlife rehabilitation license. A fawn's mother was killed by an automobile nearby, and a state biologist took the fawn to Cynthia. He also had two orphan raccoons and brought those along at the same time.

Cynthia had a long-haired terrier female who was nursing six-week-old pups. Cynthia wondered whether the dog would nurse the raccoons and the fawn. It did. In fact, the dog successfully raised her own pups, the raccoons, the fawn, and a gray fox pup. The fawn simply wandered off to the wild after 2½ months, while the raccoons and the fox stayed a little longer.

Deer are quick to recognize when a dog is chained or penned up, and in that situation they are not frightened even if the dog barks.

There are also several records of the tables being turned and the deer attacking the dogs. Charles Laux of Freeport, Pennsylvania, was hunting raccoons one night when his hounds were set upon by a buck. The buck was slashing at the dogs with his feet and the dogs were jumping in and out trying to bite the buck. The fight moved into a small stream, and here the deer really began to work the dogs over. Using a front foot, he pushed one dog underwater. Every time the dog came up, the buck pushed him under again. Laux was afraid the buck would kill his dog, so he hit the buck over the antlers with his cane. The cane bounced back, hitting Laux on the chin and badly bruising him. The buck, tiring of the sport, shook his head and trotted off. Laux and his dogs were glad to see him go.

One of the most unfortunate aspects of the dog problem is that dog owners just can't believe, or don't want to believe, that their dog would chase deer, let alone kill them. People with this attitude are not likely to cooperate by restraining their dogs unless forced to do so by the law. And then there is also the argumentative individual who feels his "rights" are being violated when his dog is not allowed to roam free to do as it likes. It is most unusual to encounter a person like one dog owner who was confronted by a Pennsylvania district game protector and his deputy. The man's dog had been caught by the deputy running deer. When the conservation officers interviewed the man about his dog's deer-chasing activities, the man picked up his gun, went outside, and eliminated the problem. This drastic "solution," of course, can usually be prevented by more responsible behavior by dog owners.

Two additional items should be mentioned regarding the relationship between the deer and its predators. The natural predators of the past, and those that still survive today, have been so efficient in controlling deer populations that the deer has never developed the intrinsic population controls exhibited by species such as the lemming, snowshoe hare, and gray squirrel. In areas where the natural predators are gone, human hunters have to take up the slack because the deer can be its own worst enemy. Its breeding potential is such that without population control, destruction of the deer's range and habitat is inevitable.

It must also be remembered that no predator ever wipes out its prey species, although the population of one, or the other, greatly affects the life of the other. The prey–predator relationship has existed for millions of years. It swings like a pendulum from one extreme to the other, and only in the precise center of the arc, where it remains for just a brief time, can it be said to be in balance.

Take the hypothetical case of a large piece of virgin forest in the North Country. Deer have expanded their range up to the edge of the forest as farmers opened the land. A few wolves inhabit the forests but do not invade the farmlands because of the concentrated presence of humans on the farms.

Suddenly, the huge forest is opened up because of pulp cutting, timbering, or perhaps a forest fire. This forest land is not suitable for farming for any of a number of reasons, such as soil infertility or too short a growing season. If it had been suitable for farming, it would have been cleared sooner. After being opened up by cutting or fire, it is allowed to remain forest land.

Once the climax tree canopy is pierced, allowing sunlight to strike what had been a shaded forest floor, a profusion of berry plants, ferns, mushrooms, and grasses start to grow. These plants will be followed in a couple of years by natural reforestation, by aspen, birch, alder, maples, pines, hemlocks, or cedars, for example. In a short time, this area is superb wildlife habitat, because it produces almost limitless foods of many types, and cover as well.

The deer are quick to respond to this situation and expand their range to fill the niche. There is very little predation at first because the wolves were driven from the area by the presence of the woodcutters and the lack of cover, or by being deprived of both food and cover if the clearing was fire-caused.

The deer herd, unfettered by predators and launched by an unlimited food and cover support system, explodes. On a high-quality, unlimited diet, the does produce more young, more of which are female. The young are larger and healthier from the start and in turn breed earlier, producing even more large, healthy offspring. Within four years, say, the deer will be reaching maximum weight, the bucks will be producing exceptionally large antlers, and the herd population will have maximized so that all available habitat is filled.

Except during the hunting season, humans seldom venture into the area, so there is really little disturbance. The wolves drift back into the area, drawn by a burgeoning prey species. The deer population is so high that the wolves do not have to go far, nor hunt very hard. In response to the easy living, the wolves, too, have larger litters, with more pups surviving. This increased predation has little effect on the deer population, because the wolves are the apex of the food pyramid and the food and habitat base is broad and solid. And the weather has been benevolent. Mature does carry twins and triplet fawns every spring.

But nothing in nature is static, certainly not habitat. Everything else in the rejuvenated forest land is also growing. Between the sixth and eighth year, the deer population has peaked, although the wolf population has not, because the predator's cycle always follows the prey's cycle by a year or two.

By the ninth year, the deer are declining in numbers because they are beginning to encounter food shortages. There is no starvation as yet, because the wolf population has now peaked and the predators are depressing the deer population. With food becoming harder to find, the does are not going into the winter in as good condition as before. They are coming into estrus later; there are no triplet fawns; some of the does are bearing a single fawn; fawn mortality is high; and almost none of the female fawns breed at seven months of age.

Food shortages become critical for the deer in the tenth year because the foliage of most of the saplings has grown beyond the deer's reach. The umbrella effect of the high foliage eliminates most of the understory vegetation. The wolf population is still high because, although the wolves have to travel farther because the deer are fewer, the deer are not as strong and are more easily captured. This is a hard winter for the wolves, but harder for the deer. The winter is extremely cold and windy, and a number of the deer die. By scavenging and taking the starving deer, the wolves have more than enough to eat and produce the same large litters.

The eleventh year repeats the pattern of the tenth. But the wolves travel farther, they expand their territories, and they encounter different wolf packs more often. Strife among the packs results in the death of a number of the wolves.

The deer population reaches an all-time low in the twelfth year, due to lack of food and fairly high predation. The few does that are able to carry a fawn full-term produce mainly bucks, so that the sexual composition of the population is radically altered. The herd is now primarily older deer because most of the young died during the first two hard winters. During a third cold winter, with little food, most of the deer die because the older deer don't have good enough teeth to cut off the woody old browse, which is all that is available. Their bodies are scavenged and, although the female wolves give birth to smaller litters, scavenging allows them to give birth. But that summer, most of the wolf pups starve because the adult wolves have few deer to hunt. There are few buffer species, because a deer-destroyed forest is almost barren of underbrush, and without underbrush there is little or no wildlife of any type.

Some good came out of all this. With the deer herd only a remnant, some of the vegetation begins to come back along the edges of the ponds, lakes, and old wood roads. It does not recover in the wooded areas because of the shade. The few deer that survived now have access to more food and quickly regain their strength and vigor. The does go into the rutting season in better condition, and the fawns they bear are predominantly female. Even so, the land cannot reach its former high productivity unless the climax habitat is again drastically altered by man or fire. The drastically reduced deer herd cannot support a wolf population at all, and those wolves that did not starve migrate to areas with more prey species. The lives of prey and predator are balanced on a seesaw, the fulcrum of which is the habitat.

THE CONSEQUENCES OF MISMANAGEMENT

The classic example of deer mismanagement that proved the destructiveness of overpopulation took place on the Kaibab Plateau in northern Arizona. The Rocky Mountain mule deer that inhabited the Kaibab were magnificent animals with exceptionally large antlers. The Navaho and Paiute Indians came into the area to hunt the deer each fall but did not live there. The plateau, which covers about a million acres (404,700 ha), is bordered on the southeast and south by the Grand and Marble canyons, on the west by Kanab Canyon, and on the north and northeast by a belt of semidesert and open plains providing very little water. Thus, the deer were effectively fenced in.

The area had been leased to ranchers for grazing since 1893, and as many as 200,000 sheep, 20,000 head of cattle, and large numbers of horses were pastured there. There were between 3,000 and 4,000 mule deer on the plateau when it was made into a preserve by President Theodore Roosevelt in 1906. It was designated as the Grand Canyon National Game Preserve.

With the creation of the preserve, all hunting was forbidden. The Forest Service, which administered the preserve, established game patrols and employed trappers and hunters to eliminate predators. With guns, traps, and poison, they killed 4,889 coyotes, 781 cougars, 554 bobcats, 20 wolves, and a few bears. Gradually the sheep were prohibited from grazing there, although cattle and horses continued to be pastured in the area.

With the elimination of all natural predators and with no hunters harvesting any of the surplus animals, the deer population erupted by 1923. Although the official government estimate of the population was a conservative 30,000, other estimates put the population at over 100,000. The experts could not agree on the population figures, but one thing apparent to all was the tremendous destruction being done to the vegetation. And it was not just the damage done by the deer. Overgrazing by cattle and horses had destroyed most of the native grasses and forced the cattle to feed on browse in direct competition with the deer. It has been calculated that one cow will eat as much food as six deer. Overgrazing by livestock denuded the soil of its protective vegetation, and thousands upon thousands of tons of topsoil were lost through erosion.

In 1919, Edward Goldman, one of the top biologists with the Bureau of Biological Survey, was sent to the area to study the deer problem. His reports were filled with warnings of imminent starvation of the deer. Goldman's concern was echoed and supported by George Shiras, III, on the basis of his personal investigations of the Kaibab.

To complicate the problem, in 1919 part of the plateau had been made into the Grand Canyon National Park and put under the jurisdiction of the Park Service instead of the Forest Service. When the Forest Service suggested that the deer herd be thinned out by shooting, stiff opposition came from the Park Service, which would not tolerate any shooting on its land. Further opposition came from Arizona's governor, George Hunt, who was opposed to any federal control as a violation of states' rights. When the Forest Service wanted some of the deer hunted, the Governor threatened to have his game wardens arrest any man who came off the plateau with a deer. While the squabbling raged on, the deer population skyrocketed—and, like a skyrocket, it was about to explode.

Zane Grey evidently believed that cowboys could do everything in real life that he had them do in his novels. He gathered 40 cowboys and 70 Navaho to herd the deer off the plateau and across the Colorado River in an effort to thin out the herd. Two futile attempts in 1924 proved once again that you don't drive deer. Not a single deer was herded off the plateau.

That winter, the deer died by the thousands. Goldman, in the spring of 1925, figured that between 30,000 and 60,000 deer had died in the previous two winters. Subsequently, limited shooting was approved. Between 1925 and 1930, 1,124 deer were killed by federal employees, 2,652 were livetrapped and removed, and 11,641 were killed by sportsmen under special federal permits. This total of 15,417 deer taken in five years did not even keep up with the annual herd increment.

In 1930, a panel of experts in forestry, livestock, and wildlife toured the area on behalf of the federal government. This panel

Management Problems

recommended that the war on predators be stopped, that all wild horses be removed from the plateau, and that livetrapping, regulated hunting by sportsmen, and controlled shooting of the deer by federal authorities be increased.

Hunters had taken deer in 1929, and in 1930 over 5,000 were harvested. Through the combination of starvation, disease, lowered birth rates due to malnutrition, and shooting, the deer herd was finally reduced to less than 20,000 by 1931. At last the range began to show signs of recovering. And so did the deer. The management program that had begun through trial and error eventually evolved into the proper game and range management, brought about by application of scientific research. The Kaibab Plateau, when I visited it a couple of years ago, still showed some scars of the range devastation that was once so common. But trophy bucks are again being taken from the Kaibab.

Let me cite a few more examples of the deer's breeding potential. In 1962, the Michigan Game Division put 6 bucks and 11 does on South Fox Island, an island of 3,000 acres (1,214 ha), or about 5 square miles (13 sq km) off the mainland in Lake Michigan. Although there had been deer on the island previously, they had all been killed off. Logging had taken most of the big timber, but the second growth that covered most of the island provided excellent habitat.

Two of the deer died from unknown causes that first winter, so 15 deer were the nucleus of the population explosion that followed. Some local landowners shot a few deer each year, but by 1969 only about 40 had been taken. In 1969 it was estimated that there were at least 500 deer on the island, and the range showed signs of deterioration. To alleviate these conditions, a special hunting season was held, and the hunters took 188 deer.

But the breeding potential was still too high, and a concentrated hunting effort was made in 1970, in which 382 deer of all ages and both sexes were taken. A population count the following spring showed that there were still 194 deer on the island, and the population level the following fall would be back up around 400. It was calculated that the deer harvest every year would have to be heavy enough to maintain that number as a peak population. Otherwise the range destruction would continue, and the island's carrying capacity would deteriorate further.

The Llano Basin of the Edwards Plateau in central Texas provides another lesson on the reproductive capabilities of deer. The eradication of the screwworm was one of the leading factors that permitted this deer eruption. There are approximately 525,000 acres (212,468 ha) in the basin. From 1954 to 1961, there was an average density of 14.4 deer per 100 acres (40.5 ha), or 75,600 deer. That is 92.1 deer per square mile (35.5 per sq km). By 1961 the deer had increased to 18.9 per hundred acres, or 99,750 in the basin. That is 121 deer to the square mile (46.7 per sq km)—one of the heaviest deer concentrations in the nation. This fantastic population took its toll; the deer became smaller each year because of malnutrition. The average field-dressed whitetail buck in 1961 weighed only 72 pounds (32.7 kg), while an average doe, field-dressed, weighed 55 pounds (24.9 kg).

Such increased density breeds stress as competition for food and cover increases. The friction generated between the deer usually lowers the reproductive rate. Although the herd population increased, the reproductive capability of each doe decreased.

The 1981 figures showed that the population on the Edwards Plateau had dropped. Bandera County averaged 9.1 deer per 100 acres (40.5 ha); Kerr County, 11; Edwards County, 9.5; Sutton County, 11.9; and Crockett County only 5.2. The Llano Basin once had what was considered to be the highest concentration of deer in the country, with Kerr County, in 1963, having an estimated 23.4 deer per 100 acres, or a staggering 149 deer to the square mile (57.5 to the sq km).

Princeton Township in New Jersey provides yet another classic example of what can happen when deer hunting is banned. The owners of many of the large estates in that lovely township did not want hunters on their lands, so they pressured the township committee to ban all hunting in the township. The ban was illegal, because the right to hunt is a state prerogative. It was made illegal to discharge a firearm within the township, a right residents had. This ban was passed in 1972, despite the fact that no firearm accident involving deer hunters had ever occurred in the township.

By 1981, the residents were seeking relief from the devastation that deer were visiting upon their shrubbery. A single deer can denude three $40 ornamental yews in a single night. Deer–car collisions had risen 342 percent, from 33 in 1972 to 113 in 1981, with a number of the accidents going unreported. Bowhunting had been encouraged, but in situations like this either sport gunhunting or control gunhunting is needed to alleviate the condition.

Drastic overpopulation of deer did not occur before man eliminated the large predators. When nature was in balance, the predators increased, as the prey species increased, and the localized decrease of both creatures was likewise synchronized. Deer do not have the built-in control mechanisms that keep the populations of lemmings, some hares, grouse, gray squirrels, some voles and mice, and other cyclic creatures under control. Cyclic animals attain a peak population density within a predictable period, after which the population is abruptly self-thinned by a combination of such factors as massive dispersion, decreased reproduction, and ailments closely related to population density. A new cycle then begins. The cyclic creatures do not destroy their range on a large scale, as deer do, before their populations are reduced. Deer populations can be kept under control only by eliminating the annual increment. This can no longer be done by predators, so the surplus must be harvested by man. And the most efficient and practical method is by regulated sport hunting.

When anti-hunting groups and preservationists voice such platitudes as "let Nature take care of its own," "get Nature back in balance," and "let's bring the big predators back," they are also proclaiming that they have lost touch with reality. Everyone would like to see nature more in balance, but to achieve this we would have to drastically reduce human population and destroy most of our civilization.

The single most destructive force in the world is our own burgeoning human population, with its resultant habitat destruction. We are currently losing a million acres of habitat to development each year. As our population increases, we need more homes, schools, factories, roads, and so on. Room for these can only be made by taking away wildlife habitat of some kind. With its habitat gone, the wildlife that formerly lived there is also gone—gone as effectively as if it were killed at the moment the development started. This concept is hard for many people to accept.

Many people prefer to think the wildlife just moves out of the area being built on. This is true, but in most cases wildlife really has no place to go. If the habitat into which the wildlife is forced to move could support more wildlife of that kind, such animals would already be there. Nature abhors a vacuum, and almost every niche is already filled at all times.

Although human population growth in the United States has slowed, all demographic projections show that the population will rise tremendously over coming decades. The destruction of habitat will continue apace, and the numbers and types of our wild creatures will continue to decline. There is nothing game managers can do about this. It is a matter of personal concern for everyone.

Unless we can control our own population, most of the wild creatures and wild places will be lost, and with them the quality of our own lives.

The events on the Kaibab, Grand Isle, South Fox Island, and others all provided a fantastic education for wildlife managers and biologists throughout the world. The price was high, but, as with any good education, it was worth the price. We have learned our lesson. Or have we?

The game managers, the biologists, and many hunters have learned from the devastation that the deer bring on themselves. But more of the general public, especially the preservationists, have not. They react out of emotionalism and ignorance—by-products of the Bambi syndrome. Such emotionalism stopped New Jersey's first proposed "doe day" back in 1958, when most of the opponents were deer hunters who knew very little about deer. Such emotionalism caused a "pocket-sized Kaibab" in New Jersey from 1970 to 1974, when such preservationist groups as Fund for Animals, Friends of Animals, Society for Animal Rights, and others stopped a deer hunt in New Jersey's Great Swamp.

The Great Swamp National Wildlife Refuge is an area of some 6,000 acres (2,428.2 ha) in Morris County, New Jersey. Biologists on the refuge proposed a limited hunt in 1970 to reduce the rapidly expanding deer herd. For four years, legal action by preservationist groups stopped any hunting in the area. The herd continued to increase and the destruction of the range was severe. Damage to the shrubbery of homes in the area surrounding the Great Swamp was also severe. And the deer began to die of starvation. Ten confirmed cases of starvation occurred in the spring of 1974, and the biologists estimated that at least four times as many cases of starvation had occurred that winter.

The first hunt in the Great Swamp finally took place in December, 1974. Busloads of protesters from the preservationist groups picketed the area, carrying placards charging collusion among the hunters, the National Rifle Association, the arms industry, and the wildlife biologists.

An interesting footnote to these demonstrations was turned up by Dr. James Appelgate, working with polling experts from Rutgers' Eagleton Institute of Politics. He found that "the greatest opposition to hunting comes from college-age females living in urban areas." One has to wonder how many of these hunting opponents have had an opportunity to observe wildlife, for that matter, know anything at all about this subject.

It was estimated that there were 554 deer in the Great Swamp area, or about 55 deer to the square mile (262 per sq km). That is more than twice the average population usually found on good range. And the habitat in the Great Swamp was in very poor condition. Each day for the six days, 106 hunters hunted the Great Swamp, harvesting a total of 127 deer.

One of the bucks taken had 7 pounds (3.2 kg) of papilloma tumors on his head and neck. The tumors covered one eye and had destroyed the cornea of the other, resulting in blindness. Two other deer also had crippling tumors. Many of the deer had heavy loads of parasites, and most were below the weight and size for the average New Jersey deer. Several deer died the spring following the hunt from a bacterial skin infection known as *dermotophilosis*. This was only the second known occurrence of deer deaths due this disease. In 1975 a parasite was found on the deer that had never been found on deer before. None of these tumors, parasites, or diseases had ever been found on the Great Swamp deer until overcrowding and starvation became rampant. It is a well-known biological fact that the incidence of disease and parasites increases as the health of the deer declines.

Hunting has continued at the Great Swamp each year, with a goal of bringing the herd down and keeping it down to about 250 animals, the carrying capacity of the land.

The biologists and wildlife managers have the hard facts on deer management. But facts do not sway the preservationists from their continued plotting of a course that will wreck the deer on the shoals of disease and starvation.

Here are two more fiascos caused by the misguided efforts of preservationists: one took place in the Florida Everglades in 1982, and the second is the continuing conflict on Angel Island, California.

In June 1982 two tropical storms caused tremendous flooding of the lowlands in Florida's Everglades Wildlife Management Area, forcing the whitetails of that area to higher ground. The deer consumed most of the vegetation on these "islands," as the weeks went by, long before the flood waters abated. The Florida Game and Fresh Water Fish Commission, in desperation, attempted to prevent the total destruction of the habitat and mass starvation of the deer, by authorizing a special two-day permit hunt in just two sections of the management area. The plight of the deer had already become a media event, and, for the first time for many, people got a chance to see the real world of wildlife, bereft of the cuteness of a Disney nature spectacular. Although they were used to seeing human deaths on their evening television shows, people were appalled to see death rampant among animals, too.

What started out as a preservationist movement to "save" the deer soon turned into an anti-hunter campaign. A court injunction was issued stopping the hunt in the northern section, from which preservationists proposed to remove 2,000 deer, alive, in eight days. Both the rescue and the relocation effort failed. At a cost of about $8,000, 18 deer were finally captured, of which 12 died when relocated. Subsequently, 66 percent of the northern area deer —992 out of 1,500 animals—died of starvation, and the habitat will take several years to recover.

Fortunately the hunt in the southern area took place. Of the estimated deer in that area, hunters harvested 723, or 17 percent, of the herd. Only 6 percent of the deer died of starvation. Because the habitat was not destroyed, both the habitat and the deer could be reestablished within a year.

Angel Island is a 750-acre (303.5-ha) park in the middle of San Francisco Bay. Before becoming a park, it had been a United States Army Base. In 1915, deer were introduced onto the island to provide hunting for the officers, who kept the deer from overpopulating the area. When the island became a state park, all hunting was banned. With no predators and no hunting, the deer population, predictably, exploded.

To prevent destruction of the range and starvation of the deer, the California Department of Fish and Game shot 50 deer. Although the herd reduction was done by the Division of Fish and Game marksmen, and not the hunting public, a hue and cry was raised by the preservationists.

In 1976, the size of the deer herd was again out of hand. Again the Department of Fish and Game planned to remove some of the deer by staging a managed hunt. The preservationists would have none of it. Their counterproposal to feed the deer carried the day.

By 1981, the deer were at a crisis stage. When the Department of Fish and Game proposed another hunt, it was stopped by a lawsuit by the San Francisco S.P.C.A., which demanded that the deer be removed and relocated. The Department of Fish and Game conceded that it would probably be less expensive to move the deer than to fight a long, costly court battle. So 215 deer were livetrapped and moved to Cow Mountain, in Mendocino County, which had good deer habitat. Researchers from the University of California at Berkeley did a year-long study of the relocated deer. The relocation had cost over $100,000. Only 13 percent of the 215 relocated deer lasted as long as a year. That came to a cost of about $4,000 for each of the 27 deer that lived.

About 40 deer were left on the island, and their population shot back up as the vegetation began to recover. In 1984, the Department of Fish and Game recommended that a hunt be allowed to reduce the herd to 50 animals, all the habitat can safely support.

Again the S.P.C.A. objected, but this time they suggested sterilizing the deer. After two months of hard work by hundreds of volunteers, 30 does had been captured and sterilized. The program is doomed to fail because only half the does were caught. The rest will soon overpopulate the island. It is not practical for the Department of Fish and Game to attempt to feed or to relocate the deer, nor is sterilizing a part of the herd the answer. Things came to an impasse between the Department of Fish and Game and the S.P.C.A. There is never an impasse among deer; they just go on being deer, destroying their habitat if they overbreed.

Fanatic preservationists are an unwitting yet tremendous threat to the future of most wildlife. Their "no hunting" attitude, if allowed to prevail, would guarantee that wildlife populations in the future would be only a fraction of what they are now. The wildlife would not only diminish in numbers but also in physical size and health as their ranges are devastated. Under natural conditions, most species produce more young each year than their habitat can support. Unless something is done to increase the carrying capacity of a species' habitat, the yearly surplus of that species, whether it be made up of old or young specimens, must die. There is absolutely no biological argument about that incontestable fact; the surplus must die. The only question is how. It can be through starvation, predation, disease, or hunting. It is this surplus that sport hunting takes, and sport hunting thereby serves as a tool of proper game management. Not a single species of animal has been put on the endangered list by sport hunting. And the success stories of the management of deer, antelope, Canada geese, most ducks, and so on have been made possible by money provided by the sport-hunting fraternity.

Preservationists are motivated by a sincere belief in their cause. But sincerity does not necessarily make a belief right. Preservationists just don't know the facts. The facts prove that most game species are their own worst enemies; that if allowed to reproduce unchecked, they destroy their food, their habitat, and ultimately themselves.

A favorite preservationist theme is that "if we bring back the wolf and the cougar—the big predators—we won't need to hunt the deer." I gave a lecture on deer before a nature club in Westchester County, New York, in March, 1977. It is amazing how many deer inhabit the little wooded enclaves of that county, just 30 miles (48.3 km) north of New York City. Sure enough, a club member advocated bringing back the predators to do away with hunting. The facts will reveal the impracticality of this motion.

In 1975, licensed hunters took 103,225 deer in New York. Biological research has proved that a cougar eats, on the average, a deer a week, or 50 per year. It has been proved that a pack of eight wolves will kill 29 deer every 63 days. To equal the number of deer legally harvested by the hunters, New York would need a huge supply of predators.

Cougars never were as plentiful as wolves but are much more efficient predators, so we can allot them 37,500 deer per year, even if New York has "only" 750 cougars. To kill the remaining 65,725 deer would take 438 packs of eight wolves each—3,504 wolves. New York is one of the chief milk-producing states. How do the preservationists propose keeping these 750 cougars and 3,504 wolves from killing the dairy cows and beef cattle, which are so much larger, "dumber," and easier to kill than deer? How many people in urban Westchester County are going to want their share of the cougars and wolves? The preservationists had a chance to prove their argument. In 1975, the controversy over reducing the wolves in three of Alaska's game-management areas reached incendiary levels. Governor Jay Hammond offered to ship some of the live wolves down to any of the states that wanted them. There were no takers.

As much as I thrill to the sight of a deer, I thrill even more to the far rarer sight of a cougar or wolf. And, yes, I would like to see these predators returned to appropriate wilderness areas in the eastern United States. There are true wilderness areas in New York's Adirondack Mountains, for example, where relatively few people hunt. The Adirondacks have a high loss of deer due to starvation, and I believe these predators could help to alleviate the loss there. Both the wolf and the cougar are wilderness animals and would be unlikely to leave the wilderness. They could help to keep the deer herd in balance with the range, and without reducing the game available to hunters.

In the 1980s, battles raged over returning the wolf and the cougar to Yellowstone National Park. The elk in Yellowstone are destroying the habitat. I can't say *their* habitat, because it is also, or *was* also, the deer's habitat. I find elk today in sections of the park and in numbers that I have never seen there before. Yet I am *not* seeing the mule deer where I always found them before. Where I find them, they are reduced in both numbers and antler size. The predators would help to get the ungulate species back in balance with each other and with their habitat. The predators were eliminated from the park through ignorance; ignorance should not be allowed to prevent their return.

Humans would have nothing to fear from either predator. In North America, there is only one authenticated record of a wolf attacking a man. The wolf was rabid. It was killed, and the man was not. Cougar attacks on humans have been authenticated, but they are extremely rare. The farms adjoining the wilderness areas would probably lose some livestock, for which the farmers would have to be compensated. I do not believe the losses would be frequent enough to make the cost of compensation prohibitive. In a few cases, where livestock predation continued, an individual predator might have to be eliminated.

In some regions, the numbers of predators must be controlled for the sake of deer and other wildlife whose habitat man has altered and diminished. In other regions, predators need all the protection they can be given, and they should be reestablished in true wilderness areas where they have been extirpated. But the problem of too high a deer population for the carrying capacity of the habitat cannot be solved by substituting predation for hunting.

19

Automobile Fatalities and Other Accidents

A ccording to figures released by the American Automobile Association—figures gathered from all the states' game and highway departments—over a million wild creatures are killed on United States highways every day. That total of over 365 million wildlife deaths each year is a shocker.

A comprehensive listing of roadkilled deer was published by the Fred Bear Sports Club for the year 1974. In that year, the number of deer actually picked up by wardens and other officials amounted to 146,229. That figure is the only one that can be substantiated, but, having worked as a deputy game warden for many years, I know that the figures reflect only a part of the kill.

For one thing, many deer that are struck by automobiles are not killed on the road but run, walk, or stagger out of sight and die. I tracked one young injured buck by following the toboggan-like trail it made in the snow with its chest. Both of the buck's front legs had been broken, and it was traveling by sliding on its chest, pushing with its hind legs. I dispatched this buck, but many fatally injured deer are never found.

Nor do the statistics reveal the many roadkilled deer that are removed by motorists or local residents for meat. In a few states, such as Pennsylvania and New Jersey, this is legal, but in most states picking up a roadkilled deer amounts to illegal possession of deer and is as much of a crime as poaching. Although I hate to see good venison wasted, it is easy to understand the states' position about the removal of roadkilled deer by anyone other than authorized personnel. There are just too many unscrupulous people who would deliberately kill deer by one means or another, then break a few bones and claim the animals were roadkills.

I estimate that at least a third of the deer that are killed on the roads do not appear on the fatality lists. The figure of 146,229 roadkilled deer may be official, but it is far too low. I would be willing to bet that a more realistic figure would be over 200,000.

There were an estimated 200,000 deer–car collisions in 1984. These collisions will continue to increase as long as deer herds continue to increase, and as long as we continue to put thousands of additional cars on our roads each and every day. Some states have repealed the 55-miles-per-hour speed limit that was imposed as an energy-conservation measure in 1977. Disregarding the pros and cons of that limit, I only note that when the limit was imposed, deer-car collisions declined. They will undoubtedly leap upward again where the limit is rescinded.

Pennsylvania leads the United States in the number of road-killed deer. In 1974 it had 26,445 deer killed on its roads. The year 1975 was a record-breaker with 29,914. As the Pennsylvania report emphasizes, "those figures include only those deer which are physically removed from the state's highways by Game Commission personnel." The report also points out that the number of deer

In springtime, what appears to be one roadkill may actually be two or three; this doe was carrying two full-term fawns when she was hit.

killed by vehicles in Pennsylvania exceeds the number of white-tails harvested by hunters in 35 other states in that year.. The highest number of roadkills usually occur during the rutting season, in October and November. More deer are killed on weekends than on weekdays because the traffic is heavier, and the peak hours are about sunrise and sunset to two hours later, because this coincides with the peak of the deer's daily movements. Weather, temperature, and moon phase seem to make no difference.

A smaller peak period of road kills occurs in April and May. As already mentioned, May is the month that pregnant does become territorial and drive off their yearlings, much to the yearlings' dismay and confusion. April is the time that the first new sprouts of grass are shooting up. The grass along the highway edges is always up and green sooner than the grasses in the nearby fields or woodlands, for a number of reasons. Most new highway shoulders are fertilized to get them established. Established roadways are usually mowed regularly; those in the East never have the cut material gathered, and this green mulch makes excellent fertilizer. The absorption of the heat and its reflection by concrete or macadam creates a thermal strip providing a greenhouse effect, and this stimulates the grass to grow earlier. Much of the rain that falls on a highway is sprayed back on the shoulders by the automotive traffic, so the vegetation gets more than its normal share of moisture. Deer prefer to feed on fertilized or rich soil and so gravitate to the highways with their heavy traffic.

Blair Carbaugh and his associates conducted research on two 6-mile (9.6-km) stretches of Pennsylvania's Route 80. It was estimated that the area surrounding these stretches had about the same deer population. One stretch of the highway ran through a heavily forested area, while the other ran through open farmlands.

Of the 6,500 deer observed during the entire research period, 79.4 percent were seen in the forested area's highway right-of-way. Only 2.9 percent of the deer were found on the agricultural area's right-of-way. The remaining 17.7 percent were not on the right-of-way. Of the deer seen, only 40 were killed in the agricultural area, while 286 deer were killed in the forested area.

Although these figures do not bear out the percentages of observations, they do point out that a highway right-of-way is an "attractive nuisance." The forested area's right-of-way offered deer the first green vegetation available in the spring.

Although its effect has not yet been determined, it is a known fact that deer that feed heavily upon highway vegetation have a higher residual lead level in their bodies. They ingest the lead with the grass they eat, which has been polluted by automobile exhaust.

Pennsylvania's estimates of the size of its deer herd and the number of roadkills per year indicates that 1 out of every 32 deer in Pennsylvania will be killed by an automobile. In 1983, New Jersey estimated a loss of 1 out of every 35 of its deer to automobiles. In the northeast we lose a much higher percentage of our deer than do most of the other states and provinces because of the tremendous volume of traffic on high-speed roads that cut through the heart of good deer country.

Surprisingly, I can find no comprehensive, nationwide compilation of human deaths in the United States caused by collisions with deer on highways. But in 1969, in Michigan alone, nine persons were killed in deer-car collisions. Nationwide, the number of fatalities must be considerable. Statistics show that there is usually a 4 percent injury rate and 1 human death in every 3,500 deer-car collisions. If so, the 1984 death rate, for example, would have been approximately 57 people.

A 1983 Michigan study by Christopher Hansen showed the average auto-repair bill from a deer-car collision in 1984 to be $569. Today the rates would be even higher, but at those rates the total damage bill for 1984 would have been $113,800,000.

In addition to the loss of human life, human injuries, damage to vehicles, loss of time due to vehicle damage, repair bills, and increased insurance costs, there is a value to the actual meat that is lost. I know that the deer killed range all the way from fawns to big bucks, and there is, perhaps, no way to average out the weight. The average deer weighs 125 pounds (56.7 kg) alive, of which 47 pounds (21.3 kg) is pure meat. Figured at a current value of $2.39 per pound, for 200,000 deer killed, we come up with a loss of $22,466,000. And the loss is even greater if the third of the deer that are hit but not counted are included in the calculations.

My advice is always to drive defensively and be particularly alert in areas of high deer concentration. Most public highways have road signs warning motorists of deer-crossing areas. Be as alert to them as you would any other hazard warning sign.

During the periods of darkness, drive with your high beams on unless another car is coming toward you. This will allow you to see a deer at a greater distance. Watch for deer and also for the eyeshine of deer along the side of the road. When a deer crosses a road ahead of you, slow down and watch for more deer that may be following. If the deer that crossed was a doe, you can almost bet that her two young are just a few jumps behind. Apply your brakes to avoid hitting a deer, if possible, but don't jeopardize your vehicle or your own life by sudden braking if traffic is close behind you. Nor should you violently swerve to avoid hitting a deer.

Many ideas have been tried to lessen the carnage on the highways. New Jersey borrowed an idea that originated in the Netherlands. The Dutch found that by using Van De Ree stainless-steel mirrors to reflect the lights of automobiles into the woods, they cut their deer kill to zero. New Jersey installed the mirrors on 5.4 miles of the Garden State Parkway. The three sections that had the heaviest deer kill were chosen. The mirrors were mounted 36 inches (0.9 m) off the ground, with 160 to 200 mirrors to the mile. Success has been only minimal, perhaps due to the fact that the light intensity of the mirrors is cut down by their distance from the center of the road. The New Jersey mirrors had to be placed much farther from the road than those in the Netherlands because space had to be left for disabled cars and for snow removal. The use of mirrors by New Jersey has been discontinued. Maine has had better results with the mirrors and continues to use them.

A device that has been meeting with great success is the Swareflex Wildlife Reflector, produced by the Swarovski Company of Austria, and imported and distributed in this country by the Strieter Corporation of Rock Island, Illinois.

A number of states have run rather extensive tests on the reflectors with very satisfying results. According to a report by the Washington State Department of Transportation, the reflectors were placed in four test sections of Highway 395, in the eastern part of the state, each of which had a mortality rate for whitetail deer. The reflectors were alternately covered and uncovered during the test period, from 1981 to 1984. "During this period, 52 deer were killed at night in test sections when the reflectors were covered and 6 deer were killed at night when the reflectors were uncovered. This difference in deer-vehicles collision rates between the covered and uncovered periods is significant, indicating that the reflectors were effective on this highway during this time period." The difference was a reduction of 88 percent in the number of deer killed.

The state of Minnesota came up with an 89.5 percent reduction in the number of deer deaths due to automobiles, in areas equipped with the Swareflex Wildlife Reflectors. Prior to the installation of the reflectors, 38 deer were killed in the 1-mile (1.6-

km) test zone. The first year after the reflectors were installed, only 6 deer were killed. During the second and third years, 3 deer were killed each year. Of the 12 deer killed, 2 were killed during the daylight hours, when the reflectors would not be working, and 3 were hit at the end of the reflector line. Although the researcher studying these results could not be positive that the entire reduction in deaths was due to the installation of the reflectors, he was willing to regard them as the probable cause.

Pennsylvania did considerable experimenting with deer-proof fences on stretches of Interstate Route 80, where the roadkill was highest. The greatest number of kills occurred where the highway had been carved through the forest with no grassland or fields to act as buffer zones. One-way gates were installed so that if a deer did get onto the highway by coming around the end of a fence or jumping over it, the animal could get back off the road. The gates worked, and the fences drastically reduced the highway kills.

In 1974 the last strip of Interstate Route 80 in New Jersey was opened, just a few miles from my home. This section of the highway, between the Hackettstown exchange and the Blairstown exchange, runs through some of our very finest deer territory. When the highway opened, carnage resulted. As many as four deer in one night were killed on this 8-mile (12.9-km) stretch, and hardly a day went by without at least one deer being killed. In 1976 the highway department started to install chain-link fence the entire length of Route 80 through New Jersey. The idiocy of this measure is that the fence is 5 feet (1.5 m) high in some sections and even lower in others. This is no hindrance to a deer. By 1986, it was clear that the deer kill along Interstate Route 80 in New Jersey has been drastically reduced. Not that the fences helped; they didn't. It is simply that all of the deer that traditionally traversed the area the highway now transects have been killed or died. The few deer killed there now are ones that are dispersing from their home range, being forced from their ranges, or expanding their range during the rutting season. The chain of habit has been broken—at a cost.

Interstate Route 70, near Mud Springs, Colorado, intersected a major migration route of the mule deer of that area. When the state built that section of highway, it installed an underpass for the deer. For a considerable distance on each side of the underpass, the highway was equipped with deer-proof fencing 14 feet high, 14 feet wide, and 99 feet long (4.3 m × 4.3 m × 30.2 m). The tunnel had a concrete floor and low-wattage lights in the ceiling. Deer passage through the underpass was monitored by an electric-eye counter, by track counts in the raked dirt on either side, and by visual observation. About 60 percent of the deer used the underpass, although it took up to three days for some of them to work up the courage to do so. Observers think that a dirt floor and no lights would make the underpass more acceptable.

Deer-proofing a highway completely is an impossibility because of the cost. Fences are most effective, but very costly, varying between $25,000 to $75,000 per mile ($15,625 to $46,875 per km) according to the type of fence constructed. The underpasses and overpasses are the most expensive of all, and need fences in conjunction. The Swareflex reflectors cost about $45 per mile (about $28 per km). A number of states have tried using additional overhead lighting to enable the motorist to see the deer sooner at night. The lights did not deter the deer, and didn't help the motorists either.

A new device that I have personally tested and found to be very effective is the Sav-a-Life Deer Alert. These deer warning devices come two to a set, and both are needed. They are mounted on the bumper of your automobile. When you drive over 30 miles per hour (48.3 km per hour), the air rushing through the devices causes one to emit ultrasonic sound waves at 16,000 hertz and the other to emit at 20,000 hertz. It is the same principal as the silent dog whistle. You cannot hear the sound from the inside of your vehicle, but the deer can hear it for several hundred yards.

It is extremely hazardous to drive at night in my area; I have had 12 deer collisions with my various cars and trucks. I don't want any more. Most of the time the deer bolt away from the side of the road as my car approaches because of the Sav-a-Life devices. If they are a short distance from the road, they usually just freeze where they are. If they are some distance from the road, I really get their attention, while other cars pass by unnoticed. The devices are rated to be 80 percent effective, but they can save a lot of lives, including, perhaps, yours.

Fences figure importantly in connection with deer and deer damage. There are areas where some crops have to be fenced to prevent deer damage, but there are also areas where fencing is impractical. There are areas where certain crops cannot be raised because of deer damage, and others where deer damage precludes planting any crops at all. Back home on my family's farm, we could not plant soybeans because the deer ate every sprout as soon as it broke through the earth.

Electric fences are seldom practical because deer jump over them. Some farmers have tried three-strand electric fences, and deer either jumped over the low ones or through the strands if they were widely spaced. A deer that touches a single strand of electrified fence is not grounded and therefore feels nothing at all. Where prolonged dry spells occur, electric fences cannot be used because there is not enough moisture in the dried soil to make a good ground.

Cyclone-type fences are best, but the cost is usually prohibitive. Most farmers use hog-wire fencing with 6- to 8-inch (15.2- to 20.3-cm) mesh. Whereas a 4-foot (1.2-m) fence is high enough to keep in livestock, a deer fence should be at least 6 feet (1.8 m) high. A 7-foot (2.1-m) fence is much better. A deer can, but rarely will, jump an 8-foot (2.4-m) fence. Various designs have been tried in which the top strands have been made of barbed wire, some with the tops inclined inward, others with the tops inclined outward. None of the designs or types of wire make much difference to the deer. The major deterrent is the height of the fence.

The statistics on roadkilled deer can also be used as a game management tool. Over a period of time, they can tell of the deer herd's age and sex composition, shifts in ranges, dispersal of tagged deer, general health, and much more. James Durell, Assistant Director of Game Management in Kentucky, deplored the increase in the number of roadkilled deer in his state, yet he was not entirely unhappy about the statistics. He declared that as long as the roadkill was increasing steadily, it meant the deer population was also increasing steadily.

Richard E. McCabe, in the *White-tailed Deer,* reports:

In the United States alone, as of December 31, 1981, there were 3,852,697 miles (6,200,147 km) of improved public roads and approximately 100,000 miles (100,930 km) of primitive and unimproved roads. Of the improved public roads, 83.6 percent were in rural areas. And, within the contiguous United States, paved roads and their rights-of-way account for about 1.1 percent of the country's land surface—equivalent to the areas of Rhode Island, New Hampshire, Vermont, Massachusetts, and Connecticut combined. At present (1980–1983), new roadways in the United States are being constructed (graded and surfaced) at a rate of more than 11,200 miles (18,000 km) per year. In addition to their interference with whitetail movements and the elimination of prime or critical deer habitat in many cases, roadways compound management difficulties by erosion-control, windrow and esthetic plantings, and winter salting that attract the deer. Fencing, overpass and underpass structures, and other means of preventing or reducing deer-vehicle accidents and accommodating deer movements are expensive, often prohibitively so. Increased attention to roadway site selection and construction planning is needed to alleviate current and future problems.

Automobiles are not the only vehicles that kill deer. Railroad trains and airplanes also kill them, though not in such great numbers. During the winter, when snow is on the ground, deer that are not yarded up will frequently walk along railroad tracks because the passage of trains keeps these lines of travel open. In 1970 a railroad that was relocated because of the Libby Dam on Montana's Kootenai River started operating in the midst of some of that state's best winter whitetail range. A survey of the carcasses along the railroad was made each spring to tabulate the damage. As many as 37 deer to the mile (23 to the km) were killed by the trains in some of the best wintering areas.

In 1976 in Clearfield County, Pennsylvania, a deer crossing the runway at a local airport was struck by an airplane. Pilots from different parts of the country have told me that they often have to "buzz" the grass airstrips at some of the little rural airports to chase deer off the runways before attempting to land. Many of these back-country strips are hewn out of wilderness, and the runways may provide the only grass available in otherwise wooded areas. Even in open country, the grass on the runways is more tender and contains more nutrients because it is constantly being mowed and the grass is always comparatively new growth.

Power lines and other wires are very hazardous to deer. Near Erie, Pennsylvania, a high-voltage line broke loose from its insulator and fell to within a foot of the ground without interrupting electrical service in the area. Three deer, a buck and two does, touched the wire and were instantly electrocuted. Two red foxes that came to eat the deer were also electrocuted. Two mongrel dogs made the same mistake and suffered the same fate.

On Cherry Creek, in Montana, two mule deer bucks were slugging it out during the breeding season. As they shoved each other around, one of the bucks' antlers got twisted up in a strand of barbed-wire fence. As the bucks continued to twist and turn, their antlers were thoroughly and fatally tied together by the wire. Eventually, one buck was strangled by a piece of the wire twisted around his neck. When the two deer were found by James R. Martin, the live one was so close to death that he had to be dispatched, too. In the course of the fight, the deer had pulled loose yards and yards of wire and had broken off 13 fenceposts.

When I visited the Gage Holland Ranch near Big Bend, Texas, in 1967, I was given a mule deer skull that had about 20 pounds of fencing wire wrapped around its antlers. Evidently the buck had run into the fence and torn loose hundreds of feet of wire. As he turned around and around, the wires were twisted into a cable. Not being able to free himself, the buck died.

Wire and deer are a bad combination. I have found deer twisted up in woven-wire as well as barbed-wire fences in both winter and summer. I have seen them break their antlers, their jaws, and their legs on cyclone fences. And there really isn't much we can do about it. It's just one more hazard of civilization that deer have to contend with.

Deer can also get into trouble in their own habitat and without human help. They have broken through thin ice on ponds and rivers, become mired in mud, and fallen off cliffs. In June 1963, near Kaibab, Arizona, a muley doe and her two fawns were electrocuted when lightning hit a huge ponderosa pine under which they were standing.

I have a number of records of deer that have been caught in tree crotches that they had tried to jump through. Some were caught by the feet and several were caught around their middles. Near Ludington, Michigan, a six-point buck reached up to get some apples that had lodged in a large hollow in the crotch of an old apple tree. When he tried to drop back to the ground, his antlers got caught in the hollow. The harder the buck tried to pull away, the more pressure he put on his neck. At least in this case death must have been quick, for the area was not torn up by the deer's hoofs.

Bounding through the woods, deer sometimes impale themselves on unyielding snags or branches. One deer, killed by a hunter, was found to contain a dead stick that had punctured a lung long before he was shot. The wound had healed, the lung had inflated itself again, and the stick had acquired a gristlelike covering that prevented it from further damaging the lung.

While on patrol, District Game Protector H. P. Goedeke of New Alexandria, Pennsylvania, got a radio call from his wife. He wasn't sure she had the story straight, but he finally decided to go check out the reports of a deer up on a roof. The deer had run up a ramp from the ground and fallen onto a roof 30 feet (9.1 m) lower down. It had then jumped down to a still-lower roof, where the warden found it. When chased, the deer jumped off that roof to the ground 10 feet (3 m) below and dashed away.

John Doebling and Lincoln Lang, game protectors in Monroe County, Pennsylvania, had to rescue a deer that had run up a trestle and fallen into a commercial coal bin. The wardens fastened a syringe to a pole and climbed into the bin to give the deer a tranquilizing injection. The deer, thrashing around in the bin, discouraged that idea. Finally, the men succeeded in lassoing the deer and then giving it the injection. Once the deer was drugged, they managed to truss it up and remove it to wild country, where it was released.

Game Protector H. T. Nolf of Fort Washington, Pennsylvania, had to remove a deer that had crashed into a supermarket. While he was subduing that deer, he got a call about a deer in a local

The Swareflex reflector in the foreground contains a prism device which reflects the headlights of oncoming cars at right angles, deep into the roadside cover, warning deer and other wildlife before they reach the highway. The local chapter of Deer Hunters of America maintains these reflectors in cooperation with the Minnesota Department of Transportation. photo courtesy of Streiter Corporation

These three-legged deer were almost certainly born in a hayfield and injured by mowing machines.

bowling alley. That one, too, was safely removed. It was the second deer that he had removed from a bowling alley. He has also had to remove deer from a grocery store and a local bank.

The president of a Freeport, Maine, bank came back empty-handed from a deer hunt in that state. He had just returned to his office when a six-point buck came crashing through the plate-glass window. The buck evidently had noticed an error in his balance because he dashed right into the bookkeeping department. From there he went down to the basement to sulk. He was removed by the local game warden.

Some deer accidents have to do with the rutting season. Robert Clark, in Colville, Washington, saw a buck that had found his equal when he saw himself reflected in a store window. The buck gave all the signs of aggression while the other mirrored his every action. Determined to prove he was the better of the two, the buck lowered his head and charged the window, instantly smashing both the glass and his opponent.

In the Siskiyou National Forest, near Sullivan, Oregon, a blacktail buck challenged his reflection in the side of an automobile. Not daunted by his rival's fierce appearance, the buck charged and proceeded to bash in the door panel and the fender. He didn't do the grill any good, either.

It must have been an embarrassment to one buck to charge into a concrete lawn deer and get his antlers locked with those of his unyielding foe.

20

Improving Deer Habitat

A s has been stressed in earlier chapters, the deer population bears a direct relation to the carrying capacity of the land. When habitat improves, the population will increase; when it deteriorates, the deer population will thin out, by dispersal or, at worst, by disease or starvation. Human modifications of deer habitat have a direct effect on deer populations. Sometimes the effect is inconvenient to humans: deer munch food crops or inadvertently prune shrubbery. But human intervention in forested areas can create good habitat in wilder places as well.

REDUCING CROP AND LANDSCAPE DAMAGE

As noted in the last chapter, deer are seldom deterred from "trespassing" by any but the highest (and thus most expensive) fences. A number of other means have therefore been tried. Deer repellents have been used with varying (and limited) success. Some areas have used propane-gas cannons or carbide flash guns to fire loud blasts at timed intervals. These can be used only away from all human habitation because the periodic explosions would not be tolerated by residents. Unfortunately, deer become accustomed to the noise and the cannons lose their effectiveness. In California in 1976, I heard devices used to keep deer and birds out of vineyards that periodically emitted high-pitched shrieks and perhaps even ultrasonic waves. These seemed to be effective, for I saw no depredation from any wildlife. The deer probably find the high-pitched sound painful.

Lights have been placed in fields to ward off deer, but, if anything, the illumination helped the animals to see what they were eating. Repellents consisting of dried fish meal and the blood of cattle from slaughterhouses have been tried, again with very limited success. The first rain usually nullifies their effectiveness.

Creosote oil and naphthalene flakes give some protection, but for a limited time only.

Goodrite Zip, a chemical deer repellent made by the Goodrich Chemical Company, has proved to be one of the best commercial deterrents available. There are other compounds that are effective repellents for use on plants during the fall, winter, and early spring—the times when deer do most of their browsing.

Tests done by Thomas P. Sullivan in British Columbia on both snowshoe hares and blacktail deer have proven that wolverine urine is effective in suppressing depredation of forest seedlings by these animals. Other predator urines should work equally well. The wolverine urine deterred destruction for up to two months. If this urine can be duplicated synthetically, it will prove a tremendous boon to the big timber companies as well as homeowners.

Any company that comes up with an effective long-range deer repellent is going to make its shareholders rich. Repellents are of great value to orchardists and inestimable value to homeowners. As more and more people from cities build homes in suburban areas, the destruction of deer habitat increases apace. The owners landscape their property with expensive ornamental shrubbery that the deer promptly eat because their normal browse has been reduced and they probably fed in that same area before the houses were built. Deer particularly love yews—those nice, big, expensive yews—and it takes about three to four yews to feed one deer for one night. No one has tabulated the cost of the destruction of ornamental shrubbery by deer, but it must run to millions of dollars yearly.

I am sometimes called in to appraise deer damage and offer solutions, especially on large estates. Repellents are one answer for some situations, some of the time. Fencing would do the job, but many estate owners feel that fences would detract from the beauty of their estates, and they are right. Often, there is no easy or complete solution. I find many of these large estates closed to hunting. Many sportsmen would love to hunt on the estates, and each deer taken would lessen the problem. When the estate owners post their land against hunting, they are actively working against the most logical solution to their problem—that of reducing the deer population.

Diversionary foods are sometimes helpful, although not practical for every landowner. Good plantings include plots of corn left standing unhusked in the fall, an alfalfa or clover patch left unmowed, rye grass seeded as a winter cover crop after the harvest, and buckwheat left unharvested. These foods will temporarily lessen damage, but they do not solve the problem and may in the long run contribute to it by attracting deer if the deer herd remains unharvested.

The same problem exists with the use of deer attractants. Tests in California have proved that spraying mixtures of molasses, salt, and trace minerals in a water base on plants will attract deer to eat foods that they might otherwise find unpalatable. As a boy on the family farm, I had the job of running the dry corn stalks through a chopper and then sprinkling a molasses solution on the fodder to make it attractive to the cows. The same idea was used in the California tests. The deer stripped the sweetened foliage from the undesirable plants and left the conifer seedlings alone. The major drawbacks are that the mixture itself and the spraying are quite expensive, that the success of the diversion depends on how much brush is available to be sprayed, and that the diversion may attract additional deer to the area. Moreover, the sprayed vegetation may sometimes be less nutritive than what the deer would normally eat, so this method could occasionally lead to management problems.

Another recourse for homeowners is to plant shrubs and flowers that are relatively deer-resistant, or unattractive to deer. It must be borne in mind that under normal conditions a deer may not eat a plant that it will feed upon when forage is scarce in the winter. Some of the shrubs and flowers that are highly unpalatable to deer are lilac, boxwood, jasmine, chinaberry tree, Hercules' club, century plant, persimmon, sea buckthorn, holly, smoke tree, butterfly bush, black locust, pepper tree, wax myrtle, narcissus, daffodil, jonquil, aloe, columbine, clematis, iris, larkspur, foxglove, and English ivy.

Some apple orchardists I know lessen the damage of deer, mice, and rabbits by piling all the yearly pruned branches in low piles around the perimeter of their orchards. The bark and the tips of these twigs are more nutritious and more palatable than anything these creatures could reach in the orchards. The twigs have to be pruned anyway, so both the orchardists and the wildlife benefit. Of course, most orchardists welcome hunters, in order to reduce the deer herd. In most cases especially in the Eastern part of the country the deer in these areas cannot be overhunted, and they are drawn to apple trees like iron filings to a magnet.

Improving Deer Habitat

CUTTING TIMBER TO IMPROVE HABITAT

Most of the large lumber companies do everything possible to attract hunters to their lands, even providing maps and maintaining the backcountry roads so that the hunters can get into as many sections as possible. This practice is good public relations and good business.

Clear-cutting large blocks of timber or pulp is the most practical method of harvesting the wood that grows there, and it creates excellent wildlife habitat. Many people oppose clear-cutting because it denudes large areas of trees all at one time, and on steep slopes it can cause erosion due to fast water runoff. I decry the erosion and the temporary silting of streams as much as anyone, but the timber must be harvested if we are to continue to live in wooden houses and use paper and countless wood by-products.

I'll never forget first seeing the vast clear-cutting operations of the pulp companies in Canada, where I guided wilderness canoe

This clear cut tract shows excellent deer food sprouting in the second year after the lumber was harvested.

trips for many summers. When the huge tracts of virgin spruce went down before the saws and machines, I was appalled. The land looked like photographs taken during World War I that I had seen of "no-man's land." I thought the practice of clear-cutting was criminal.

But the following year, the entire area was clothed in the greenery of bracken ferns and berry bushes. Within two years, the spruce seedlings were starting to show, if you looked carefully. The logging roads that had been bulldozed were soon covered with white-pine seedlings. The seeds of the white pine are so large that the tree cannot reseed itself through the duff and debris of a forest floor; they must fall on bare earth. Only when these areas were "logged off" was the white pine able to come back. Within five years a new forest was well on its way.

Those virgin spruce forests had held little wildlife; the newly emerging forests abounded with wildlife, and the deer expanded their range northward. And so it is with many clear-cut areas.

When the mature forests go and the second growth sprouts, the wildlife populations explode. Therein lies a problem for many of the lumber companies. The Douglas fir is probably the most important tree to the lumber companies of Washington and Oregon. The Douglas fir is also one of the most important foods of the mule deer and blacktail deer. It is easy to see why the lumber companies woo the hunters. They want to have as many deer harvested as possible.

One of the greatest problems facing the whitetail deer today is that over much of its range very little, if any, lumbering or timber cutting is being done. The conditions that allowed the fantastic explosion of whitetails in the early 1900s no longer exist, nor are they apt to be duplicated again on this continent. The whitetail population is continuing to increase, on the whole, because the deer are still being introduced to areas from which they had been wiped out.

The whitetail population was not high in early colonial days in those sections of the East that were heavily forested. When the virgin timber was taken out by the settlers, the deer population increased, but so did the hunting pressure. That pressure was eventually reduced, mainly because the deer's very existence was threatened toward the end of the 1800s. At that time the United States was still a nation of woodburners, and forest growth was constantly being stimulated by the felling and regrowth of the trees. When I was a boy, my family had three wood stoves. All the

Some habitat is better than it appears to the layman's eye: very little browse is apparent in this hardwood forest, but the deer have been pawing for acorns, and there is brush nearby.

farm families of my area constantly harvested firewood from their fencerows and woodlots. Where wood was taken out, more was constantly growing back in, and the deer benefited from the abundance of browse.

Today, although I live in an all-electric house, I use the electricity for heating only as a reserve. I am back to heating the entire aboveground level of the house, my living quarters, with heat from my wood stove. The wood I burn I get by thinning my own woods or cutting up blowdowns, and by doing the same on the land of some of my neighbors. Even if I clear-cut my land, I could not get new growth started, because the local deer population is too high to allow forest regeneration. The mountains behind my home have thousands upon thousands of acres of trees that are about 30 to 40 years old. And thousands of these acres are old, abandoned farm fields or woodlots that are no longer cut. This is the basic cause of the chronic food shortage for the deer in my area. The trees have grown beyond the deer's reach, and this climax forest will remain useless to them until it is cut again, or burned over, or until disease wipes out enough trees to open the forest canopy.

Gypsy-moth caterpillars were first brought to Cape Cod from Europe for experimental work in producing stronger silk. Since then, they have become a scourge in many areas, completely denuding and destroying some forests. Yet in some areas, the consequences of their destruction have not been entirely bad. In some places they have opened up the climax forest cover and have started massive plant succession, which will be a bonanza for wildlife.

Deer are faced with the dual calamity of losing habitat to urbanization—about 10,000,000 acres (4,047,000 ha) per year—and losing the use of much remaining habitat because of the problem of the forests growing beyond their reach. One of the basic tenets of wildlife management is that mature forests and wildlife do not go together. In mature forests there is very little wildlife of any kind, because of a lack of food. This principle is as hard for many hunters to grasp as it is for nonhunters. Many hunters go up to the North Woods or other forested areas to hunt deer, but they find very few deer in extensive areas of mature forests.

The early Indians were efficient hunters, and they were the first to practice game management in this country. The colonists found extensive burned-over areas in the midst of the virgin forests. Some of these fires had been set by the Indians. The Indians knew that the number of deer and other animals in an area increased in direct proportion to the food available. Deer might run through the mature forests and would go there in the autumn to feed on the mast crops, but they didn't live there. They were and are creatures of the "edge."

Edge is a word that describes excellent wildlife habitat. Edge is where grassland or an open area meets or abuts a forested area. Edge usually means both food and shelter, and that is what habitat is all about.

Wild turkeys need mature forests for mast, but they also need open grassland where their poults can feed on insects. About 80 percent of all ruffed grouse nests are located along old wood roads or along the edges of the woodland. A single wood road creates edge through a forest by allowing sunlight to reach the forest floor. Pheasants and quail venture into croplands and open areas, but they nest in the edge. Every brush row is edge favored by cottontail rabbits.

The game birds and animals, and most of the nongame species, will be found in the transitional stage where grassland or old farmland is being reclaimed by the forest. This stage of plant succession is edge on a grand scale—until the forest matures. One of our major problems is that we are running out of edge because we are no longer creating any. We are not cutting wood in the eastern forests as we once did; we are not clearing homesteads; we are not cutting woodlot roads through our forests. In the late 1800s, only about 20 percent of Vermont's land was still covered with forests; today, because of farm abandonment, the state has reverted to 83 percent forest. Although the mature forests produce cover, they produce little in the way of deer food. Often where new suburban enclaves or industrial developments rise, the manicured lawns and asphalt strips and blankets end abruptly where the mature forest is left standing—without any real edge. Even in some of our most important farming country, much of the edge has been removed. The small, traditional "patchwork" farms have given way in many regions to huge "agribusiness" corporations that practice modern "clean" farming—enormous single-crop tracts unbroken by woodlots, windbreaks, or natural fencerows of shrubs, small trees, or brush.

Today, we are aware of the importance of wilderness, not only to wildlife but to our own physical and mental well-being. Many of us are better able to cope with the stresses of civilization if we can escape periodically to some of the wild spaces. Many people benefit just by knowing that there are some wild places to escape to, even if they never actually get to them. I, for one, could not live if all the "wildness" were taken out of life.

Nevertheless, many people have a reverence for trees that is not necessarily good for the forest or for wildlife. Most people are against cutting trees. In the western states this is not a problem because most of the forested tracts are owned by lumber companies or are on National Forest land. Multiple land use is the major theme of our national forests.

Multiple-land-use programs are sometimes a cover for abuses such as overgrazing and poor timber-cutting practices. I am very much against spending millions of dollars tearing apart virgin wilderness areas to facilitate the cutting of timber that is not needed and that is being sold at a tremendous loss. Nevertheless, the multiple-use idea is a sound one, and the hunting and lumbering must be continued. It is the abuses that must be halted. Our forests must be managed to provide the greatest benefits to the largest number of people. This means that the bulk of our forests should be thinned or cut off systematically, by sections, at a prescribed rate, for a sustained yield. When this is done, both wood and wildlife are renewable resources.

In the eastern states, most of the timbered land is in forest preserves or state parks, or privately owned, so multiple-use management—including timber cutting—is often opposed. Where the land is in forest preserves or state parks, a legislative act is frequently required before any cutting can be done. It is vital that our state foresters and game departments work in closer harmony, getting together to inventory the forested land: its trees, game, and habitat. Then sound decisions can be made about which lands to hold inviolate, which can be managed, and when and how to manage them. These two divisions ought never again to act independently, as they so often did in the past. The soil, the water, the vegetation, and the wildlife are not separate entities but integral parts of a whole. Only when a united front is maintained can a legislative body be approached with reasonable hope that our lands and wildlife can be properly managed for the best interests of all.

The woodlands that are in the private sector are another situation. Many of the old mountain woodlots are held by absentee owners who don't know what they own. Most of these people could be persuaded to have at least some of the timber on their land cut if they knew that their interests would be protected and that wildlife would benefit. This, too, requires a sales job by a team of foresters and game biologists. In my lecture appearances, I have found that many people will abandon misguided preservationist attitudes if the biological facts are presented to them. More timber must be harvested and more forests opened up if we are to maintain healthy deer and halt the diminishing of deer populations—for although we have a high overall deer population, the herds are suffering needless privation and reduction in many regions.

Almost 75 percent of all hunting for whitetail takes place on private land, so landowners' cooperation is most essential. The state of Vermont has worked out a very extensive and successful program with the wildlife division, the forestry division, and landowners. All three are cooperating to improve both the deer herd and their habitat.

The Pennsylvania Game Commission, a leader in deer re-

Improving Deer Habitat

search and management, had the foresight to provide a place for the deer to live and a place for hunters to hunt the deer. Over a period of many years, the commission acquired State Game Lands, millions of acres held in perpetuity for wildlife of all kinds and outdoor recreation of all kinds. These lands—and the taxes paid on the lands—all come from hunters' license fees, permits, and game-violation fines.

The hunting seasons in Pennsylvania are carefully tailored to keep the deer herd in balance with the land and food available. The herd is stabilized at about 800,000, with an annual harvest of about 140,000. In 1976, to provide better habitat and more winter food, the commission sold the commercial timber on 8,300 acres (3,359 ha). This timber must be cut in the winter so that the tree tops can be used as deer food. In addition, the commission has crews cutting brush on 2,800 acres (1,133 ha) where there is no timber that can be commercially harvested. These brush-cutting operations slow down the forest's race to maturity and improve the habitat enormously for many kinds of wildlife.

Over the years, much of Maine's wintering deer areas were growing up into even-age stands of trees that were beyond the reach of deer. In the 1950s the Department of Inland Fisheries and Wildlife devised a cooperative, voluntary winter-habitat management program between the state and some private landowners. Although the program was sound, it was not done on a scale large enough to benefit enough of the state's deer.

To expand the program, the state made aerial surveys of all major deer wintering areas. Maine has over 10 million acres (4 million ha) of unorganized township land that comes under the jurisdiction of the Land Use Regulation Commission. On that land, 903 deer wintering areas were located. A law was passed requiring private owners of such land to obtain a permit before harvesting any of their timber. The idea was to get the owner, the forestry department, and the game department together before any cutting was done, to ensure that the interests of all three are served.

Blocks of trees are being cut on a 15-year basis. This maintains mature tree stands, which serve as winter shelter for the deer, while the cut areas provide the sprouts needed for food. And the blocks that are cut systematically provide a sustained yield of timber. Through such cooperation, everyone benefits, especially the deer.

Deer are quick to take advantage of the tops of trees felled in winter unless deep snow prevents the animal from moving about.

Frank Brochu, supervisor of a hardwood-cutting operation at Bingham, Maine, reported that 11 deer came out of the woods and began to feed on the tree tops while the cutters were still working. The next day all 11 of the deer were bedded down in the immediate area, waiting for the saws to start up. Deer have learned that chain saws mean food. In the Upper Peninsula, the Cusino deer herd is often subjected to starvation. While the lumbermen there cut up the felled trees, the deer often feed on the tops.

I saw the same thing happen in my own area when large-scale timbering was done for the first time in many years. Fortunately, the year of the largest operations coincided with the year of the deepest snow. The deer had become accustomed to feeding on downed tree tops even before the snow came. The dense forest and the numerous gulleys gave good protection from the wind, so the deer did not have to yard up. Over 50 spent the winter in the immediate vicinity. The sound of a chain saw was like the gong of a dinner bell. This is the most practical way of providing supplemental winter food. Feed the deer with a saw and an axe.

CONTROLLED BURNS AND OTHER HABITAT BOOSTS

Smokey the Bear died in late 1976. For a quarter of a century, he was a very successful symbol for forest-fire prevention. The message that he conveyed reached untold millions. But even before Smokey died, a lot of what he stood for had already died, though there is still no excuse for negligence that could start a forest fire. In May 1986, a forest fire devastated 75,000 acres (30,352.5 ha) in the coastal forests of North Carolina and wreaked havoc upon the wildlife when it swept through the refuge there. Fire prevention is still of tremendous importance in most areas, in most situations—but not in all of them. Fire is gaining acceptance as a tool of forestry and wildlife management. That area in North Carolina will have more wildlife four years from now than it ever did. It will be high-quality wildlife habitat for the next 20 years and so will fire-ravaged Yellowstone Park.

Game managers and foresters have long understood that some forested areas are scrub-covered and their trees, grown too high to provide food for the hoofed big-game animals, will not yield harvestable timber. Much of this scrub would be too expensive to cut. Fire is the logical answer. Fire has always been one of nature's ways of keeping the woodlands under control. The fires that used to sweep across the prairies kept the trees from invading the grasslands. The fires that swept through much of the forests were even more beneficial, because they opened the solid forest canopy to sunshine and started anew the cycle of subclimax, or transitional, plant succession.

Before the 1900s, forest-fire control was almost unheard of. Huge, destructive fires ravaged miles and miles of timber throughout the last century. These fires, coupled with the timber harvesting, set the stage for the eruption of the deer population that took place in the early 1900s. Since then, the control of forest fires has become almost a science, and this control has coincided with the decline of lumbering. The combination of fire control and reduced lumbering contribute to the chronic food shortages that plague our deer herds in many sections of the country, particularly in the Northeast, by allowing forest vegetation to grow beyond the reach of the deer.

Of course, accidental fires still occur. In 1975 a forest fire burned almost 2,000 acres (809.4 ha) of Worthington State Park in New Jersey. All of the local fire companies and hundreds of volunteers fought the fire for days before bringing it under control. These fire fighters performed heroic work. Yet it would have been better for wildlife if the fire had covered more area and had been hotter. There was very little harvestable timber in the burned area, and no homes or personal property were threatened. On almost half the burn, the blaze was only a ground fire that consumed the dead leaves and downed wood, and killed off some underbrush. It did not kill most of the trees or open up the forest canopy, so as wildlife habitat the forest was in worse condition afterward than it had been before. The trees continued to shade out the understory, and almost no regenerative browse grew in.

In the areas that *were* burned clean, the trees that were killed were primarily scrub oak and rock oak. It is true that in good mast years these trees might produce 200 to 400 pounds of acorns per acre (224.1 to 448.2 kg per ha). In poor years they might produce only 10 to 50 pounds per acre (11.1 to 56.3 kg per ha). The first year after the burn, I observed that such an area grew up with snakeroot, berry bushes, ferns, and many forbs and grasses. That fall many of the deer did not come down off the ridges at all but stayed up in the burn area and fed on the new growth. Within two more years, the new sprouts that shot up all over the area produced up to a ton or more of browse per acre (2,241.6 kg per ha) and continued to do so for at least 10 years afterward.

Worthington State Park is one of New Jersey's public hunting areas and is subjected to intensive deer-hunting pressure. The number of deer there was low for the amount of land, not only because of the hunting pressure but because the area was a mature forest. When a species' population is below the carrying capacity of the land, the population grows at the highest rate possible for that species. This is known as the law of inversity. The more dense a population becomes, the more the reproduction rate declines. Because of the hunting pressure, the Worthington deer population will be kept in line with its food supply. Hunting in the area should be excellent for the next couple of decades.

The fire gave the grouse population, which had steadily declined in that area, the chance to make a strong comeback, as did the snowshoe hare. At different times the New Jersey Game Department has tried to restock the extirpated snowshoe hare. These stockings were doomed because of competition with the deer for the available food. The burned area could produce food enough for both species. Even more good would have resulted had the fire been hotter and more extensive.

That big burns have been exceedingly beneficial to most kinds of wildlife has been proved time after time by such examples as the Tillamook burn in Oregon, which resulted in more and bigger blacktail deer in the area, and the Kenai burn in Alaska, which benefited moose. Management plans in both areas call for the continued use of fire, on a rotating basis, to keep the habitat productive.

One of the largest controlled burns in which the federal government actively participated was the fire in Grand Teton National Park in 1975. Thousands of acres of scrub were allowed to burn to improve the browse for elk and mule deer. After years of publicizing the evil of all forest fires, the government was the target of tremendous criticism, even more during the fires in Yellowstone in the summer of 1988. There is a compulsion in man to put out a fire. Most of the time that compulsion is laudable, but we now know that a *controlled* fire can yield marvelous benefits. We now have the job of convincing the general public.

The greatest benefits could be gained from controlled burns in mature forests by burning swaths no more than ½ mile (804.7 m) wide. Deer do not like to be more than 400 yards (365.8 m) from cover at any time.

Ideally, the forest managed for the maximum benefit of all would be divided into strips further divided into thirds. The mature, harvestable timber would be cut and removed from one third

Improving Deer Habitat

of the land. The deer could feed on the tops that winter. The following spring that third would be burned, providing good deer food for about 20 years. At the start of the second 20-year period, the next third would have the timber harvested and then would be burned. At the start of the third 20-year span, the last third would be harvested and then burned. In the northern sections of the country it takes about 60 years to produce good, harvestable timber. In the South, where timber grows faster, the time cycles could be shortened to 12 or 15 years or whatever is required to grow harvestable timber. On this rotating basis, we would produce a maximum population of many kinds of wildlife and a sustained yield of lumber.

The deer could feed on the new sprouts that come in after a burn. And because of the profusion of sprouts, the foresters would have a better chance of growing the kinds of trees they wanted on these tracts. In some areas, the destruction of seedlings by deer is so great that for every 500 seedlings planted to the acre, no more than 8 grow to be harvestable trees. The maturing strips of trees would provide excellent cover for the deer as well as acorns and other types of mast crops.

This plan would benefit the forests as well as the wildlife, because all too frequently nonproductive trees are allowed to grow on unmanaged lands. Although we do need some large, dead trees to provide homes for cavity-nesting birds, unmanaged forests have too high a percentage of trees that are crooked, deformed, or rotten on the inside. These trees benefit nothing.

Do I expect this plan to be adopted? No. At least not until all of the departments concerned can devote the necessary years to educating the public.

A mixture of browse plants is always desirable. Deer do not feed on any single plant species exclusively, no matter how nutritious it happens to be. For a deer's rumen to function properly, the animal must ingest a variety of foods whenever possible. A mixed stand of vegetation offers more in the way of food and shelter than a solid stand of any one species. In the northern states it is common to find unbroken stands of hardwoods, especially in areas that at one time were farm fields. These hardwoods offer food but no shelter. Where possible, blocks of conifer trees (up to an acre) should be interspersed throughout the hardwood trees. This would greatly benefit the deer because it would break up the wintering concentrations. On level terrain, it does not matter where these blocks are planted. Where the terrain is hilly or rolling, the blocks of evergreens should be planted on the south or southeast slopes or down in the hollows. These are the areas favored by the deer in winter.

As a naturalist and conservationist, I am often asked by ecology groups to endorse some project or program. I often do, but not until I have satisfied myself as to the merits of each endeavor. One program I will not endorse—in fact, I fight against it—is the underground installation of high-tension power lines.

To some people the towers and their sweeping wires are esthetically offensive. Personally, I am often fascinated by the form and composition of the towers, but I will admit that in some areas the proliferation of these towers is an eyesore. In urban areas, underground installation may be a good idea, but I don't want to see it happen out in the country. Where the big power lines run through a forested tract, as they do in my area, the swaths cut

The powerline cut is a clear example of an "ecotone", or edge habitat, which is marked by rich growths of browse and grazing forage near cover.

through the woods are often the only steady source of good, nutritious deer food. Like the backcountry roads I mentioned earlier, they provide excellent "edge" habitat.

The right-of-way beneath the power lines is usually 250 to 300 feet (76.2 to 91.4 m) wide. Periodically, the brush on the right-of-way is cut to prevent it from growing into trees that could interfere with the power lines. The new brush that is constantly sprouting up makes excellent deer food. The brush that is cut and piled along the edge of the right-of-ways creates excellent habitat for cottontail rabbits and nesting birds. The electric companies do this cutting in their own interest, but inadvertently they create miles and miles of edge and excellent wildlife habitat.

In September 1976 I visited the University of California's Hopland Field Station, operated by the university's Division of Agriculture. The director, A. H. Murphy, provided me with a knowledgeable guide, a young graduate student named Mike Frey. I was interested in the work being done to improve conditions for the blacktail deer in the dense stands of chaparral. Much of the chaparral country is on rolling hillsides that are too steep for crops. Erosion is a problem, too. This area was formerly used very heavily for sheep ranching, but sheep in California have been reduced from over 3 million head to about a million. Many of the areas are blanketed with the high bushes of chamise, manzanita, madrona, and the like that make up the chaparral. These dense stands are so extensive in some areas that there would be no food for the sheep, which prefer grasses and forbs. Many California ranchers, like their counterparts in Texas and elsewhere, are now engaged in game ranching—managing their land as hunting clubs and charging the hunters a fee. (In Texas, many ranchers are surviving only because they are leasing out their land for deer hunting.) The dense stands of chaparral brush have to be managed for the deer, because mature, unmanaged stands offer only cover. Most of the brush grows beyond a deer's reach.

Controlled burning is one of the best methods of opening up the chaparral brush. Studies show that perhaps 100 pounds of browse per acre (28.4 kg per ha), per year, may be all that is available to the deer in the mature stands. After a fire there is far more browse per acre on the same land. At the Hopland Station, I saw an area that had been burned about a month earlier. The bases of the mature brush had already begun to put out millions of sprouts. After a burn, the deer will often feed on some of the new sprouts of brush types that they ordinarily find unpalatable.

Crushing is another method of opening up the chaparral where the hillsides are not so steep that they preclude the use of the heavy equipment needed to do the job. Crushing, used alone, also allows the deer to feed on the flattened tops of the chaparral. Frequently, brush that has been crushed is later burned. In this fashion, the deer gain the greatest benefits from the existing plants, because fire not only opens up space for new sprouts but also revives the soil—literally fertilizes it.

The crush-burn areas are often planted with grass seed and then, two years later, are crush-burned again. This second burning in two years usually kills the chaparral brush and keeps the opened areas in grass. This benefits the sheep as well as the deer. The deer perhaps benefit most, because the second burning gives them access to grasses and forbs as well as the new brush sprouts. These grassy areas create lots of edge.

The main drawback to controlled burning in the chaparral is that California is so often a tinder box. The regulations now require such an outlay of equipment and manpower to make sure the fire remains under control, that the cost is becoming prohibitive.

The creation of browseways is also gaining in popularity. This approach is less efficient than the controlled burn, but the cost is far less. A huge heavy-duty tractor is used in conjunction with a heavy-chain flailing instrument. This is used to beat or cut the shrubs off about 6 to 12 inches (5.2 to 30.4 cm) above the ground. The advantage is that it doesn't create a fire hazard; nor does it cause any erosion, since the soil is not disturbed. Browseways facilitate both the deer's and the hunter's movements through the chaparral, while providing new brush sprouts in all the areas that have been cut. Some browseways still provide good browse 15 years after their creation. I am against the use of chemicals.

California has also experimented with chemical sprays to kill brush. If there are no long-lasting effects of the herbicides, and they are not harmful to wildlife, this may prove to be the most inexpensive method of all. Such sprayings will be of greater benefit to agriculture than to wildlife unless only small areas of brush are opened up to create edge and to plant limited amounts of grasses. Whereas burning, crushing, and flailing do not kill the brush—and the deer benefit from the new sprouts—the chemicals do kill the brush.

Minnesota has been using a phytocide (2,4 dichlorophenoxyacetic acid) as an aerial spray. This spray is used on areas where the existing vegetation is not used by the deer. In a very short

Improving Deer Habitat

The chaparral country of eastern California: even inhospitible-looking country like this can be managed for productive, stable deer populations. Crushing can create "browseways" of excellent forage.

time grasses, forbs, and herbs grow profusely on the sprayed areas, providing excellent deer food.

Maine has been experimenting with nitrogen fertilizer to increase the growth of brush and to boost the protein rate. Investigations have shown that on areas tested, the normal protein content of browse such as aspen, sarsaparilla, and bracken fern is about 7½ percent in April, 10 percent in early summer, and down to 6¾ percent in December. When nitrogen was applied to this browse, the protein content shot up to 15 or 16 percent, which botanists think is about optimum for the plants. Nitrogen fertilizer is expensive, however, and will probably never be applied on a large scale. Nevertheless, the use of the fertilizer doubled the benefits of this browse to the deer.

Studies in New York showed that on plots heavily fertilized with nitrogen the deer browsed 80 percent of the flowering dogwood. On untreated plots, they ate only 3⅗ percent of the shrubs. Apparently, the deer's taste for mineral-rich forage is the factor.

Studies in Oregon and New York showed that the chemical composition of soils changed dramatically after fires. Nitrogen, which had been converted from an inorganic substance to an organic one after being assimilated by the plants, was once again available to the new forest growth. Potash from the burned wood produced enormous amounts of potassium. These two fertilizer components greatly enriched the soil and the plants. Deer will feed even more avidly on the new sprouts that spring up after a burn than on those that follow a lumbering operation.

Arthur Einarsen, in his work on blacktail deer, found that not only did the chemical composition of the soil change on burned-over land, but so did the protein content of the plants. An analysis of the blacktails' favorite foods, such as vine maple, thimbleberry, blackberry, and salmonberry, showed an average protein rate of 12 percent on a new burn. The same plants on a six-year-old burn showed only 6⅘ percent protein. This is to be expected because with time the chemicals in the soil are leached out by rain and sun. The chemicals are also removed by deer eating the plants that had assimilated the chemicals from the soil.

Eventually the chemical value of the soil on burned-over land drops below that of cut-over land. Fire makes the chemicals available immediately, whereas the decomposing tops of vegetation release the same chemicals at a slower rate, making them available for a much longer period.

THE PROBLEMS OF ARTIFICIAL FEEDING

Whenever we have a winter with deep snow and cold weather, concern about the deer's well-being mounts. Game departments brace themselves for a flood of letters and telephone calls. Concerned citizens suggest (or demand) that the deer be fed at once. These people always seem very sure that supplemental feeding will solve the starvation problem.

There are many reasons why the artificial feeding of deer doesn't work. The main one is that instead of solving the deer problem, it compounds it. The second major drawback is that the cost of a feeding program is prohibitive. Feeding deer during emergencies does far more for the human psyche than it does for the deer. At such times people feel they should do *something* for wildlife.

When the number of deer in a given area does not exceed the carrying capacity of that land, there is no need for supplemental feeding. But, given the breeding potential of deer, this utopian situation does not exist except where 35 to 40 percent of the deer herd is harvested each year. Most deer harvests are far lower than this, so the deer population soon exceeds the carrying capacity of the land, and starvation is imminent. Where deer overpopulate their range, a herd reduction of at least half is needed to bring the balance back. If overpopulation becomes a long-term proposition, a 50 percent reduction may not be enough. It depends on the damage to the range. *But artificially feeding the deer is never the answer.*

Such feeding permits the survival of more starving deer. The food may mean that a starving doe not only will survive but bear young. If she is well fed, she may bear the normal set of twins instead of being barren or having a single fawn. Unless that doe or her young can be harvested, there will then be three deer struggling to survive on land that did not have food enough to support one deer. Now three times as much supplemental feeding is needed, and the problem grows progressively worse. Experimentally, an anti-fertility drug was tried on deer, the objective being to lower the birthrate where such overpopulation developed. Unfortunately, it did not work.

I am not suggesting that the deer be allowed to starve. I am stating that the wildlife managers and the biologists should be allowed to manage the deer. They should be allowed to regulate the proper harvesting of the deer so that starvation is not a problem and supplemental feeding is not needed.

Feeding deer on an individual basis or in special situations is not wrong. What is wrong is to allow a situation to develop in which a large part of a state's deer population is threatened, and then to expect the game department to feed the deer. In the past, some game departments have attempted to feed deer on a large scale. No state has the manpower or the money needed for the job, nor have the results ever been truly beneficial.

For a feeding program to be successful, it must be started *before* there is snow, so that the deer are freely moving—coming to the food. This has the drawback of concentrating the deer. High concentrations intensify the problem and increase the danger of disease and the spread of parasites. If the deer are getting sufficient food, they can withstand the cold, and if there is heavy cover, such as a stand of conifer trees, nearby, they will not be forced to leave the food to yard up elsewhere. Most feeding programs are started only after the deer are on the verge of starvation. Then they may be getting too little too late. After deer are yarded up, the food has to be delivered to the yard, because the deer will not leave the protected area. Studies in Maine have shown that deer food, either natural or artificial, that is 100 yards (91.4 m) away from the yard will be left untouched.

A major drawback to feeding deer in their yards is that unless the food is dropped in by helicopter, its delivery produces compacted trails to the yards. Dogs and other predators then follow the trails right to the deer.

When the snows pile up, many sportsmen's clubs and other groups and individuals attempt to feed the deer in their areas. They often go to the local supermarkets and get stale bread and discarded greens—lettuce, cabbage and the like. This is hauled out into the woods for the deer. Tame deer will eat bread. Wild deer will not. If the greens are placed where the deer can get to them, they may be eaten, but the small amount of nutrition in these discards is worthless.

People often ask me what *is* good food for the deer. I repeat that the best way to feed them is with saws and axes. Cut good nutritious browse for the deer. Keep in mind, however, that you are not legally free to cut browse except on your own land.

Alfalfa or clover hay is also good food, and most deer will readily eat it. Timothy hay is not good deer food, although the deer will pick through it to eat the dried weeds and forbs mixed in it. Deer that have access to corn in the fall can be fed corn in the winter. This is an excellent deer food because it is high in carbohydrates, which produce body heat. Wilderness deer are not likely to eat corn, however. They don't seem to know what it is. Commercial deer pellets, manufactured by many of the large milling companies, are excellent deer food and are usually readily eaten. In emergencies, the pellets fed to goats or rabbits also make good deer food.

It has been found that supplemental feeding of protein foods is actually better for the deer than is the feeding of energy foods such as corn. Corn has a tendency to reduce cellulose digestibility, and most of whatever foods the deer are eating in the winter will be very high in cellulose.

For years, hundreds of acres of corn were raised near my home. That corn helped to create a deer problem. When the farming was discontinued, the deer herd was much too large to be supported by the natural food available. We lost a great many deer to starvation because at that time we could not harvest does.

When enough time had passed, our deer no longer knew what corn was. Each fall they came to feed on the dropped apples in an orchard near my home. One year the apple crop was exceedingly poor. Because I like to watch the deer, I put corn out for them. But no corn had been grown in the valley for about five years. The deer did not know what it was. They smelled it and were not afraid of any odors on the corn, but they would not eat it. None of the deer that came in had been alive when corn was a mainstay; the link was broken and the corn remained untouched.

Whether a private feeding program is good or harmful depends on the situation. My friend Helen Whittemore loved to watch deer, and she fed them regularly for many years. Helen fed several hundred pounds of cracked corn each day, as well as dropped apples and apple pumice when it was available. I have seen over 50 deer at her feeding station at one time. Although they are fully protected on her land, they are hunted when they leave. So, although the deer do come in to feed, they remain healthy and wild and they do not build up an excessive population. In Helen's case, the feeding was therefore beneficial.

A number of the neighboring farms here in Hardwick Township, New Jersey, have banded together and control perhaps a thousand acres (about 400 ha). Their hunting club usually takes more than two dozen bucks during the first two days of the season and they also harvest some of the does, but their limited membership takes far less than the annual increment of deer. In the past I have seen as many as a hundred deer in some of their fields. They have long had an efficient winter feeding program. They take good care of their deer, but they have just too many.

Their deer do not die of starvation, but the feeding concentrates them and thereby makes them vulnerable in other ways. Two years ago the deer in New Jersey were hit by epizootic hemorrhagic disease (described in the Life in Summer chapter). Almost every occurrence of disease is the result of overpopulation. One of the hardest-hit areas was that of the neighboring farms—the scene of the feeding program.

There are times, of course, when feeding may be necessary for the deer's survival. When the winter is excessively severe, or if it is essential to preserve the nucleus of a deer herd, then feeding may be the only solution. To be brutally practical, under such circumstances it is cheaper to feed the deer—and risk any later consequences—than to restock the deer if they are wiped out.

21

Population Management

K nowing the number of deer in given areas in each state is of the utmost importance to biologists, wildlife managers, game departments, and the general public. Management requires a knowledge of how many deer are in a herd, the age-and-sex composition of the herd, and the quantity and quality of the range.

TAKING THE CENSUS

Livetrapping allows the biologists to check on individual deer, but the traps catch only a fraction of the animals. Radio transmitters pinpoint the exact whereabouts of a particular deer, thus accumulating much data on daily activities and travels. Tagging provides

A radio-collared deer is helping managers make telemetry studies of winter movements.

a basis for long-range records on weights, ages, and dispersal. Still, when we want to know how many deer are using any given area, we have to count them by some means. The most effective and accurate method is by large-scale drives. This can be prohibitively costly because of the manpower involved, though most gun clubs are more than willing to volunteer the help of their members for such undertakings.

The areas to be driven must be tracts that can be thoroughly and completely surrounded by the number of people involved. Those who are designated as counters are placed in position first, downwind where possible. All of them must be able to see the counter on either side. If they are placed on a road, they can be as far apart as such sightings allow. Since counters count only the deer that pass on their right sides, there is no duplication. The drivers must be close enough together to move all the deer within the area. The drivers move slowly and with a minimum of noise. It is of the greatest importance that the drivers' line be kept as straight as possible and that all drivers be able to see the driver on either side. The drivers count only the deer that run back through their line, and they, too, count only the animals on their right side. This is the most accurate method of censusing deer in a specific area.

We once took a census of the deer on Coventry Hunt Club land by going out at night and driving along, or through, all of the fields adjoining the woodlands to count the deer with the aid of automobile headlights and spotlights. When this was done for four consecutive nights, we had a good estimate of how many deer were on that land. Because I knew the areas intimately, I did not need to drive each one. Instead, I could do sample areas and then make projections. This method works well on appropriate terrain and is comparatively inexpensive.

By doing the census in late summer or early fall, we obtained not only a total count but the sex ratio and an accounting of the fawn crop. So long as the ratio of fawns to does was high, we knew the range was still adequate and the herd was increasing. When the deer had reached the carrying capacity of the range, our fawn count dropped.

Mule deer are usually tallied when they are concentrated in February or March. A count is made by two persons, walking on opposite ridge tops of each drainage. Each person counts only the deer seen on the opposite slope. This is much more effective than for each to try to count the deer on the near slope. The number of deer seen is tallied against the estimated acreage covered, to give a count per square mile. This figure, calculated against the total square miles of the mule deer's winter range, gives an estimated herd population.

Where a crew of census-takers is unavailable, other methods can be used that require only one or two people if the area to be covered is not too large. One such method is a track count, which is best made after a rain or snow. Old tracks can be raked over. If the tracks are to be counted on a dirt road, dragging heavy brush or a piece of chain-link fence behind a car will usually wipe the old tracks out.

Counts of droppings are also made. Deer defecate an average of 13 times each 24 hours. Some researchers advocate picking up all old pellets before counting the new ones. That's too time-consuming. I have found it much faster to mix up some bright water-paint and use a pump oil can to spray a few drops of paint on the old defecations. In a short time you complete your counting of the heaps of new pellets, and the water-paint has washed away. This method, like the track count, is useful because it requires no crew of census-takers.

In areas that have snow, the easiest way is to count the defecations on top of the snow. The number of heaps of pellets divided by the number of days since the snow fell, divided by 13, gives you the number of deer using the area censused. Pellets decompose very rapidly in April and May because of the frequent rainstorms during those months.

Counting deer by airplane has never been very accurate, because even the smallest planes fly too fast and the counters miss too many deer. The use of helicopters is far more accurate, since helicopters can hover in one spot to check cover thoroughly, and they can be lowered over dense conifer cover to flush the deer out. New Jersey makes extensive use of helicopters to census the deer when there is snow on the ground. Elsewhere, too, helicopters are being used increasingly.

Infrared photography is also gaining wider usage in censusing wildlife. The heat of an animal's body affects the film and shows up as white spots. For deer, this method has been more efficient in summer than winter because a deer's winter coat is such an effective barrier to heat loss that the body heat fails to register on the film.

In addition to these methods, statistical analysis and computers make use of the data gathered from the return of the hunter-kill cards. This data would be even more accurate if all hunters, *whether successful or not,* were required to turn in a report card at the end of each license year. Human nature being what it is, I suppose this is an impossibility. Even states that require the cards to be sent in have a large percentage of hunters who just don't bother. More data could be gathered if each hunter were required to fill out a report card when getting a new license. Unfortunately, many hunters don't buy their licenses until just prior to the hunting season, so the data would be over a year late and of no use in setting seasons and limits for the current. year. Another drawback is that many states do not require landowners to have a license while hunting on their own lands.

New Jersey, like a number of other states, requires that every deer killed be presented at a checking station within a day. This very effective regulation gives the state the opportunity to gather and tabulate a wealth of data very soon after the season ends. It is an inconvenience to the hunters, but some inconveniences are necessary.

New Jersey's deer-management program was computerized in 1974. The state was divided into 40 management zones. This new system divided the state by prominent boundaries such as major highways. Most of our major highways follow natural geographical divisions—for instance, the valleys between mountain ranges. Prior to this, information was gathered by counties—along political lines that were not recognized by the deer and couldn't be recognized by most humans, either. The new system allows data to be collected from smaller units, where the deer and range characteristics are similar. Data about the number of deer inhabiting each zone, their reproductive rate, age, sex and weight classes, and general condition, as well as the range conditions, are tabulated and fed into a computer at Rutgers University.

The 72 mandatory checking stations provide a wealth of information on about 11,000 deer in the one week of the hunting season. Analyzing this data manually had been costly and time-consuming. Using the computer, the deer managers now have an overall view of most aspects of New Jersey's deer herd and are better able to manage the deer by increasing or decreasing the hunting pressure in each of the deer-management zones.

Despite the fact that the law mandates that each deer harvested be taken to a checking station for verification, many New Jersey hunters still don't do so. If such hunters were caught, they could be prosecuted; but the state does not have the staff needed for such enforcement. Several years ago, New Jersey hunters were given new incentive to report. Any hunter who presented a deer at a checking station was rewarded with a free license entitling the hunter to take an additional deer. Immediately the total recorded harvest went up, although the hunters who took a second deer made a negligible impact on the herd.

RESTOCKING AND LIVETRAPPING

Restocking was the salvation of the North American whitetail herds. This is the quickest way to get deer back into an area where they have been wiped out. Deer populations swing like a pendulum between boom and bust. The goal of deer management is to shorten the swing, and if possible, stabilize it in the center of its arc. In most areas, the pendulum is still swinging toward the boom for whitetails.

Today, there are not many areas in the United States that are actually in need of restocking. Most of the good deer country already has deer. In 1966, Roy Anderson, Chief of Game Management for the Tennessee Game and Fish Commission, imported 27 young blacktail deer from Oregon. The deer were released on the grounds of the Volunteer Army Ammunition Plant, on the outskirts of Chattanooga. In 1967, three dozen more blacktails were brought in to bolster the new herd. Free-running dogs killed 6 of the deer and automobiles killed 11. But the herd thrived, and by 1971 it numbered 115. These deer have grown larger in size than their western counterparts because of the abundance of food. A 3½-year-old blacktail buck may weigh 140 to 170 pounds (63.5 to 77.1 kg) and have a rack that sports 8 or 10 points. If all goes as planned, some of the surplus will eventually be released into the wilds of Tennessee.

One of the major problems of livetrapping and removing surplus deer from an area is that there are so few places that need to be restocked. Several states estimate that it costs about $200 per deer in materials, time, equipment, and manpower to livetrap and relocate one deer. Generally speaking, livetrapping is successful only in times of food shortages, because the deer must be baited into the traps with food. (Where the traps are to be used in one area over a long period, they are frequently baited with salt instead of food. This often works when natural food is plentiful.) Most livetrapping projects are carried out during the winter, when hunger lures the deer. A drawback is that the cold and snow may force the deer to yard up, and most deer yards are not accessible to roads. The snow makes it difficult to get trucks into the areas where the deer are located.

Public ignorance and vandalism also cause problems. There are some people who inadvertently trip a trap while they are just looking at it, and there are others who trip it "just for the heck of it." There are also misinformed people who trip the traps to prevent deer from getting caught and release those that are caught. Sometimes these people even include misinformed hunters who don't want any of "their" deer taken away.

Almost all deer that are livetrapped are tagged. Metal or plastic tags or streamers are placed in their ears. Some are prepared for monitoring by placing transistorized radio transmitters around their necks. They are also aged and weighed. By the use of the numbered and coded tags, biologists can obtain ready information on a deer's age, travels, and so on. The radios allow the biologists to locate the deer at any time of day or night and to study their daily activity patterns and the areas they frequent, their escape routes, dispersal, and so on. This radio monitoring is relatively new. It is a valuable tool for biologists and wildlife managers.

The trapping of deer has its lighter side, too. Biologists near Ashland, Maine, caught one adult doe 8 times in one year, in the same trap. A puzzle to the biologists is that, although she has been caught so often and is conspicuously marked with ear tags, she has never been seen except when she has been caught in the trap. With regard to the number of times she was caught, the Ashland doe set no record. One state livetrapped a whitetail doe 50 times in three months.

Deer can become so used to being trapped that they practically take up residence in the trap. A few years ago, when New Jersey was trapping and tagging deer in the mountains above my home, one trap became the exclusive property of one doe. She was so eager to get the apples used for bait that she ran right in every time the trap was set. She finally had to be removed from the area.

The traps most commonly used are modified versions of the venerable box trap with dropping end doors, tripped by a treadle. New Jersey uses the old-style all-wooden trap. Some states are using wooden traps with metal doors and guides because they don't warp and jam when wet. Some of the newest traps are made of aluminum. Their advantage is lightness. They weigh about 180 pounds (81.6 kg), whereas wooden traps weigh about double that.

In California I saw the lightest type of all. The framework was made of pipe, and sides, top, and even the gates were made of heavy fish netting. There was no bottom to the trap. The four guide pipes for the gates were driven into the earth to give the trap extra rigidity and prevent its being knocked over. The entire trap weighed about 35 pounds (15.9 kg).

The Pisgah trap has an added chute at the end to facilitate boxing the deer for transportation. Most deer-transport boxes are made of plywood and designed to keep the deer in as near-total darkness as possible. Almost all wildlife remains much calmer if it cannot see what is going on.

These traps provide researchers with obvious advantages, since the deer can be handled, aged, sexed, weighed, tagged, marked, and examined. Blood samples can be taken, and the animals can be moved to other areas if need be. The major drawback is cost. Modified snares are more economical if tagging is the main intent.

Several types of collar snares have been developed. These snares, placed in deer runways, allow the deer to tag themselves. When a deer puts its head into the snare, it pulls the collar about its neck to a predetermined size and locks it. As the deer continues to pull, the collar snaps free from the snare wire. Biologists have found that the snares are about 50 percent effective, allowing them to collar and tag large numbers of deer at a fraction of the cost of other methods.

The use of immobilizing drugs, delivered by the use of an arrow or a dart gun, has gained wide acceptance. This method is used mainly on animals in holding pens. The range and accuracy of the darts leave much to be desired.

22

Poachers

Wherever there is illegal money to be made, somebody is always ready to make it. Just as organized crime has invaded legitimate business, the poachers have stepped up their operations against legal hunting.

A recent and comprehensive survey on the poaching situation in the United States was conducted by John Cartier, a field editor for *Outdoor Life* magazine. He divided the continent into seven major sections. Questionnaires that he sent to all of the states and provinces provided his statistics. Cartier found that deer poaching, over the past decade, has increased by 19 percent in the Prairie States, 34.3 percent in the Southeast, 59 percent in the South Central region, 60.6 percent in the Northeast, 88 percent in the Inter-Mountain States, 99.1 percent in the North Central region, and 112.1 percent in the Far West. The major increase in deer poaching is due to the increasing disregard on the part of the general public for all law and order. Another important factor is that the explosion of the deer population in most places coincided with the human population explosion. There are more people to shoot the deer and more deer for people to shoot. The rise in deer poaching is also the result of a decline in most areas of small and upland game and in the waterfowl populations, with a resultant decline in the poaching of these species. Then, too, the poaching of deer gives a much greater return for the time and effort expended. One average 125-pound (56.7-kg) deer yields the poacher as much edible meat as 50 rabbits or ducks.

During the Depression, many backwoods people and farmers took a deer now and then to feed their families. Many of these people still take a deer for the same reason, even though their financial situation may be vastly improved today. Some farmers feel that, although the deer belong to the state, the crops the deer eat and the land the deer live on belong to the landowners. I'm not about to argue the merits or morality of that issue because the landowners and farmers have a good point. In any event, these people do only a small fraction of the poaching.

It is the professional poacher, the outlaw, shooting the deer to sell the meat, who makes the heavy inroads on deer herds. Many such poachers conduct their operations with the precision of military maneuvers. Many use scope-sighted, highpower rifles and are equipped with four-wheel-drive vehicles that can negotiate the roughest terrain. Some are even equipped with CB radios with which they keep in touch with cohorts to help to spot game and with lookouts who watch out for conservation officers. Such organized groups move into a back area, kill all the game possible, and are on their way in a couple of hours.

For example, while hunting with a friend in Wyoming, John Cartier discovered 13 dressed-out antelope and mule deer hung up in an old shack on the friend's ranch. By the time the friend and Cartier got back with a conservation officer, the shack was empty. That was no subsistence operation.

A couple of years ago, four professional poachers from Rhode Island were arrested and convicted of shooting 25 deer near Chelsea, Vermont. When that many animals are involved, there can be no doubt that the poachers are professionals, engaged in selling meat and skins, and sometimes trophy heads.

A friend of mine told me that on several occasions, while hunting in Maine, he had been approached by poachers who offered him a buck with a rack of whatever size he could pay for. The larger the rack, the higher the price. The bucks had been shot before the season opened. My friend was a real sportsman who wouldn't dream of buying his deer—and wouldn't need to. Many hunters, after hunting all week without shooting a deer, will buy deer from poachers rather than go home empty-handed. The hunters who patronize these poachers evidently lack the intelligence to realize that they would have a better chance to take their own deer if the poachers hadn't "skimmed off the cream" before the legal season. The easiest prey for these poachers is the nonresident hunter who happens to be an urbanite, has not done much hunting, knows little about it, is easily discouraged, and—to borrow a phrase of current jargon—is more interested in an "ego trip" than a real hunting trip.

The practice of selling trophies has been common in many states, including my own. Today New Jersey has sharply curtailed this practice by requiring that no deer be moved without being tagged by the hunter and that every deer killed be taken to a checking station within 24 hours. A number of other states now employ similar means to combat this type of poaching.

A deer's eyes turn green half an hour after death, rigor mortis sets in about 3 to 4 hours later, and it takes at least 10 or 12 hours for a carcass to freeze solid, even where the weather is extremely cold. The inside of the body cavity turns dark as the meat is exposed and tends to dry out. With all of these clues, it is fairly easy to tell how long ago the deer was killed. Trophy deer can still be sold, but the deer cannot be stockpiled before the season. Trophy poachers are most apt to be local professionals.

Most commercial poachers concentrate on taking deer out of season and selling the venison—mainly to individuals and to a lesser extent to restaurants. They care little about trophies. They find it much more profitable to shoot any deer they can get in their sights.

The greatest difficulty in stopping all this is apprehending the violators. Having worked as a deputy conservation officer, I can attest to the huge number of hours of waiting or patrolling that it takes to catch a single poacher. And sometimes those hours turn out to be fruitless; that poacher escapes. Conservation officers have to be at about the top of the list of unappreciated heroes. This situation is changing and will continue to change as more hunters and the general public come to realize that the deer the poachers take are *their* deer. The active cooperation of indignant citizens is at last beginning to make life harder for the poachers.

No state has enough conservation officers to do the job properly, however hard they try. The area of Minnesota, for example, is about 84,000 square miles (217,560 sq km). To patrol that area, the state has 129 conservation officers, or one to every 650 square miles (1,683.5 sq km). Minnesota had 359,432 licensed hunters in 1975, or one officer to every 2,786 hunters. New Jersey has 35 conservation officers, or one to every 229 square miles (593.1 sq km). We had 159,475 licensed hunters in 1975, or one officer to about 4,555 hunters. Deputies help, and the conservation officers are now also getting help from the Game Protective Association. This association is made up of members of organized gun clubs who volunteer their time to assist the conservation officers whenever needed. They are entitled by state law to carry badges and make arrests. The results have been very good, and other states would find it advantageous to establish similar organizations.

Unfortunately, conservation officers can devote only a portion of their time to apprehending deer poachers. The officers in most states have to enforce all of the game laws, help with game releases, enforce the fishing laws, help with fish stocking, give protection to all protected nongame species, appear in court to prosecute cases, write up reports, and do public relations work for the division. It is a wonder that any deer poacher is ever caught.

How many deer are poached? How many poachers are caught? Naturally, there are no concrete statistics to answer the first question, but many of the states report some astonishing figures.

One state estimated that 41,000 deer were poached each year in just that state. A state in the Northeast calculated that the illegal deer kill equaled 95 percent of the legal hunter kill. A north-central state reported a 400 percent increase in deer poaching in the past decade. Most states require that a poacher be caught in the act or be found in possession of illegal game. The odds are definitely in favor of the poacher.

George Dahl, Chief of Michigan's Department of Natural Resources Law Enforcement Division, stated that in the 1973–1974 big-game season 2,763 violators were arrested. But according to Michigan statistics, 181,183 violations were probably committed, which means that arrests were made for only about 1½ percent of the violations.

To compound the problem, after a poacher is apprehended, too many lenient judges make a mockery of the laws. Unless the fines are mandated by law and are high enough to be a deterrent, being arrested is merely an inconvenience for a poacher and being fined just part of the cost of doing business. At one time, Pennsylvania confiscated any vehicle involved in poaching. *That* was a deterrent. Most states confiscate the poachers' guns, but some do not. Some states revoke the poachers' hunting privileges, but some do not. Revocation of hunting privileges is a very important deterrent and should be more widely applied, although I realize that this will not stop professional poachers any more than the registration of guns would stop professional criminals from possessing them.

Many of the older, rural judges, who remember the Depression years, are inclined to be lenient and seldom impose more than the minimum fine. As these older judges retire and younger, more conservation-minded judges are presiding at the courts, the fines are mounting. Judges who are elected locally are also reluctant to impose stiff fines on local poachers. Petitions have been circulated and publicity given to many unfair decisions of these judges. The public that elected lenient judges can vote them out of office again—if public apathy to deer poaching can be overcome. State-mandated fines now take most of this power of decision away from local judges, so the proper action should be instigated by the hunters and general public at the state level.

It is no coincidence that the states with the highest fines have the lowest rate of increase in poaching. The north-central state that had a 400 percent increase has an average $69 cost for illegally killing a deer. New Jersey has a minimum $200 fine, and deer poaching has not increased in proportion to the population.

Missouri has one of the lowest rates of poaching increase in the nation, a mere 15 percent in the past decade. It imposes an average fine of $300 per deer.

A few years ago, during a meat shortage, Missouri had an outbreak of cattle rustling as well as deer poaching. All of the Missouri law-enforcement agencies cooperated in their expenditures of money and manpower to blanket the state with posters and with notices in the papers and over the airwaves, alerting the public to this illegal killing. The public responded enthusiastically by refusing to buy illegal meat and by reporting the sellers to the authorities. The outbreak was stopped and the trend even reversed.

You can't appeal to a poacher's heart or sense of honesty, but you can hit where it hurts—in the pocketbook. Stiffer fines, confiscation of weapons, loss of hunting privileges, and public cooperation can and will reduce the poaching of deer.

The National Rifle Association supplies free wallet-sized cards that can easily be carried around. They should be filled out by anyone witnessing a game violation and then forwarded to the proper state authorities. All bona fide hunters are eager to rid their woods and fields of the unsavory characters who kill game illegally, and increasing cooperation is being shown. Through the education of all hunters and the general public, the poachers can be deprived of a market. If poachers can't sell the deer they shoot, they will soon stop shooting deer.

PROFILE OF A POACHER

Many states and provinces have done extensive research on known game violators, and a profile of the typical poacher has emerged.

The average poacher is male, 24 years old, and still single, separated from his wife, or recently divorced. He lives in a rural area, most likely a wooded area. He belongs to, or is anxious to join, a subculture in which the taking of game, by any method, at any time, is condoned. Instead of carrying a stigma with it, the act of poaching is rewarded with recognition or acceptance among peers. Poaching may be a family tradition. The typical poacher is a better hunter than the average deer hunter because he spends more time hunting.

The deer poacher is most often a "blue-collar" worker who works on a farm or a road crew, as a logger or a truck driver, or in construction. Three of these five occupations put the poacher outdoors in the vicinity of the game he poaches and on or near the land on which he carries on his poaching activities. Three-quarters of these poachers graduate from high school, but have no further education; the rest are dropouts.

Alcohol or, to a lesser extent, drugs is associated with about three-quarters of all violations. The poacher had been drinking, was drunk, or had started out from a bar. "Whiskey courage" lessens the fear of being caught, or gets the poaching started as a lark.

Although poaching is done year-round, it greatly increases from September to February, peaking in October and November. About 54 percent of all poaching occurs on the three days of the weekend, with Fridays accounting for 18.9 percent; Saturdays, 20.1 percent; and Sundays, 14.6 percent. Most of the poaching takes place between 5 P.M. and midnight, 11 P.M. being the peak.

Approximately 47 percent of those arrested for deer poaching have also been arrested for traffic violations. Some 20 percent had been arrested for drunkenness, arson, assault, breaking and entering, mugging, attempted murder, negligent homicide, manslaughter, rape, armed robbery, or drug charges. This latter group had a total disdain for any and all laws. Common feelings were that the laws were made for other people, were made to be broken, or were undue restrictions on their personal freedom. Poachers do not see themselves, much less deer, as part of a greater whole.

Not all poachers are poor in money and education. There is a certain class of poacher who is wealthy, often owning large tracts of land, and thinks he is simply above the law.

Nevertheless, the average poacher, when he worked, earned less than $14,000 a year. Many were unemployed for a part of the year; many held no regular job. Most drove pickup trucks; almost half had four-wheel-drive vehicles. Their firearms were good to better-than-average. Over 50 percent said they poached to provide meat for their families, although 48 percent had only themselves to support. Others poached for the money derived from the sale of the entire deer to sportsmen, the venison to friends and restaurants, or from the sale of trophy antlers.

Poaching Is Big Business. I decry the emphasis that has been placed on trophy antlers; we are creating a madness that is going to make it next to impossible for the average law-abiding citizen ever to have the opportunity to shoot a trophy buck. At one of my deer seminars in Wisconsin, a man showed me photographs of a set of antlers that he had just sold for $12,000. Prices of $15,000 to $20,000 are not unheard of, and an unscored first-quality whitetail head was reported in a number of sporting magazines to have been sold for $35,000. Deer with antlers like those, deer with a value like that, are just not going to be around when the legal hunting season arrives, not if a poacher has anything to do with it.

An acknowledged poacher, interviewed by *The Detroit News*, said, "This is big business. Some guys are making in excess of $50,000 each year while drawing unemployment." In 1981 Federal officers in the Midwest broke up a ring of poachers from four different states, confiscating tons of game of all kinds and game parts. The take had been in the hundreds of thousands of dollars per year.

Poaching Is Widespread. All of the states contacted reported that poaching was one of their biggest problems, that it was widespread, and that often the illegal take was much larger than the legal one.

New Mexico figured the legal take of deer in 1977 was 20,000, but estimated that the poachers had taken 34,000. Michigan figured the 1984 illegal kill by poachers topped 120,000, while the legal take was 129,200. Poaching became so heavy in the Shenandoah National Park, Virginia, that the roads had to be blocked to all after-dark traffic at certain times of the year.

In an effort to combat this criminality, Missouri set up the system described earlier, employing a concentrated effort by their law-enforcement agencies, but relying principally upon the general public. Now more than 40 states have similar programs, featuring toll-free telephone hot lines manned around the clock. Informants' identities are kept secret, and there is a cash reward if a call leads to an arrest or conviction. The success of these programs has less to do with the financial rewards than with the guaranteed anonymity. Many of the informants donate their reward money right back to the game departments.

Stiffer fines and penalties are needed. It is estimated that only 1½ to 2 percent of all poachers are apprehended. This makes the risk exceedingly small for such high financial rewards. Wisconsin has raised its fine for poaching a deer to $1,000. Other states should do likewise. As the United States becomes increasingly urbanized, as Depression-era judges retire, as the public becomes more aware of the enormity of the problem, more states will set the stiffer penalties—both in cash and in jail time—that are needed to deter violators.

Poaching Is Dangerous to All Concerned. All law officers know that they are laying their lives on the line each time they go on duty. Criminals may be armed; game protectors *know* that all deer poachers are armed. Game protectors, or conservation officers, have the most dangerous of all law-enforcement jobs, and the statistics prove it. William B. Morse, who studied the statistics on assaults made upon conservation officers over a period of 16 years, states: "The overall assault rate was 2.72 per one hundred officers during 1980–1984. At that rate, almost 82 percent of them would be assaulted during a 30-year career." And many of the assaults have ended in the death of the conservation officers.

23

Wildlife Management
and Politics

O ne of the foremost needs of the majority of fish and game departments is to be severed completely from all political connections. I do not have statistics on how many top jobs in conservation and fish and game departments are filled with political appointees, but the number is far too large. Many of the departments cannot manage their game properly because the major decisions are made not by game biologists but by the legislatures. Many of the legislators have little or no knowledge of the basics of wildlife management, and without this knowledge they are vulnerable to pressure groups.

When the top jobs in the department are filled by political appointment, qualified, competent biologists trained in game management cannot advance to positions of authority where their knowledge could be put to the best use. Instead, the qualified men are relegated to lesser jobs. Chafing under the impossibility of advancement, they often quit working for government, and for wildlife, and seek jobs in the private sector.

Wildlife-management policies must be flexible because many variables, such as weather, often change the conditions affecting the wildlife, and decisions must be made at once. Often there is no time to wait for these decisions to arrive by the slow route of legislative procedure. There are times when bad weather prevents a deer kill large enough to harvest the proper number of animals. A quick decision to extend the season will allow for this contingency. Some game departments, such as Pennsylvania's, having no political strings pulling them, are able to manage their wildlife for the benefit of the wildlife, the hunters, and the general public. But many other states are not so fortunate.

Let me quote a few examples of how politics can thwart wildlife management. The first quotation is from the August 1976 issue of *Michigan Out-of-Doors*:

"Biopolitics" killed a proposed antlerless deer hunting season in a sizable portion of the eastern Upper Peninsula this year. Biologists had recommended issuing 8,930 permits to take antlerless deer in a large area in the eastern Upper Peninsula, a proposal that would have sparked vitriolic charges at the commission meeting by doe-hunting foes. The issue was defused, however, by Department of Natural Resources Director Howard Tanner's recommendation that the Commission reject the biologists' proposal.

"The eastern Upper Peninsula biological recommendation is based on a harvestable surplus of 2,000 animals," Tanner said, "but we as a department have paid a tremendous price for pushing for antlerless seasons in the face of stiff opposition there. A harvest or not of 2,000 animals cannot be that important biologically."

In *Illinois Wildlife* for November 17, 1976, an article entitled "Capers in Conservation," by John Warren, had this to say:

As this installment was written before the November 2nd election, no conclusion can be given herein as to the course conservation may take in Illinois during the next two years.

One thing is certain, however, Illinois will very shortly have its seventh conservation director in eight years.

With this kind of political maneuvering, any real advance in resource preservation is impossible—which is exactly what those who plunder and despoil the resources for personal or corporate gain wish.

Conservation administration in Illinois has finally now sunk so low that the top executive no longer need know anything at all about fish, wildlife, forestry, law enforcement, parks, outdoor recreation, nature preserves, etc. All he need be, is a product of the Governor's office, with no background in natural resources either from an educational or an experience standpoint.

Michigan and Illinois are by no means the only states where wildlife and wildlife biologists have been victimized by politics.

Observers have frequently cited Vermont as a classic, and tragic, example. In 1968, *Vermont's Big Game and Waterfowl Review and Forecast* tersely declared that "the waste continues under legislative control." Two years later *Vermont's Game Annual* reported that "in most sections of the state, the 1970 fall deer population will exceed the size of the 1969 fall population. . . . Physical deterioration of bucks and does will certainly continue with further reductions in the already sharply retarded reproductive rate." In 1971, *Vermont's Game Annual* reviewed what had happened in the previous five years—and was continuing to happen:

> In the Review of the 1966 Big Game Seasons, Fish and Game Department Commissioner Edward F. Kehoe stated that it was "obvious" our deer herd was at a "turning point." No matter how obvious the situation was at that time, perhaps no one really knew how swiftly that point turned and how disastrous and far-reaching the results would be.
>
> That a turning point was unavoidable should have been no surprise to Vermonters. In 1963, the "crash" of the deer herd was predicted by the Department as an absolute certainty if immediate steps were not taken to correct the situation. In each of the subsequent three years, the warning was repeated in many Department publications, at hundreds of speaking engagements and through all of the news media. . . .
>
> By March of 1966, the newly reapportioned Vermont Legislature felt it was time to do something about the increasing crop and forest damage complaints and to lessen the incredible numbers of deer being wasted on the highways, to free running pet dogs and to illegal hunting. The resulting bill, effective until March 31, 1971, created an advisory Interim Committee (consisting of the Chairman and Vice Chairman of both the House and Senate Fish and Game Committees) and directed the Fish and Game Board to hold public hearings in those sections of Vermont which were overpopulated with deer. Further limitations, not included in the bill but mutually agreed upon between the Committee and the Fish and Game Board, restricted antlerless deer removal to no more than eight percent of the available antlerless population and required that proof of a seriously depraved situation must exist before such token controls could be applied.
>
> By November of 1966, Vermont had crowded more deer per square mile within its borders than had ever before been

recorded by any state. The saturation point had been achieved. . . .

An average December, followed by a mild January, lulled most of us into thinking that the winter of 1966–67 was to be another mild one. But February brought us some of the most bitter winter temperatures ever recorded in Vermont. These extremely low temperatures, coupled with deep snows, caused winter mortality to rise sharply. March dumped more than normal amounts of snow on southern Vermont and continuing subnormal temperatures brought further statewide hardship to our deer herd. In March of 1967, as Commissioner Kehoe was issuing his "turning point" statement to Vermonters, the inevitable had begun.

From March of 1967 to March of 1971, the Vermont white-tailed deer herd suffered winter mortality that, in all likelihood, exceeds any ever documented in any other state or province, or for that matter any similar land mass in the world.

Losses during a single winter in some yarding areas exceeded 200 carcasses counted per square mile and this does not include fall illegal or crippling losses. . . .

The decline of the deer herd beginning in February and March of 1967 has continued to the present day. Depending on winter severity, it could continue downward for several more years.

Some people blame the demise of the herd on "shooting does and fawns." The fact is that we have never taken more than eight percent of the antlerless herd by lawful hunting while the harvest level to stabilize herd size should be about 20 percent. . . .

Over widespread areas of Vermont, 50 percent of all fawns entering the winter yards never live to see spring. Damage in three out of four winter yards is so severe on all species of palatable and nutritious deer food that at least a decade will pass before meaningful improvements in food availability could possibly be seen. . . .

Vermont's buck kill has declined sharply since 1966, but the Department has never had control of the deer herd in Vermont.

All of the deer harvested during antlerless seasons held from 1966 to 1970 were only slightly more than the number of deer killed on the highways over the same period. Another comparison to keep things in perspective is that the deer killed by dogs during the past *two* winters of 1969–70 and 1970–71 exceeded the total antlerless harvest for the *five-year* period between 1966 and 1970. This statement is not a reflection of the effectiveness of the Department's enforcement efforts on free running pet dogs. It is the inevitable result of carrying too many half starved wintering deer that are unable to escape from this wasteful form of predation.

The only real argument that opponents to herd management have is the decline in buck kill. True, the buck kill did go down, but a closer examination of this decline will reveal that *the decline was greatest in areas where antlerless deer have not been harvested for 50 years.* . . .

Even with mild winters, the buck kill may never recover to levels enjoyed during the middle sixties. A return to higher harvest levels of deer can only be attained after a recovery in the quality of the habitat. Recovery of the habitat will take several years of intensive herd management. . . .

The Vermont Legislature now bears absolute responsibility for the future of Vermont's white-tailed deer herd. Legal authority for the Vermont Fish and Game Department to declare even token antlerless seasons has expired. . . .

Unfortunately, our deer herd will continue to suffer until we all decide to face the distasteful facts. Our overbrowsed winter yards continue to decline and will not support the numbers of deer we have known in the past. Man has removed Nature's large deer predators and *he* must assume their role in checking and balancing Vermont's deer population. . . .

Still the warnings went unheeded, and in 1972 the same publication characterized Vermont's whitetails as "one of the most underharvested and poorly managed deer herds in the country." The reason was beginning to sound like an old refrain:

Sincere Department efforts to correct the problems and meet its objectives of perpetuating a healthy deer herd met with a number of obstacles. Limited public appreciation of the situation and public acceptance of the recommended corrective measures certainly were major obstacles. The Legislature has not been willing to support the Department to the extent necessary to implement sound management practices. . . . Legislators should be asked to justify their decisions and supply evidence supporting their actions.

Did the legislators at last do so? The answer can be found in the 1975 edition of *Vermont's Game Annual*:

Findings of the spring dead deer surveys coupled with all the other studies of the Deer Project continue to dramatically point up the need for well-regulated, antlerless deer harvests in most sections of the state. The deer herd and the hunter are both getting short-changed by a lack of proper management.

The last sentence strikes home. Need anything more be said on the subject of politics in wildlife management?

The severe winter of 1977–78, after a succession of moderate winters, again took a tremendous toll on Vermont's deer herd. The decline could no longer be blamed on the game department's policy of harvesting antlerless deer, because this had been prohibited for the previous eight years.

By 1979, even the politicians realized that the Vermont deer herd was sick, and that if it was to have a future, they would have to allow it to be given the proper medicine. The biologists had the medicine, and the politicians finally had the courage to allow the medicine to be administered. In the face of a steadily declining buck harvest, the game department was allowed to increase the number of antlerless deer that could be taken in order to reduce the number of deer to the carrying capacity of the land.

In 1982, the entire deer harvest for the state was still only 9,946 animals, compared to highest post-1953 harvest of 17,000. By now it was acknowledged that the state's deer range could be improved only by reducing the entire deer population. Continued antlerless deer seasons are helping Vermont achieve this goal.

24

The Value of Our Deer

I n 1984, the estimated population of whitetail deer was 18,299,350 in the United States and 1,017,000 in Canada, for a total of 19,316,350. It was estimated that there were 3,443,900 mule deer and 1,510,000 blacktails in the United States and Canada, totaling 4,953,900. That gives us a 1984 grand total of 24,270,250 deer of all species.

What are those deer worth? There is no way to put a value on them that would be accepted by everyone, or even by most people. Deer mean many things to many people.

To people like me, their value is immense. In practical terms, they are a large part of my "bread and butter." As a wildlife photographer, I make a good part of my living from the sale of photographs of deer. I sell at least a hundred deer photos for every photo of an African lion or elephant. As a lecturer, I find that deer are one of my most popular subjects. As an author, I have chosen to make deer one of the main topics I write about.

More than my livelihood is involved. I have taken over 150,000 photos of deer, and I am just as thrilled and anxious to take my next as I was to take my first. I started taking photographs of deer as a hobby, just because I loved to watch and study deer. Gradually my hobby became my profession. If, for some reason I never sold another deer photo, I would go right on taking pictures of deer.

I am not alone in my love of photographing deer and other wildlife. In fact, I have to be careful not to be crowded out. Glen Chambers, coauthor with me of "Photographing the Whitetail" in the Wildlife Management Institute book *White-tailed Deer,* uncovered some very interesting statistics.

In 1970, there were about 4.9 million wildlife photographers nine years of age or older in the United States—2.9 percent

L.L. Rue photographing deer; note the subject's reflection in the lens. photo courtesy Len Rue, Jr.

of the national population—whose photography accounted for more than 40 million recreation days (U.S. Fish and Wildlife Service 1972). Five years later, approximately 15 million wildlife photographers—8 percent of the population—spent 156.7 million recreation days (U.S. Fish and Wildlife Service 1977).

The latest national survey (U.S. Fish and Wildlife Service 1982) revealed that in 1980, nearly 128.7 million persons older than six years of age in the United States participated in nonconsumptive outdoor activities, including more than 34.1 million who made trips specifically to observe, photograph, listen to or otherwise enjoy wildlife. In addition, almost 89 million participated in at-home activities of which the primary nonconsumptive purposes were wildlife-related. For more than 104.6 million individuals who made trips away from home for a primary purpose *other than* enjoyment of wildlife, 85.7 million reported that wildlife proved to be an important secondary nonconsumptive component of the event. And about 102.3 million Americans indicated enjoyment of unplanned opportunities to hear or see wildlife at home. There was, of course, overlap (94 percent) of numerical participants among the various primary and secondary categories. Nevertheless, it is quite obvious that, to a majority of the United States population, nonconsumptive appreciation for and use of wildlife are valuable recreational, aesthetic and/or educational experiences.

A total of 377.4 million days in 1980 were spent on trips to observe, feed or photograph wildlife, including 74.5 million days for wildlife photography alone. Approximately 5 billion days were spent by Americans observing and/or photographing wildlife around the home or neighborhood. The primary and secondary nonconsumptive users of wildlife spent about $14.7 billion in 1980, of which $10.7 billion was spent on equipment. Of the latter, $2.6 billion of the equipment expenditures was exclusively for nonconsumptive wildlife-related use. And $1.17 billion (45 percent) of the $2.6 billion was for binoculars and photo equipment. Wildlife photography, therefore, not only is a popular recreation, it is a big and growing business.

A study by Ducks Unlimited (1982) showed that, of that organization's 465,000 member/subscribers, 93.1 percent owned still-cameras and 26.1 percent owned movie cameras. Of all camera owners, including those who were deer hunters, each had used on the average about 15 units of color film and 8 units of black and white in the preceding 12-month period.

What is the esthetic value of deer to people who just want to watch them? It is beyond calculation. There are no records of the number of people who go out for a drive on a Sunday afternoon chiefly in the hope of seeing deer. Some of the backroads in my area on Sundays—and on evenings throughout the week—are crowded with cars carrying people who are looking for deer. Six times more people visit wildlife refuges to merely watch the wildlife than to hunt in those areas.

Professor Sidney Wilcox's 1976 study at Arizona State University assigned the value of $40 per day for the recreational value of hunting. This is what a hunter would be willing to pay. This is not the actual cost per day, because in some cases the figure would be high and in others far too low. The states reported that each firearm hunter averaged 5.41 days of hunting, and each bowhunter averaged 8.75 days of hunting. Multiplying 9,098,771 licensed gunhunters (in the United States in 1975) by 5.41, we can calculate 49,224,351 man/days of hunting. The man/days times $40 per day comes to $1,968,974,040. The 1,186,049 bowhunters (in 1975) multiplied by 8.75 equals 10,377,928 man/days of hunting. That, at $40 per day, comes to $415,117,120 worth of recreation. The two figures give us a total of $2,384,091,160 of recreational value to the United States for the year 1975, if we consider only hunting in the estimate.

I realize that not all of these licensees hunted for deer. Many hunted for other game. But I also realize that there are no statistics on the numbers of unlicensed landowners who hunt deer. We know that 21.4 percent of those who hunt do so legally without licenses. If we trade off the unlicensed hunters for the licensed hunters who don't hunt deer, we might well come up with approximately the same statistics given. There is no way of knowing for sure, but the estimates we do have are impressive.

Taking the known figures a step further, it can be theorized that if six times as many nonhunters as hunters use the refuges and other wild areas, not counting national parks, the total is about 62,000,000 nonhunters. There is no season on hiking, birdwatching, nature study, nature photography, or similar activities, and these people spend more days engaged in their hobbies than do hunters. If we allot only 10 days afield to each nonhunting participant in outdoor recreation (an extremely conservative estimate, since a government survey indicates that the actual figure may triple that) we can safely calculate at least 620,000,000 man/days of nonhunting outdoor recreation—plus a total of 59,602,279 man/days spent afield by hunters using both firearms and bows. If we allow the same value of $40 a day for the 620,000,000 man/days, the total value is astronomical.

I have been quoting sport-hunting statistics from only one source—the 1975 Arizona State University study—and readers may wonder if the figures are inflated. On the contrary, they are extremely conservative. At the end of 1977, the Fish and Wildlife Service of the U.S. Department of the Interior released an analysis of its own 1975 survey. In that year, according to the government survey, more than 20 million Americans hunted. Over 13 million were big-game hunters who spent over 126 million man/days engaged in their sport, and most of those people—more than 12 million—hunted deer whether or not they hunted any other game. *Of more than 2½ billion dollars spent by big-game hunters on equipment, transportation, food, and lodging, the deer hunters contributed over a billion.* And this does not include expenditures for licenses and other hunting fees.

Of course, those hunters who bring home venison get food as well as sport for the money spent, though in dollars and cents the food procured does not compare with the outlay. The ratio of successful to unsuccessful hunters varies widely with locale, but in the United States as a whole perhaps two hunters out of ten will harvest a deer in any given year. Professor Wilcox's tabulations showed that 2,269,848 mule deer were harvested in the United States in 1975. He figured that an average mule deer or whitetail would yield about 56 to 57 pounds (25.4 to 25.9 kg) of boneless meat, while an average blacktail would yield 42 pounds (19.1 kg). The total harvest in boneless meat came to 122,816,330 pounds (55,709,487.3 kg). Premium-grade ground beef was then averaging $1.15 per pound in the United States. Using that price, the total value of the harvested deer meat was calculated at $133,846,931.

As to the skins, many are wasted, many are used by the hunters themselves, and many are sold. Since the average sale price is $2 a skin, we can add something in the vicinity of $4,500,000 for deer hides. Yet the total meat and skin values are irrelevant compared with the esthetic and recreational value of deer. Even in purely practical terms, a better indication of the true value of the resource is the willingness of hunters to pay $40 per man/day without a guarantee of *any* return in meat or hide values.

In 1986, I updated Professor Wilcox's figures. In May 1986 the price of premium ground sirloin was $2.39 per pound, up from 1975's $1.15. Multiplying that by the 122,816,330 pounds (55,709,487.3 kg) gives an updated figure on a year's venison at $293,531,028.70. All other figures also had a correspondingly higher value.

But let's look again at Professor Wilcox's $40-a-day estimate. Would the *non*hunting public be equally willing to pay $40 a day for wildlife-related recreation? I hope so, but I doubt it. Up to now the nonhunting public certainly has not paid their way in supporting wildlife. They have taken a free ride, paid for by hunters in license fees, excise taxes on hunting equipment, special-permit fees, and so on.

A major factor differentiating European and American game conservation is that in Europe the game belongs to the landowner, while in the United States it belongs to each of the states—that is, to the public. Although the game belongs to the general public, the game departments that administer the protection and harvest of the game are not supported by tax dollars from the general public. Most of the wildlife refuges, most of the state game lands, most of the wildlife research and management, and most of the protection provided for any wildlife, both game and nongame species, come from the hunters' license fees. It is true that *commercial* hunting came close to pushing some of our species over the brink to oblivion, but that was not true of *sport* hunting. It was the sport hunters who brought back the deer, the ducks, the wild turkey, the pronghorn antelope, and other creatures. It was also the hunter who introduced the ringneck pheasant, chukar, gray partridge, and other foreign species that have now become part of our native fauna.

In 1975, according to Lynn Greenwalt, Director of the U.S. Fish and Wildlife Service, U.S. hunters paid $154,919,581.69 for hunting licenses. They paid an additional $51,100,000 in excise taxes on guns and ammunition. This excise is a self-imposed tax put on these items at the request of the hunters, who actively lobbied—both privately and through their various organizations—for the passage of the federal law requiring the tax. All the money thus derived goes to the individual states for wildlife work.

The hunters also paid $11,800,000 in taxes on handguns and archery equipment, of which 70 percent went to the states for research and other wildlife purposes. The remaining 30 percent went to the states for hunter-education programs.

Since 1935, when the duck stamp program was first instituted, hunters have bought over $150,000,000 worth of federal migratory-waterfowl stamps. All of this money has gone into the purchase, rental, and maintenance of refuge lands for the ducks, geese, and swans of North America, as well as countless other game and nongame species that utilize the refuges.

In 1975 alone, hunters contributed more than $228,000,000 for wildlife restoration. Since 1923, they have donated more than $2.1 billion for wildlife.

When preservationists advocate bringing back the predators to control the game so that hunting will no longer be needed, they

The Value of Our Deer

overlook the crucial fact that predators don't buy licenses, pay taxes, or make charitable donations. Without the revenue from the hunters, most of the state conservation departments would be out of business. There would be no wildlife protection, no wildlife research, no wildlife refuges—and in a short time there would not be much wildlife.

Customarily, the states' fish and game boards, which govern the direction of the fish and game departments, have been made up of officials elected by farmers and other landowners, sportsmen's clubs, and conservation groups. These were the people who were sufficiently interested in wildlife to "put their money where their mouth was." They are the ones who have always picked up the tab. Today the preservationists are demanding a voice in the affairs of wildlife and a seat on the governing boards, and they are getting what they want. That is democracy at work, but why not let them pay for the privilege? A rallying point before the American Revolution was the issue of "taxation without representation." In effect preservationists are trying to perpetrate another injustice: representation without taxation.

A number of states experimented with the sale of nonhunting wildlife stamps, but the returns were dismal. A much better approach has been for the states to retain $2–$4 from an individual's state tax refund when the appropriate box on the tax form was checked. Most of this money is earmarked for nongame research and protection. New Jersey has taken in as much as $400,000 a year. Programs like this must be encouraged. It is patently unfair for hunters to carry the burden alone.

Many misguided individuals among the general public contribute money to preservationist organizations. Before donating money to one of these groups, people should scrutinize its activities.

For example, the Los Angeles Times exposed the activities of Belton P. Mouras, founder and president of the Animal Protection Institute. Mouras's organization is the one that for a number of years flooded newspapers and magazines with photos of the worst examples of trapped animals that he could locate. The ads stated that tax-free contributions would put a stop to wildlife's suffering, but the main beneficiary of the contributions was Belton P. Mouras. In 1974, he collected about $100,000 in salary and fees from his group. According to the Times, "The Animal Protection Institute provides no direct relief to animals and does relatively little lobbying of Congress to enact animal protection laws."

In 1971, during the period when Friends of Animals stopped the deer hunting in the New Jersey Great Swamp, that tax-free organization collected $468,166. Alice Herrington, the driving force behind F.O.A., got a salary of nearly $20,000, while administration and general expenses ran about $80,000. The bulk of the money—$330,604—went to veterinarians for spaying cats and dogs. Friends of Animals did a lot of lobbying and ran many advertisements about wildlife, but I find no record of direct contributions for the purchase or maintenance of wildlife refuges or anything else of that sort.

I once was hired by a group of wealthy people who had banded together in a group called the Deer Protection Association. I went to their meeting prepared to lecture on basic deer biology and management. No! That's not what they wanted. They wanted to know how they could get rid of the deer! The deer were eating the expensive shrubbery on their estates. They didn't want to put up fences because fences were unsightly. They wouldn't allow hunters to come in to reduce the tremendous deer herd because they didn't want anyone on their property. No, I didn't have the answers to their problems. I didn't have any sympathy for them, either.

25

Public Education
and the Future

The greatest and most difficult task still confronting wildlife managers is that of educating the general public, hunters included. The state and provincial game departments have to develop better public-relations departments. The general public is not being reached; the reports and recommendations of the game biologists are not being read and understood. The game departments have to develop better liaison with the soil conservation departments, the county agricultural agents, and the forest departments. These departments have to coordinate their efforts because they are not separate entities; they are all part of the whole, and the public must be made aware of it.

I take my programs and seminars across the nation in an effort to do just that. I am proud of the fact that my seminar on deer, based on this book, is suitable for people of all ages; it is family fare. I try to share my love for and knowledge of the deer with all.

To keep hunters and the general public informed, most game departments publish game magazines. In many states these magazines are now combined with the conservation magazines of the natural resources division. This is a step in the right direction, since the combined magazines reach a much wider audience. The sorry situation is that not enough hunters and far too few other people get their states' magazines. Every game department should see that its publications get into every school in its state. Schoolchildren must be reached.

Many of the general sporting magazines are devoting additional space to the proper management of game, all game, not just deer, and this is good. Specialized magazines, such as *Deer & Deer Hunting* and *Whitetail,* concentrate on good deer management. An educated hunter is a better hunter.

More game departments should develop interesting programs and lectures that can be given in the schools and to church groups, women's clubs, garden clubs, service clubs, and fraternal organizations. Such groups always need speakers, and they are the groups that must be reached, because they probably would not seek out programs on wildlife on their own. These are the people who are constantly bombarded by preservationist groups, some of which do more harm than good, as I have pointed out. The preservationist groups are, for the most part, adept fund-raisers, and they make far better use of television, radio, magazines, and newspapers than do any of the game departments.

Spokesmen for the game departments often speak at meetings of sportsmen's clubs. That's fine. The sportsmen need to be informed too, but having game departments speak only to sportsmen's groups is like having Democrats speak only to Democrats. Far greater results can be obtained by talking to people who may not share the same viewpoint. The object of talking to more diverse groups is to give them information they did not have before and expose them to a viewpoint they may never have considered.

I know, from personal experience, that these groups can be reached, that they are interested in wildlife, that they can be taught the genuine concepts of wildlife management. I know it can be done, because I make a good part of my living doing it. I used to deliver an average of 200 lectures a year and conduct about 2 months of outdoor-education classes, mostly for school groups of all age levels. Young people, in particular, loved these programs; they soaked up all the facts about wildlife that I could give them. I never tried to make a hunter out of anyone, for that is a personal decision. But I did get students and others to understand that hunting is a most important wildlife-management tool. After all, the youth of today are the adults of tomorrow—the taxpayers and policymakers of tomorrow. To get the message across to the widest possible audience, I produce a series of family-oriented outdoors-and-nature television programs and videotapes. It is up to everyone interested in wildlife to see that the public gets the facts about wildlife. Only when the public has the facts and understands them will wildlife be managed in its own best interest and that of the general public.

Many wildlife agencies send out newsletters to all members of the Outdoor Writers Association of America and to all the newspapers within their states. But more game-department bulletins should be distributed to all the radio stations in each state. Almost all radio stations have a community bulletin board for free public-service announcements. I have never heard a game-department announcement in any state, and the game departments are missing the opportunity to reach a vast audience. Most states also have public-education television. These networks always need new material. Since the game departments need more media outlets, cooperation could be mutually beneficial. The killing of wildlife does not have to be presented. Most hunters themselves find the killing of game the least of the reasons for hunting, contrary to preservationist accusations of bloodthirstiness. Certainly the kill is necessary. It is how we harvest the game. But there is no reason to emphasize it dramatically. I think that the *American Sportsman* television series did sportsmen a disservice by sometimes emphasizing this aspect of hunting. Television programs should be factual, basic, life-history studies of wildlife and wildlife problems —and should show game managers' solutions to those problems. Hunting should be shown because hunting is one of the most important tools a game manager has.

The prestigious National Wildlife Federation has published a leaflet entitled *Should We Hunt?* in which the question is asked, "What is the role of hunting?" And it is answered:

> Thinning out of game populations has been accomplished for years by hunting.
>   Why hunting?
>   Because the federal government, all 50 state governments, all of the nation's major conservation organizations and reputable wildlife biologists recognize regulated hunting as an efficient means of reducing surplus wildlife populations. As the United States Council on Environmental Quality stated in its *Fifth Annual Report,* "Under American law and custom, sport hunting—properly regulated and based on scientific principles—is considered a legitimate management technique as well as a form of recreation."

This is the kind of basic message that must be broadcast and elucidated so that the game departments can get on with their job of managing deer and other wildlife.

In several of the previous chapters, and again in this one, I have stressed the problems of game management. Yet it would be misleading—and it is certainly not my intention—to leave you with the impression that America's deer are in a sorry plight. They are, in general, better off than they have ever been. As I stated much earlier, we probably have at least as many deer now as were present when this continent was first explored. And we certainly have more, far more, than we had in the supposedly "good old days" of the 19th century. My reasons for stressing the problems are, first, to present an accurate picture and, second, to do what I can to encourage the solution of those problems. Having done that to the best of my ability, I want to end by summarizing the prospects—the future—of America's deer.

I believe it is a bright future. Management problems notwithstanding, deer are America's most managed, and most manageable, big-game animals. They are also our most common, most widely distributed, most adaptable, most hunted, most loved, best known, and perhaps most fascinating animal. The strong universal interest in deer assures that their well-being will always be accorded a top priority.

Of course, there are and will continue to be both localized and regional setbacks. In my own area of New Jersey, a tremendous influx of new homes, with the resultant destruction of habitat, will certainly diminish the local deer population. Yet my state—the whitetail population of which was once little more than a memory—has been recording a larger deer harvest than ever before. Like most whitetail states, it should continue to support a large, healthy herd in spite of any local setbacks.

Prospects for our mule deer are not quite so bright as for whitetails and blacktails. The muley populations are declining in many regions. In Canada's prairie provinces, mule deer have been nearly wiped out. A major part of the trouble lies in the expansion of the whitetail's domain. This is not a matter of competition only but of accompanying problems—for example, the scourge of brainworms brought by whitetails. Mule deer have not developed sufficient tolerance of these devastating parasites, as whitetails have. In addition, the mule deer herds are affected by changing habitat as well as sharply increased hunting pressure in much of the West. These problems are compounded by permissive public-land policies that condone the overgrazing of public lands by domestic livestock. But mule deer are becoming much more wary animals than they once were, and are gradually evolving more adaptable traits. Through it all, they are surviving and will continue to. Moreover, as the public becomes better educated and the game departments are enabled to carry out their duties more effectively, muleys will probably increase once again in some areas and be reestablished in others.

Blacktails are already showing signs of increasing. Since the number of hunters is at an all-time high, these deer furnish proof that properly regulated sport hunting is a help rather than a hindrance to healthy deer herds. In Alaska, the blacktails are not only increasing their numbers but are also slowly expanding their range.

Since whitetails are far more widespread, they are hunted by far more sport hunters. Whitetails, too, are increasing their numbers. They are doing so in almost all parts of their range, and that range is steadily growing. I am sure that at no time have America's deer been more numerous or healthy, and in no times have their prospects been brighter.

APPENDIX I

Favored Foods

In Part II of this book, describing the life cycle and habits of deer, I discussed browsing and grazing behavior as well as some pronounced food preferences during each season. This appendix presents fuller lists of preferred foods, although no such enumeration can be truly complete. A particular species of plant may be abundant and strongly preferred in a particular locale, yet the same plant may be scarce or absent in some or most other locales. If a plant suddenly proliferated in a new locale, it might or might not quickly achieve favored status. For example, in wilderness areas where the animals have never learned what corn is, even very hungry deer tend to ignore corn if it is put out for them. But in farm country, corn is a favorite deer food. I have tried to list the foods in an order of preference, but the order is very rough indeed, since food preferences vary so sharply from one locale to another. A number of foods, such as cottonwood, mesquite, and Douglas fir, obviously have only a regional availability, though they are extremely important where abundant. Certain plants are favorable in all seasons, while others are not. In some cases (particularly with regard to mule deer, which migrate seasonally between the high country and the lowlands) this may reflect the elevations at which a plant grows. In other cases, a plant may be dormant or absent during the colder part of the year. And in some cases, though a plant is available throughout the year, it is more nutritious or tastier to deer at one stage of growth than at another.

COMMON SPRING FOODS

Whitetail	Mule Deer	Blacktail
May hawthorn	sunflower	grasses
clover	fescuegrass	sedges
alfalfa	bluegrass	horsetail
cinquefoil	bromegrass	bracken
crabapple	ragweed	trailing blackberry
teaberry	kohleria	fireweed
trailing arbutus	needlegrass	red elder
greenbrier	wheatgrass	thimbleberry
dandelion	ricegrass	salmonberry
plantain	goldeneye	salal
corn	mountain mahogany	willow
wild strawberry	silktassel	cedar
trefoils	oak	deerbrush
lespedeza	manzanita	chamise
aster	wild oats	live oak
sunflower	dogwood	scrub oak
pokeweed		mountain mahogany
jewelweed		Douglas fir
poison ivy		
New Jersey tea		
bitterbush		
serviceberry		
red maple		
Japanese honeysuckle		
sassafras		
willow		
speedwell		
blueberry		
big and little bluestem		
curly mesquite		
tall dropseed		
magnolia		
Yaupon holly		
bigleaf gallberry		

COMMON SUMMER FOODS

Whitetail	Mule Deer	Blacktail
red maple	deerwitch	salal
striped maple	fescuegrass	black raspberry
blueberry	pine	red alder leaves
blackberry	eriogonum	willow
greenbrier	serviceberry	bracken
alfalfa	oak	grasses
corn	bluegrass	sedges
dogwood	kohleria	thimbleberry
swamp ironwood	needlegrass	vine maple
ferns	wheatgrass	salmonberry
wild rose	ricegrass	fireweed
mushrooms	wild cherry	red elder
bluegrass	mountain mahogany	sow thistle
bearberry	silktassel	figwort
wheatgrass	gramagrass	mushrooms
sassafras	mushrooms	trailing blackberry
wild grape	lupine	huckleberry
chestnut oak	knotweed	plantain
pokeweed	thimbleberry	clover
sunflower	elderberry	pearly everlasting
blackeyed Susan	dogwood	yarrow
crabapple	mesquite	rose
soybean		interior live oak
wild hydrangea		buckeye
elderberry		chokecherry
jewelweed		chamise
aster		honeysuckle
sumac		poison oak
cabbage palm		foothill ash

COMMON AUTUMN FOODS

Whitetail	Mule Deer	Blacktail
acorns	creeping barberry	acorns
oxalis	bearberry	filaree
plains lovegrass	snowberry	bromegrass
whorled nodviolet	snowbush	manzanita
mat euphorbia	jack pine	chamise
arrowleaf sida	sunflower	scrub oak
maple	sagebrush	deer brush
sweetfern	pine	wavyleaf ceanothus
willow	cedar	buckeye
wintergreen	mountain mahogany	wild grape
grasses	cliffrose	ferns
oak	poplar	toyon
wild cherry	needlegrass	poison oak
lespedeza	gramagrass	western chokeberry
snowberry	paintbrush	chaparral pea
greenbrier	bitterbush	wormwood
blackgum	rabbitbrush	trailing blackberry
creeping blueberry	wild cherry	plantain
holly	fescuegrass	vine maple
live oak	manzanita	annual agoseries
persimmon	eriogonum	red alder
snakeweed	quaking aspen	huckleberry
wheatgrass	sedge	salmonberry
honeysuckle	serviceberry	clover
aster	hackberry	
goldenrod	arrowleaf sida	
pussytoes	acorns	
palmetto berries		
mushrooms		
teaberry		
sumac		
blueberry		
coralberry		
sassafras		
witch hazel		
crabapple		
dogwood		
wild rose		
wild grape		
clover		
elderberry		
bittersweet		
red raspberry		

COMMON WINTER FOODS

Whitetail		Mule Deer	Blacktail
red maple	bearberry	creeping barberry	Douglas fir
striped maple	wild rose	bearberry	trailing blackberry
witch hazel	aspen	snowberry	red huckleberry
sumac	Oregon grape	ceanothus	yew
blueberry	spruce	sagebrush	madrona
hemlock	white birch	jack pine	salal
willow	sassafras	Douglas fir	sword fern
white pine	crabapple	rabbitbrush	vine maple
viburnums	Japanese	cedar	manzanita
yellow birch	honeysuckle	mountain	chamise
ash	apple	mahogany	red cedar
wintergreen	coralberry	bitterbrush	usnea
fir	honey locust	fendlera	moss
white cedar	lady's-tobacco	Sierra juniper	bracken
poplar	plantain	scrub oak	yerba santa
oaks	strawberry	fescuegrass	scrub oak
lespedeza	speedwell	sedge	buckbrush
snowberry	hawthorn	wild oats	toyon
blackgum	poison ivy	bromegrass	chaparral pea
greenbrier	mints	creek dogwood	California laurel
dogwood	goldenrod	mesquite	live oak
swamp ironwood	pussytoes	cliffrose	coffeeberry
live oak	aster	holly-leaf buckthorn	filaree
persimmon	teaberry	turbinella oak	acorns
snakeweed	acorns	mountain misery	wavyleaf
		stonecrop	ceanothus
		elderberry	
		redberry	
		California buckeye	
		antelope brush	
		black oak	
		Pacific serviceberry	
		bitter cherry	
		western chokecherry	
		tesota	
		velvet elder	
		sunflower	
		cottonwood	

APPENDIX II

Deer Activity Cycles

Although deer are almost as individualistic as humans, they tend to follow a fairly clear pattern of activity each day. The pattern revolves around their times of feeding and movement between places where they bed down and where they browse. By keeping these patterns in mind, photographers and deer watchers have much better luck in finding their quarry.

DAILY ACTIVITY CYCLES

In general, deer are crepuscular animals; that is, their periods of greatest activity are at dawn and dusk. Yet deer have become much more nocturnal than in centuries past because of pressure from humans. This is important to remember, because deer in completely protected areas are much more active in daytime. In fact, under normal circumstances, deer have five main periods of activity each day.

The accompanying chart illustrates these periods graphically. Both the chart and the comments that follow are based on decades of close observation. Unless otherwise specified, patterns of activity are for deer that are not completely protected. These are the deer the reader is most likely to encounter.

The chart shows deer activity periods (standard time, not daylight saving time), including the approximate duration of each. These should be taken only as rough guides and are appropriate for the period from late March through October. Horizontal straight lines in the "valleys" between activity peaks show the time deer spend resting, sleeping, chewing their cuds, and moving to and from their feeding areas.

The times shown apply especially to does, fawns, and one- and two-year-old bucks. Mature bucks with large racks tend to appear at just about sunset, so that there is only half an hour of daylight left. In the morning the older bucks are heading for their beds at first light and are usually gone by sunrise, 30 minutes later. During hunting season, the biggest trophy bucks tend to move only at night, unless they are forced from cover.

Most people know about the dawn and dusk feeding periods but are unaware of the other three active periods. These three periods are hard to detect because activity is usually near the deer's bedding area. But deer are definitely moving around and feeding. This means that there are usually three additional opportunities to watch and photograph deer, once a bedding area has been located.

Deer activity is most readily seen and photographed when deer are in open areas, such as orchards, fields, or open woodlands. But that does not guarantee that the deer will always be in an open area during active periods shown on the chart. The distance that deer bed from their feeding areas will determine when they show up. In addition, deer usually browse as they move to feeding areas. If the distance is over a half-mile (1 km), it may take the deer a half hour to complete the journey.

Daily Activity Graph

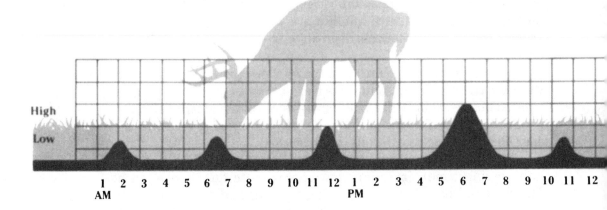

MONTHLY LOG OF ACTIVITY

Deer follow a pattern throughout the year, each month being a little different from the previous month. Bucks have a different pattern from that of does. Deer have periods when they are active only in the woods, where they are seldom seen, which will also be noted. The times given are for northwestern New Jersey and must be altered for your particular area.

My grandson, Nicolas Rue, with a whitetail fawn; I'm starting him out young. Most people retain their curiosity about wildlife throughout life.

January. By now, hunting seasons in most states are over. In states that allow an early January bow season, deer remain wary until the end of the month. Where the season ends in December, deer feed in fields until 8:00 A.M. Most bucks have lost their antlers, and so are indistinguishable from does at a distance. In the middle-latitude states, cold weather drives deer to the south-facing slopes, where the sun warms them and allows less snow buildup. In the afternoon the deer will be out about 5:00 P.M., but darkness settles in 30–45 minutes later. There is usually about a half hour of light before sunrise and after sunset.

February. This is usually the hardest month on deer. Food shortages force deer to spend more time feeding in the daytime. It is not unusual, now, to see deer feeding as late as 10:00 A.M. They may also be out in the afternoon as early as 4:00 P.M. In the northern tier of states, deer will yard up if the snow is deep. This means they will not be seen at all, except in the yards.

March. March can go either way; some years temperatures are low, while in others the weather moderates and green sprouts appear along stream banks and on hillsides. Deer may stay in fields and open areas during midday because there is little pressure from humans. In all areas deer may be seen at any time of the day because food is still scarce. Even in the north, the snowmelt is usually sufficient to allow deer to leave yarding areas.

April. This month brings the first real warmth and the first real growth of vegetation. The deer are still ravenous: they feed in the open till at least midmorning and start again in the afternoon at about 3:00 P.M. The bucks' antlers start to grow about the middle of the month. This makes the bucks more reclusive, and they restrict their activities to the smallest possible area to avoid injuring their developing antlers.

Daylight saving time usually starts in April, so although the deer don't change their times, we see them one hour later each morning. Sunrise now occurs about 5:45 A.M. and the sun sets about 8:15 P.M. (EST).

May. The deer are still in their winter coats and often stay in the shade to keep cool during warm days. Bucks continue to be inactive. But the middle of May to the middle of June is the peak of the birthing activity. Does may be seen at almost any time of the day. Prior to giving birth, the does drive off or lose their yearlings. These frantic young are frequently seen, and often killed, on highways.

June. This is a poor month for deer observation and photography. Bucks are still in seclusion. Does that have given birth become secretive. Yearling bucks move off on their own, and yearling does temporarily go off on their own and then rejoin the older does when the new fawns are about a week old. The sun rises about 5:30 A.M. and sets about 8:35 P.M., and from the summer solstice on, the days gradually become shorter. Deer shed the last of their winter hair early in June and are now in their red coats.

July. Deer observation picks up again. When the fawns are one week old, they begin to follow the doe around. Yearling females now often rejoin their mothers. If they, too, have given birth, they bring their young to join the female groupings, definitely a matriarchal society. The weather is hot and deer will be out from about 5:30 A.M. until shortly after 7:00 A.M. Deer seek north-facing slopes in the daytime for coolness. The bucks' antlers are almost full grown, and the bucks become more active. None of the deer venture out in the afternoon until shadows stretch across the fields and the air becomes cooler.

August. This is the hottest month. Days are getting shorter: the sun rises about 6:15 A.M. and sets about 8:00 P.M. All midday activity takes place in the deepest woods, where it is too dark for photography unless you are using the faster films. Most fawns lose their spots by the end of the month. Bucks' antlers are full grown and have solidified; they no longer have to be careful about antler injury. Evenings may cool off, and deer may go out to feed about 6:00 P.M.

L.L. Rue wearing the photo vest he designed for field photography.

September. This month brings warm days with cool nights, and all of the animals respond to the change. Deer can now be seen in the morning from sunrise at 6:30 A.M. until a little after 7:00 A.M. In the evening they feed from about 5:00 P.M. until dark. Between the fifth and the fifteenth, the bucks rub the velvet from their antlers, and now "buck rubs" on saplings and small trees are increasingly common. Deer are active in daylight hours until the bowhunting season opens. When there is hunting pressure, deer feed later in the morning and earlier in the evening on Wednesday and Thursday nights. That is because weekends bring out hunters in the greatest numbers.

October. During the last week in September and the first two weeks in October, all of the deer seem virtually to disappear. They abandon their regular haunts and trails; they stop feeding on corn, soybeans, and alfalfa, or in the orchards. That is because over most of North America, wherever oaks are found, the acorns are dropping. In many states the gunhunting season opens, and this increased pressure makes the deer generally more secretive. Where the gun season has not opened, most bowhunters have either gotten their deer or are hunting a lot less. Sunrise is now about 7:15 A.M. and sunset is about 5:15 P.M. At the end of the month, daylight saving time is over. About the time the clocks change, so do the deer's activities.

November. The last week and a half in October, all of November, and the first two or three weeks in December are the rutting season, when bucks are most active. Their necks are now swollen, and the mating urge causes them to expand their range. Except for the biggest, oldest trophy bucks, bucks throw caution to the wind and may be seen at any time of the day. Small-game season opens in almost all states in November, and deer-hunting season starts in many. This pressure keeps does and their young out of sight during most of the daylight hours.

The breeding season peaks from November 10 to November 20, although mating takes place for several weeks before and up to a month and a half after those dates. As the does' estrus approaches, they become more active in order to attract bucks. If you are ever fortunate enough to photograph bucks fighting, it will probably be during this peak period.

December. This is usually a poor month to see deer and is even worse for photography. Days are the shortest of the year, with the sun rising about 7:15 A.M. and setting about 4:35 P.M. The deer population has been greatly reduced, with 20 to 30 percent having been harvested by hunters. Deer that haven't been shot have been shot at; so almost all deer activity now takes place under the cover of darkness.

BIBLIOGRAPHY

Abell, David H., and Fredrick F. Gilbert. "Nutrient Content of Fertilized Deer Browse in Maine." *Journal of Wildlife Management,* vol. 38, no. 3, p. 517, 1974.

Alcock, John. *Animal Behavior.* Sunderland, Massachusetts: Sinauer Association, Inc., 1975.

Alexander, Bobby G. "Movements of Deer in Northeast Texas." *Journal of Mammalogy,* vol. 32, no. 3, p. 618, 1968.

Allen, Ross E., and Dale R. McCullough. "Deer-Car Accidents in Southern Michigan." *Journal of Wildlife Management,* vol. 40, no. 2, p. 317, 1976.

Alsheimer, Charles J. "The Oldest Whitetail on Record." *Deer and Deer Hunting Magazine,* vol. 8, no. 4, p. 50–53, November 1985.

Anderson, Dennis. "Duluth Man's Record Buck Something to Shoot For." *Pioneer Press* [St. Paul, Minnesota], November 12, 1982.

Arizona Fish and Game Department. *Arizona Deer Management Information Performance Reports.* June 30, 1975.

Armstrong, Edward, David Euler, and Gerald Racey. "Winter Bed-site Selection by White-tailed Deer in Central Ontario." *Journal of Mammalogy,* vol. 47, no. 3, p. 880–883, 1983.

Armstrong, Ruth Allison. "Fetal Development of the Northern White-tailed Deer." *The American Midland Naturalist,* vol. 43, no. 3, p. 650, May 1950.

Atkeson, Thomas D., and R. Larry Marchinton. "Forehead Glands in Whitetailed Deer." *Journal of Mammalogy,* vol. 63, no. 4, p. 613–617, 1982.

Bahnak, B. R., et al. "Seasonal and Nutritional Influences on Growth Hormone and Thyroid Activity in White-tailed Deer." *Journal of Wildlife Management,* vol. 45, no. 1, p. 140–147, 1981.

Bailey, William, Jr., George Schildmam, and Phillip Agee. *Nebraska Deer.* Lincoln, Nebraska: Nebraska Game, Forestation and Parks Commission, 1957.

Behrend, Donald F., and Robert A. Lubeck. "Summer Flight Behavior of White-tailed Deer in Two Adirondack Forests." *Journal of Wildlife Management,* vol. 32, no. 3, 1968.

Bellis, E. D., and H. B. Graves. "Deer Mortality on a Pennsylvania Interstate Highway." *Journal of Mammalogy,* vol. 35, no. 2, p. 232, 1971.

Benoit, Larry, and Peter Miller. *How to Bag the Biggest Buck of Your Life.* Duxbury, Vermont: The Whitetail Press, 1975.

Benton, Allen H. "A Three-Horned Buck." *New York Fish and Game Journal,* vol. 25, no. 2, p. 181, July 1978.

Bersing, Otis S. *A Century of Wisconsin Deer.* Madison, Wisconsin: Wisconsin Conservation Department, 1956.

Biehn, Earl R. *Crop Damage by Wildlife in California.* Sacramento, California: California Department of Fish and Game, 1951.

Bolte, John R., Jakie A. Hair, and Joe Fletcher. "Whitetailed Deer Mortality Following Tissue Destruction Induced by Lone Star Ticks." *Journal of Wildlife Management,* vol. 34, no. 3, p. 546, 1970.

Boone and Crockett Club. *Records of North American Big Game.* 9th ed. Dumfries, Virginia: 1988.

Brokx, P. A. "The Superior Canines of Odocoileus and Other Deer."*Journal of Mammalogy,* vol. 53, no. 2, p. 359, 1972.

Brothers, Al, and Murphy E. Ray, Jr. *Producing Quality Whitetails.* 2nd ed. Laredo, Texas: Wildlife Service, 1982.

Brown, Ellsworth Reade. "The Black-tailed Deer in Western Washington." *Biological Bulletin #13,* Washington State Game Department, 1961.

Brown, Robert D., ed. *Antler Development in Cervidae.* Kingsville, Texas: Caesar Kleberg Wildlife Research Institute, Texas A. and I. University, 1983.

Bubenik, George A. "The Endocrine Regulation of the Antler Cycle." In *Antler Development in Cervidae,* edited by Robert D. Brown. Kingsville, Texas: Caesar Kleberg Wildlife Research Institute, Texas A. and I. University, 1983.

Burkhardt, Dietrich, Wolfgang Schleidt, and Helmut Altner. *Signals in the Animal World.* New York: McGraw-Hill, 1966.

Cadieux, Charles L. *Wildlife Management on Your Land.* Harrisburg, Pennsylvania: Stackpole Books, 1985.

Calhoun, John, and Loomis, Forrest. *Prairie Whitetails.* Springfield, Illinois: Illinois Department of Conservation, 1974.

Carbaugh, B., J. P. Vaughan, E. D. Bellis, and H. B. Graves. "Distribution and Activity of White-tailed Deer Along an Interstate Highway." *Journal of Wildlife Management,* vol. 39, no. 3, p. 570, 1975.

Cartier, John O. "How Poachers Make a Fool of You." *Outdoor Life,* pp. 62, December 1975.

Caton, John Dean. *The Antelope and Deer of America.* New York: Hurd and Houghton, 1877.

Cheatum, E. L., and Glenn H. Morton. "Breeding Season of White-tailed Deer in New York." *Journal of Mammalogy,* vol. 10, no. 3, p. 249, 1946.

Cook, R. S., Marshall White, D. O. Trainer, and W. C. Glazner. "Mortality of Young White-tailed Deer Fawns in South Texas." *Journal of Wildlife Management,* vol. 35, no. 1, p. 47, 1971.

Cowan, Ian McTaggart, and Valerius Geist. "Aggressive Behavior in Deer of the Genus Odocoileus." *Journal of Mammalogy,* vol. 42, no. 4, p. 523, 1961.

————. "Hybridization Between the Blacktailed Deer and the White-tailed Deer." *Journal of Mammalogy,* vol. 43, no. 4, p. 539, 1962.

Dahlberg, Burton L., and Ralph C. Guettinger. "The White-tailed Deer in Wisconsin." *Technical Wildlife Bulletin #14,* Wisconsin Conservation Department, 1956.

Daniel, Walton S. "Travels of Post Oak Whitetails." *Texas Parks and Wildlife Magazine,* October 1973.

Darwin, Charles. *The Expression of the Emotions in Man and Animals.* New York: Philosophical Library, 1955.

Dasman, William. *Deer Range: Improvement and Management.* Jefferson, North Carolina: McFarland and Co., Inc., 1981.

————. *If Deer Are to Survive.* A Wildlife Management Institute Book. Harrisburg, Pennsylvania: Stackpole Books, 1971.

Davidson, William R., et al. *Diseases and Parasites of the Whitetailed Deer.* Athens, Georgia: College of Veterinary Medicine, University of Georgia, 1981.

Davis, Jerry W. "Deer Ked Infestation on White-tailed Deer in East Texas." *Journal of Wildlife Management,* vol. 37, no. 2, p. 183, 1973.

Dean, Donald J. "Streptothricosis—A New Deer Disease Transmissible to Man." *The Conservationist,* October–November 1961. •

Dean, R. E., et al. "Reticulo-Rumen Characteristics of Malnourished Mule Deer." *Journal of Wildlife Management,* vol. 39, no. 3, p. 601, 1975.

DeCalesta, David S., Julius G. Nagy, and James A. Bailey. "Some Effects of Starvation on Mule Deer Rumen Bacteria." *Journal of Wildlife Management,* vol. 38, no. 4, p. 815, 1974.

————. "Starving and Refeeding Mule Deer." *Journal of Wildlife Management,* vol. 39, no. 4, p. 663, 1975.

de Nahlik, A. J. *Wild Deer.* London: Faber and Faber, Ltd., 1959.

————. *Deer Management.* London: David and Charles, 1974.

Dickinson, Nathaniel R. "Observations on Steep-Slope Deer Wintering Areas in New York and Vermont." *New York Fish and Game Journal,* vol. 23, no. 1, 1976.

Dickson, John D., III. "An Ecological Study of the Key Deer." *Technical Bulletin #3,* Florida Game and Fresh Water Fish Commission, 1955.

————. *Deer Management.* London: David and Charles, 1974.

Dietz, Donald, R., and James R. Tigner. "Evaluation of Two Mammal Repellents Applied to Browse Species in the Black Hills." *Journal of Wildlife Management,* vol. 32, no. 1, p. 109, 1968.

Doutt, J. Kenneth, and John C. Donaldson. "Antlered Doe Study." *Pennsylvania Game News,* November 1961.

Duvendeck, Jerry P. "The Value of Acorns in the Diet of Michigan Deer." *Journal of Wildlife Management,* vol. 26, no. 4, p. 371, 1962.

Ellisor, John E. "Mobility of White-tailed Deer in South Texas." *Journal of Mammalogy,* vol. 33, no. 1, p. 220, 1969.

Elman, Robert. *All About Deer Hunting in America.* New York: Winchester Press, 1976.

———. *The Hunter's Field Guide to the Game Birds and Animals of North America.* New York: Knopf, 1974.

Fay, L. D. "A Two-Headed White-tailed Deer Fetus." *Journal of Mammalogy,* vol. 41, no. 3, p. 41, 1960.

Forand, Kenneth J., R. Larry Marchinton, and Karl V. Miller. "Influence of Dominance Rank on the Antler Cycle of White-tailed Deer." *Journal of Mammalogy,* vol. 66, no. 1, p. 58–62, 1985.

Forbes, Stanley E. "Diseases and Parasites of the Pennsylvania White-tailed Deer." *Pennsylvania Game News,* December 1961.

Forbes, Stanley E., Lincoln M. Lang, Stephen A. Liscinsky, and Harvey A. Roberts. *The White-tailed Deer in Pennsylvania.* Harrisburg, Pennsylvania: Pennsylvania Game Commission, 1971.

Freddy, David J. "Heart Rates for Activities of Mule Deer at Pasture." *Journal of Wildlife Management,* vol. 48, no. 3, p. 962–969, 1984.

Freddy, David J., Whitcomb M. Bronaugh, and Marten C. Fowler. "Responses of Mule Deer to Disturbances by Persons Afoot and Snowmobiles." *Wildlife Society Bulletin,* vol. 14, no. 1, p. 63–70, Spring 1986.

French, C. E., et al. "Responses of White-tailed Bucks to Added Artificial Light." *Journal of Mammalogy,* vol. 41, no. 1, p. 23, 1960.

Garland, Lawrence E. *A Summary of Known Deer Losses to Causes Other Than Legal Hunting in Vermont 1938–1974.* Montpelier, Vermont: Vermont Department of Fish and Game, 1975.

Gill, Gerald B. *Montana Deer Nutrition Report Study Number 48.02.* Helena, Montana: Montana Fish and Game Department, 1972.

Glazener, W. C. "An Unusual Antler-Drop Schedule in the White-tailed Deer." *Journal of Mammalogy,* vol. 50, no. 1, p. 156, 1969.

Golley, Frank B. "Gestation Period, Breeding and Fawning Behavior of Columbian Black-tailed Deer." *Journal of Mammalogy,* vol. 38, no. 1, p. 116, 1957.

Goss, Richard J. *Deer Antlers: Regeneration Function and Evolution.* New York: Academic Press, 1983.

Gray, James. *Animal Locomotion.* New York: Norton, 1968.

Gruell, George E., and Nick J. Papez. "Movements of Mule Deer in Northeastern Nevada." *Journal of Wildlife Management,* vol. 27, no. 3, p. 414, 1963.

Gunderson, Harvey L. *Mammalogy.* New York: McGraw-Hill Book Company, 1976.

Hair, J. Alexander. "Ticks Can Kill." *Outdoor Oklahoma,* vol. XXIV, no. 11, December 1968.

Hall, E. Raymond. "The Deer of California." *California Fish and Game,* vol. 13, no. 4, October 1927.

Hall, Michael H., and Edward C. Hall. "An Early-born Fawn in New York." *New York Fish and Game Journal,* vol. 27, no. 1, January 1980.

Halls, Lowell K., ed. *Whitetailed Deer Ecology and Management.* A Wildlife Management Institute Book. Harrisburg, Pennsylvania: Stackpole Books, 1984.

Hamlin, Kenneth L., et al. "Relationships Among Mule Deer Mortality, Coyotes, and Alternate Prey Species During Summer." *Journal of Wildlife Management,* vol. 48, no. 2, p. 489–499, 1984.

Hansen, Christopher S. "Costs of Deer-Vehicle Accidents in Michigan." *Wildlife Society Bulletin,* vol. 11, no. 2, Summer 1983.

Harder, John D., and Tony J. Peterle. "Effect of Diethylstilbestrol on Reproductive Performance of White-tailed Deer." *Journal of Wildlife Management,* vol. 38, no. 2, p. 183, 1974.

Harrison, Paul D., and Melvin I. Dyer. "Lead in Mule Deer Forage in Rocky Mountain National Park, Colorado." *Journal of Wildlife Management,* vol. 63, no. 4, p. 613–617, 1982.

Hartman, Fred E. "Hunting is Big Business." *Pennsylvania Game News,* February 1976.

Haugen, Arnold O. "Reproductive Performance of Whitetailed Deer in Iowa," *Journal of Mammalogy,* vol. 56, no. 11, p. 151, 1975.

Haugen, Arnold O., and L. S. Davenport. "Breeding Records of White-tailed Deer in the Upper Peninsula of Michigan." *Journal of Wildlife Management,* vol. 14, no. 3, p. 290, 1950.

Hawkins, R. E., and W. D. Klimstra. "A Preliminary Study of the Social Organization of White-tailed Deer." *Journal of Wildlife Management,* vol. 34, no. 2, p. 407, 1970.

Healy, William M. "Forage Preferences of Tame Deer in a Northwest Pennsylvania Clear-cutting." *Journal of Wildlife Management,* vol. 35, no. 4, p. 717, 1971.

Hesselton, William. "A Wooly-coated White-tailed Deer from New York State." *Journal of Mammalogy,* vol. 47, no. 1, p. 154, 1966.

————. "The Incredible White Deer Herd." *The Conservationist,* October–November 1969.

Hickman, Cleveland P., Jr., Larry Roberts, and Frances M. Hickman. *Integrated Principles of Zoology.* 7th ed. St. Louis, Missouri: Times Mirror/Mosby, 1984.

Hickman, Cleveland P., Sr., Cleveland P. Hickman, Jr., and Frances M. Hickman. *Integrated Principles of Zoology.* 5th ed. St. Louis, Missouri: C. V. Mosby Company, 1974.

Hine, Ruth L., and Susan Nehls, ed. *White-tailed Deer Population Management in the North Central States.* Urbana, Illinois: North Central Section of The Wildlife Society, 1980.

Hirth, David H. *Social Behavior of White-tailed Deer in Relation to Habitat.* Wildlife Monographs, no. 53, The Wildlife Society, Washington, DC: April 1977.

Hlavachick, Bill. "Wanderlust." *Kansas Fish and Game Magazine,* vol. 29, no. 3, p. 8, 1972.

Hoff, G. L., S. H. Richards, and D. O. Trainer. "Epizootic of Hemorrhaging Disease in North Dakota Deer." *Journal of Wildlife Management,* vol. 37, no. 3, p. 331, 1973.

Hoffman, Roger A., and Paul F. Robinson, "Changes in Some Endocrine Glands of White-tailed Deer as Affected by Season, Sex and Age." *Journal of Mammalogy,* vol. 47, no. 2, p. 266, 1966.

Horner, Kent. *Art and Science of Whitetail Hunting.* Harrisburg, Pennsylvania: Stackpole Books, 1986.

Hornocker, M. G. "An Analysis of Mountain Lion Predation, Wildlife Mongraphs, no. 21, The Wildlife Society, D.C.,: March 1970.

Hudson, Paul, and Ludvig G. Browmam. "Embryonic and Fetal Development of the Mule Deer." *Journal of Wildlife Management,* vol. 23, no. 3, p. 295, 1959.

Huegel, Craig N., Robert Dahlgren, and H. Lee Gladfelter. "Mortality of White-tailed Deer Fawns in South-central Iowa." *Journal of Wildlife Management,* vol. 49, no. 2, p. 377–380, 1985.

Husar, John. "Short Fuse of Deer Population Bomb." *Chicago Tribune,* December 2, 1984.

Jackson, Lawrence W., and William T. Hesselton. "Breeding and Parturition Dates of White-tailed Deer in New York." *New York Fish and Game Journal,* vol. 20, no. 1, p. 40, January 1973.

Jacobsen, Nadine K. "Alarm Bradycardia in White-tailed Deer Fawns." *Journal of Mammalogy,* vol. 60, no. 2, p. 343–349, 1979.

James, M. R. *Successful Bowhunting.* Fort Wayne, Indiana: Blue-J Books, 1986.

Jenkins, David H., and Ilo H. Bartlett. *Michigan Whitetails.* Lansing, Michigan: Michigan Department of Conservation, 1959.

Kammermeyer, K. E., and R. Larry Marchinton. "Notes on Dispersal of Male White-tailed Deer." *Journal of Mammalogy,* vol. 57, no. 4, p. 776, 1976.

Karns, Patrick D. "Pneumostrongylus Tenuis in Deer in Minnesota and Implications for Moose." *Journal of Wildlife Management,* vol. 31, no. 2, p. 295, 1967.

Kie, John C., Marshall White, and D. Lynn Drawe. "Condition Parameters of White-tailed Deer in Texas." *Journal of Wildlife Management,* vol. 47, no. 3, p. 583–594, 1983.

Klein, David R., and Sigurd T. Olson. "Natural Mortality Patterns of Deer in Southeast Alaska." *Journal of Wildlife Management,* vol. 24, no. 1, p. 80, 1960.

Knowlton, Frederick F., and W. G. Glazener. "Incidence of Maxillary Canine Teeth in White-tailed Deer from San Patricio County, Texas." *Journal of Mammalogy,* vol. 46, no. 2, p. 352, 1965.

Korschgen, Leroy J. "Foods of Missouri Deer with Some Management Implications." *Journal of Wildlife Management,* vol. 26, no. 2, p. 164, 1962.

Kramer, August. "Interspecific Behavior and Dispersion of Two Sympatric Deer Species." *Journal of Wildlife Management,* vol. 37, no. 3, p. 288, 1973.

Krefting, Lauritsu, and Henry L. Hansen. "Increasing Browse for Deer by Aerial Applications of 2,4-D." *Journal of Wildlife Management,* vol. 33, no. 4, p. 784, 1969.

La Gory, Kirk E. "Diurnal Behavior of a White-tailed Deer Neonate." *Journal of Wildlife Management,* vol. 44, no. 4, p. 927–929, 1980.

La Ha, Art. "Don't Waste Wounded Deer." *Outdoor Life,* p. 74–77, December 1978.

Lambiase, J. T., Jr., R. P. Altmann, and J. S. Lindzez. "Aspects of Reproductive Physiology of Male White-tailed Deer." *Journal of Wildlife Management,* vol. 36, no. 3, p. 868, 1972.

Lay, Daniel W. *The Importance of Variety to Southern Deer.* Nacogdoches, Texas: Texas Park and Wildlife Divison, 1964.

Leopold, A. Starker, Thane Riney, Randal McCain, and Lloyd Tevis, Jr. "The Jawbone Deer Herd." *Game Bulletin #4,* California Division of Fish and Game, 1951.

Linsdale, Jean M., and P. Quentin Tomich. *A Herd of Mule Deer*. Berkeley and Los Angeles: University of California Press, 1953.

Liscinsky, Stephen A., Charles T. Cushwa, Michael J. Puglisi, and Michael Ondik. "What Do Deer Eat?" *Pennsylvania Game News*, May 1973.

Loft, Eric R., and John W. Menke. "Deer Use and Habitat Characteristics of Transmission-Line Corridors in Douglas-Fir Forest." *Journal of Wildlife Management*, vol. 48, no. 4, p. 1311–1316, 1984.

Lolenosky, George B. "Wolf Predation on Wintering Deer in East-Central Ontario," *Journal of Wildlife Management*, vol. 36, no. 2, p. 357, 1972.

Loveless, Charles M. "The Everglades Deer Herd Life, History and Management." *Technical Bulletin #6*, Florida Game and Fresh Water Fish Commission, 1959.

Madson, John. *The White-tailed Deer*. East Alton, Illinois: Olin Mathieson Chemical Corporation, 1961.

Madson, John, George H. Haas, Chuck Adams, Dwight Schuh, and Leonard Lee Rue, III. *The Outdoor Life Deer Hunter's Encyclopedia*, Danbury: Outdoor Life Books, 1987.

Maguire, H. F., and C. W. Severinghaus. "Wariness as an Influence on Age Composition of White-tailed Deer Killed by Hunters." *New York Fish and Game Journal*, vol. 1, no. 1, p. 98–109, January 1954.

Mangold, Robert. *Analysis of Reproductive Data From Does Collected During Hunter's Choice Deer Season*. Trenton, New Jersey: New Jersey Bureau of Wildlife Management, January 31, 1962.

Martin, Alexander C., Herbert S. Zim, and Arnold L. Nelson. *American Wildlife and Plants*. New York: McGraw-Hill Book Company, 1951.

Martinka, C. J. "Habitat Relationships of White-tailed and Mule Deer in Northern Montana." *Journal of Wildlife Management*, vol. 32, no. 3, p. 558, 1968.

Matschke, George H. "Microincapsulated Diethylstilbestrol as an Oral Contraceptive in White-tailed Deer." *Journal of Wildlife Management*, vol. 41, no. 1, p. 87, 1987.

Mautz, William W., and Jeffrey Fair. "Energy Expenditure and Heart Rate for Activities of White-tailed Deer." *Journal of Wildlife Management*, vol. 44, no. 2, p. 333–342, 1980.

McCulloch, Clay Y., and Philip J. Urness. "Deer Nutrition in Arizona Chaparral and Desert Habitats." *Special Report #3*, Arizona Game and Fish Department, 1973.

McCullough, Dale R. *The George Reserve Deer Herd*. Ann Arbor, Michigan: University of Michigan Press, 1979.

————. "Sex Characteristics of Black-tailed Deer Hooves." *Journal of Wildlife Management,* vol. 29, no. 1, p. 210, 1965.

McLean, Donald D. "The Deer of California, with Particular Reference to the Rocky Mountain Mule Deer." *California Fish and Game,* vol. 26, no. 2 Sacramento, California: California Fish and Game, 1940.

Menzel, Karl. *The Deer of Nebraska.* Lincoln, Nebraska: Nebraska Game and Parks Commission, 1975.

Michael, Edwin D. "Birth of White-tail Deer Fawns." *Journal of Wildlife Management,* vol. 28, no. 1, p. 171, 1964.

Miller, Frank L. "Distribution Patterns of Black-tailed Deer in Relation to Environment." *Journal of Mammalogy,* vol. 51, no. 2, p. 248, 1970.

Miller, Gerrit S., and Remington Kellogg. *List of North American Recent Mammals.* Washington, D.C.: Smithsonian Institution, 1955.

Miller, Karl V., et al. "Variations in Density and Chemical Composition of White-tailed Deer Antlers." *Journal of Mammalogy,* vol. 66, no. 4, p. 693–701, 1985.

Milne, Lorus J., and Margery Milne. *The Senses of Animals and Men.* New York: Atheneum, 1962.

Mireau, Gary W. "Studies on the Biology of an Antlered Female Mule Deer." *Journal of Mammalogy,* vol. 53, no. 2, p. 403, 1972.

Missouri Department of Conservation Newsletter. *Missouri's Biggest Whitetails.* Jefferson City, Missouri: Missouri Department of Conservation, December 1975.

Moen, Aaron N. "Seasonal Changes in Heart Rates, Activity, Metabolism, and Forage Intake of White-tailed Deer." *Journal of Wildlife Management,* vol. 42, no. 4, p. 715–738, 1978.

Monson, Keith. *Remember: The Deer Do.* Monson, Cooperstown, North Dakota: 1985.

Montgomery, G. G. "Nocturnal Movements and Activity Rhythms of White-tailed Deer." *Journal of Wildlife Management,* vol. 27, no. 3, p. 442, 1963.

Mosby, Henry S., ed. *Wildlife Investigational Techniques.* Washington, D.C.: The Wildlife Society, 1963.

Muller-Schwarze, Dietland. "Responses of Young Black-tailed Deer to Predator Odors." *Journal of Mammalogy,* vol. 53, no. 2, p. 393, 1972.

Murphy, Dean A. "Deer Range Appraisal in the Midwest," *White-tailed Deer in the Midwest Symposium,* Columbus, Ohio, 1968.

Murphy, Dean A., and Hewlette S. Crawford. *Wildlife Foods and Understory Vegetation in Missouri's National Forests.* Missouri Department of Conservation, 1970.

Murphy, D. A., and J. A. Coates. "Effects of Dietary Protein on Deer." *Transactions of the North American Wildlife and Natural Resources Conference,* 31: 129–139, 1966.

Mustard, Eldie W., and Veron Wright. *Food Habits of Iowa Deer.* Des Moines, Iowa: State Conservation Commission, 1964.

Nagy, Julius G., Harold W. Steinhoff, and Gerald M. Ward. "Effects of Essential Oils of Sagebrush on Deer Rumen Microbial Function." *Journal of Wildlife Management,* vol. 28, no. 4, p. 785, 1964.

Nagy, Julius G., G. Vidacs, and G. M. Ward. "Previous Diet of Deer, Cattle and Sheep and Ability to Digest Alfalfa Hay." *Journal of Wildlife Management,* vol. 31, no. 3, p. 443, 1967.

Neff, Don J. "Forage Preferences of Trained Mule Deer on the Beaver Creek Watersheds." *Special Report # 4,* Arizona Game and Fish Department, 1974.

Nellis, Carl H. "Antler from Right Zygomatic Arch of White-tailed Deer." *Journal of Mammalogy,* vol. 46, no. 1, p. 108, 1965.

Nelson, Michael E., and L. David Mech. "Home-range Formation and Dispersal of Deer in Northeastern Minnesota." *Journal of Mammalogy,* vol. 65, no. 4, p. 567–575, 1984.

———. "Observation of a Swimming Wolf Killing a Swimming Deer." *Journal of Mammalogy,* vol. 65, no. 1, p. 143, 1984.

———. "Observation of a Wolf Killed by Deer." *Journal of Mammalogy,* vol. 66, no. 1, p. 187–188, 1985.

New Jersey Division of Fish, Game and Shellfish. "New Jersey's White-tailed Deer." *Deer Report #3,* 1976.

Newson, William Monypeny. *Whitetailed Deer.* New York: Charles Scribner's Sons, 1926.

Nixon, Charles M., Milford W. McClain, and Kenneth R. Russell. "Deer Food Habits and Range Characteristics in Ohio." *Journal of Wildlife Management,* vol. 34, no. 4, p. 870, 1970.

Owen, Wilbur B., Jr. *A Survey of the Helminth Parasites of White-tailed Deer in Five Regions of Arkansas.* Fayetteville, Arkansas: University of Arkansas, 1974.

Ozoga, John J. "Aggressive Behavior of White-tailed Deer at Winter Cuttings," *Journal of Wildlife Management,* vol. 36, no. 3, p. 861, 1972.

———. "Some Longevity Records for Female White-tailed Deer in Northern Michigan." *Journal of Wildlife Management,* vol. 33, no. 4, p. 1027, 1969.

Ozoga, John J., and Louis J. Verme. "Activity Patterns of White-tailed Deer During Estrus." *Journal of Wildlife Management,* vol. 39, no. 4, p. 679, 1975.

———. "Comparative Breeding Behavior and Performance of Yearling vs. Prime-age White-tailed Bucks." *Journal of Wildlife Management,* vol. 49, no. 2, p. 364–372, 1985.

———. "Effect of Family-bond Deprivation on Reproductive Performance in Female White-tailed Deer." *Journal of Wildlife Management,* vol. 48, no. 4, p. 1326–1334, 1984.

———. "Initial and Subsequent Maternal Success of White-tailed Deer." *Journal of Wildlife Management,* vol. 50, no. 1, p. 122–124, 1986.

———. "Winter Feeding Patterns of Penned White-tailed Deer." *Journal of Wildlife Management,* vol. 34, no. 2, p. 431, 1970.

Ozoga, John J., Louis J. Verme, and Craig S. Riemz. "Parturition Behavior and Territorality in White-tailed Deer: Impact on Neonatal Mortality." *Journal of Wildlife Management,* vol. 46, no. 1, p. 1–11, 1982.

Parker, Katherine L., Charles T. Robbins, and Thomas A. Hanley. "Energy Expenditures for Locomotion by Mule Deer and Elk." *Journal of Wildlife Management,* vol. 84, no. 2, p. 474–488, 1984.

Patton, Art. *1975 Deer Mortality Report.* Truro, Nova Scotia: Nova Scotia Department of Lands and Forests, 1975.

Peabody, Bill. "Of Teeth and Time." *Kansas Fish and Game Magazine,* vol. 29, no. 5, p. 23, 1975.

Pearson, Henry A. "Rumen Organism in White-tailed Deer from South Texas." *Journal of Wildlife Management,* vol. 29, no. 3, p. 493, 1965.

Petraborg, W. H., and V. E. Gunvalson. "Observations of Bobcat Mortality and Bobcat Predation on Deer." *Journal of Mammalogy,* vol. 43, no. 3, p. 430, 1962.

Pratt, Jerome J. *White Flags of Apacheland.* New York: Vantage Press, 1966.

Pruitt, William O., Jr. "Behavior of the Barren-ground Caribou." *Biological Papers of the University of Alaska. #3.* Fairbanks, Alaska, 1960.

———. "Rutting Behavior of the Whitetail Deer." *Journal of Mammalogy,* vol. 35, no. 1, p. 129–130, 1954.

Quay, W. B., and Dietland Muller-Schwarze. "Functional Histology of Integumentary Glandular Regions in Black-tailed Deer." *Journal of Mammalogy,* vol. 51, no. 4, p. 675, 1970.

———. "Geographic Variation in the Metatarsal Gland of the White-tailed Deer." *Journal of Mammalogy,* vol. 52, no. 1, p. 1, 1971.

Reed, Dale F., Thomas N. Woodard, and Thomas M. Pojar. "Behavioral Response of Mule Deer to a Highway Underpass." *Journal of Wildlife Management,* vol. 39, no. 2, p. 361, 1975.

Rees, John W., Robert A. Kainer, and Robert W. Davis. "Histology, Embryology and Gross Morphology of the Mandibular Dentition in Mule Deer." *Journal of Mammalogy,* vol. 27, no. 4, p. 640, 1966.

Reynolds, Hudson G. "Mule Deer Killed by Lightning." *Journal of Mammalogy,* vol. 46, no. 4, p. 676, 1965.

Richardson, Larry W., et al. "Acoustics of White-tailed Deer." *Journal of Mammalogy,* vol. 64, no. 2, p. 245–252, 1983.

Roberts, Harvey A. "Why Check the Deer Kill?" *Pennsylvania Game News,* December 1966.

Robinette, W. Leslie, and Dale A. Jones. "Antler Anomalies of Mule Deer." *Journal of Mammalogy,* vol. 40, no. 7, p. 96, 1959.

Robinette, W. Leslie, Dale A. Jones, et al. "Effects of Nutritional Change on Captive Mule Deer." *Journal of Wildlife Management,* vol. 37, no. 3, p. 312, 1973.

Robinson, William L. "Social Dominance and Physical Condition Among Penned White-tailed Deer Fawns." *Journal of Mammalogy,* vol. 43, no. 4, p. 462, 1962.

Roosevelt, Theodore, et al. *The Deer Family.* New York: The Macmillan Company, 1902.

Rue III, Leonard Lee. "A Deer In A Dishpan." *Outdoor Life,* October 1985.

———. *Sportsman's Guide to Game Animals.* New York: Outdoor Life Books-Harper and Row, 1968.

———. "Whitetails of Vermont." *Deer and Deer Hunting Magazine,* vol. 8, no. 4, p. 44–49, March–April 1985.

———. *The World of the White-tailed Deer.* Philadelphia and New York: J. B. Lippincott, 1962.

Russo, John P. "The Kaibab North Deer Herd." *Wildlife Bulletin #7,* Arizona Game and Fish Department, 1964.

Samuel, William M. "Parasites of Pennsylvania Deer." *Pennsylvania Game News,* November 1967.

Sauer, Peggy R., John W. Tanck, and C. W. Severinghaus. "Herbaceous Food Preferences of White-tailed Deer." *New York Fish and Game Journal,* vol. 16, no. 1, p. 16, July 1969.

Saunders, Barry P. "Meningeal Worm in White-tailed Deer in Northwestern Ontario and Moose Population Densities," *Journal of Wildlife Management,* vol. 37, no. 3, p. 327, 1973.

Schemnitz, Sanford D. "Maine-Island-Mainland Movement of White-tailed Deer." *Journal of Mammalogy,* vol. 56, no. 2, p. 535, 1975.

Schiedermayer, David L. "Car-Deer Collisions: Close Encounters of the Worst Kind." *Deer and Deer Hunting Magazine,* vol. 8, no. 5, May–June 1985.

Sealander, John A., Philip S. Gipson, Michael Cartwright, and James M. Pledger. *Behavioral and Physiological Studies of Relationships Between White-tailed Deer and Dogs in Arkansas.* Fayetteville: Department of Zoology, University of Arkansas, 1975.

Seton, Ernest Thompson. *Lives of Game Animals.* Boston: Charles T. Branford, 1953.

Severinghaus, C. W. "Deer Weights as an Index of Range Conditions on Two Wilderness Areas in the Adirondack Region." *New York Fish and Game Journal,* vol. 2, no. 2, p. 216–18, July 1955.

———. "Overwinter Weight Loss in Whitetailed Deer in New York." *New York Fish and Game Journal,* vol. 28, no. 1, p. 61–67, January 1981.

———. "Some Observations on the Breeding Behavior of Deer." *New York Fish and Game Journal,* vol. 2, no. 2, p. 239–241, July 1955.

Severinghaus, C. W., and Aaron N. Moen. "Prediction of Weight and Reproductive Rates of a White-tailed Deer Population from Records of Antler Beam Diameter Among Yearling Males." *New York Fish and Game Journal,* vol. 30, no. 1, p. 30–38, January 1983.

Severinghaus, C. W., and Benjamin F. Tullar, Jr. "Wintering Deer vs. Snowmobiles." *The Conservationist,* June–July 1975.

Shafer, Elwood L., Jr. "The Twig-Count Method for Measuring Hardwood Deer Browse." *Journal of Wildlife Management,* vol. 28, no. 3, p. 428, 1963.

Shaw, Harley. "Insectivorous White-tailed Deer." *Journal of Mammalogy,* vol. 44, no. 2, p. 284, 1963.

Shoener, Tom. "Fifteen Years of Big Bucks." *Maine Fish and Game,* Fall 1973.

Shope, Richard E., L. G. MacNamara, and Robert Mangold. "Epizootic Hemorrhagic Disease of Deer." *New Jersey Outdoors Magazine,* November 1955.

Short, Cathleen. "Morphological Development and Aging of Mule and White-Tailed Deer Fetuses." *Journal of Wildlife Management,* vol. 34, no. 2, p. 383, 1970.

Short, Henry L. "Rumen Fermentations and Energy Relationships in White-tailed Deer." *Journal of Wildlife Management.* vol. 27, no. 2, p. 184, 1963.

———. "Postnatal Stomach Development of White-tailed Deer." *Journal of Wildlife Management,* vol. 28, no. 3, p. 445, 1964.

Siegler, Hilbert R., Helenette Silver, David L. White, and Henry A. Laramie, Jr. "The White-tailed Deer of New Hampshire." *Survey Report # 10,* New Hampshire Fish and Game Department, 1968.

Silver, Helenette. "Deer Milk Compared with Substitute Milk for Fawns." *Journal of Wildlife Management,* vol. 25, no. 1, p. 666, 1961.

———. "An Instance of Fertility in a White-tailed Buck Fawn." *Journal of Wildlife Management,* vol. 29, no. 3, p. 634, 1965.

Silver, Helenette, N. F. Colovos, J. B. Holter, and H. H. Hayes. "Fasting Metabolism of White-tailed Deer." *Journal of Wildlife Management,* vol. 33, no. 3, p. 490, 1969.

Sinclair, Sandra. *How Animals See: Other Visions of Our World.* New York: Facts on File, 1985.

Smith, Richard P. "Oldest Buck on Record?" *Deer and Deer Hunting Magazine,* vol. 8, no. 43, p. 96–98, September–October 1984.

———. "Winter Severity and Its Impact on Deer." *Deer and Deer Hunting Magazine,* vol. 7, no. 4, p. 14–21, March–April 1984.

Smith, Robert Leo. *Ecology and Field Biology,* 2d ed. New York: Harper and Row, 1974.

Snider, Carl C., and J. Malcolm Asplund. "Invitro Digestibility of Deer Foods from the Missouri Ozarks." *Journal of Wildlife Management,* vol. 38, no. 1, p. 20, 1974.

Stadtfeld, Curtis K. *Whitetail Deer, A Year's Cycle.* New York: Dial Press, 1975.

Stone, Ward B., and John R. Palmateer. "A Bird Ingested by a Whitetailed Deer." *New York Fish and Game Journal,* vol. 17, no. 1, p. 63, January 1970.

Stone, Ward B., Stuart L. Free, William T. Hesselton, and Lawrence W. Jackson. "Polydactylism in a White-tailed Deer from New York." *New York Fish and Game Journal,* vol. 17, no. 2, July 1970.

Stout, Gene G. "Effects of Coyote Predation on White-tailed Deer Productivity in Fort Sill, Oklahoma." *Wildlife Society Bulletin,* vol. 10, no. 4, p. 329–332, Winter 1982.

Strung, Norman. *Deer Hunting.* Philadelphia and New York: J. B. Lippincott Company, 1973.

Sullivan, Thomas G. "Influence of Wolverine (*Gulo gulo*) Odor on Feeding Behavior of Snowshoe Hares." *Journal of Mammalogy,* vol. 67, no. 2, p. 385–388, 1986.

Swank, Wendell G. "The Mule Deer in Arizona Chaparral." *Wildlife Bulletin #3,* Arizona Game and Fish Department, 1958.

Sweeney, J. R., R. Larry Marchinton, and J. M. Sweeney. "Responses of Radio-Monitored White-tailed Deer Chased by Dogs." *Journal of Wildlife Management,* vol. 35, no. 4, p. 707, 1971.

Taber, Richard D., and Raymond F. Dasmann. "The Black-tailed Deer of the Chaparral." *Game Bulletin #8,* California Department of Fish and Game, 1958.

Taft, Edgar B., Thomas C. Hall, and Joseph Aub. "The Growth of the Deer Antler." *The Conservationist,* February–March 1956.

Taylor, Walter P., ed. *The Deer of North America.* A Wildlife Management Institute Book. Harrisburg, Pennsylvania: The Stackpole Company, 1956.

Teer, James G., Jack Thomas, and Eugene Walker. "Ecology and Management of White-tailed Deer in Llano Basin of Texas." Journal of Wildlife Management, Monograph #15, October 1965.

Thomas, D. C., and I. D. Smith. "Reproduction in a Wild Black-tailed Deer Fawn." *Journal of Mammalogy,* vol. 54, no. 1, p. 302, 1973.

Torgerson, Oliver, and William H. Pfander. "Cellulose Digestibility and Chemical Composition of Missouri Deer Foods." *Journal of Wildlife Management,* vol. 35, no. 2, p. 221, 1971.

Tori, Gildo. "Deer's Biological Clock Off." *The Parkersburg* [West Virginia] *Sentinel,* February 13, 1985.

Trefethen, James B. *An American Crusade for Wildlife.* New York: Winchester Press and Boone and Crockett Club, 1975.

Trodd, L. L. "Quadruplet Fetuses in a White-tailed Deer from Espanola, Ontario." *Journal of Mammalogy,* vol. 43, no. 3, p. 414, 1962.

Tullar, Benjamin F., Jr. "A Long-lived White-tailed Deer." *New York Fish and Game Journal,* vol. 30, no. 1, p. 119, January 1983.

Ullrey, D. E., H. E. Johnson, W. G. Youatt, L. D. Fay, B. L. Schdepke, and W. T. Magee. "A Basal Diet for Deer Nutrition Research." *Journal of Wildlife Management,* vol. 35, no. 1, p. 57, 1971.

Ullrey, D. E., W. G. Youatt, H. E. Johnson, L. D. Fay, and B. L. Bradley. "Protein Requirement of White-tailed Deer Fawns," *Journal of Wildlife Management,* vol. 31, no. 1, p. 679, 1967.

Ullrey, D. E., W. G. Youatt, H. E. Johnson, P. K. Fu, and L. D. Fay. "Digestibility of Cedar and Aspen Browse for the White-tailed Deer." *Journal of Wildlife Management,* vol. 28, no. 4, p. 791, 1964.

United States Fish and Wildlife Service, U.S. Department of the Interior. *1975 National Survey of Hunting, Fishing and Wildlife-Associated Recreation.* Washington, D.C.: U.S. Fish and Wildlife Service, 1977.

Verme, Louis J. "Fecundity in a Michigan White-tailed Deer." *Journal of Mammalogy,* vol. 43, no. 1, p. 112, 1962.

———. "Progeny Sex Ratio Relationships in Deer: Theoretical vs. Observed." *Journal of Wildlife Management,* vol. 49, no. 1, p. 134–136, 1985.

————. "Reproduction Studies on Penned White-tailed Deer." *Journal of Wildlife Management,* vol. 29, no. 1, p. 74, 1965.

————. "Sex Ratio Variation in *Odocoileus:* A Critical Review." *Journal of Wildlife Management,* vol. 47, no. 3. p. 573–582, 1983.

Verme, Louis J., and John J. Ozoga. "Effects of Diet on Growth and Lipogenesis in Deer Fawns." *Journal of Wildlife Management,* vol. 44, no. 2, p. 315–323, 1980.

————. "Sex Ratio of White-tailed Deer and the Estrus Cycle." *Journal of Wildlife Management,* vol. 45, no. 3, p. 710–715, 1981.

Wallmo, Olaf C., ed. *Mule and Black-tailed Deer of North America.* A Wildlife Management Institute Book. Lincoln, Nebraska: University of Nebraska Press, 1981.

Walther, Fritz. R. *Communication and Expression in Hooved Animals.* Bloomington, Indiana: Indiana University Press, 1984.

Wambold, H. R. *Bowhunting for Deer.* Harrisburg, Pennsylvania: Stackpole Books, 1978.

Warner, Roger. "Ticks and Deer Team Up to Cause Trouble for Man." *Smithsonian,* vol. 17, no. 1, p. 131–146, April 1986.

Wegner, Rob. "Acorns and Whitetails." *Deer and Deer Hunting Magazine,* vol. 4, no. 6, p. 12–19, August 1981.

————. "Stalking the Deer Poacher." *Deer and Deer Hunting Magazine,* vol. 9, no. 3, p. 10–29, January–February 1986.

————. "Whitetails and Alfalfa." *Deer and Deer Hunting Magazine,* vol. 6, no. 33, p. 8–17, January–February 1983.

White, David L. *Deer Kill Summary: Losses to Cars and Dogs, 1947–1976.* New Hampshire Fish and Game Department, 1976.

White, Marshall. "Description of Remains of Fawn Killed by Coyote." *Journal of Mammalogy,* vol. 54, no. 1, p. 291, 1973.

White, Marshall, Frederick F. Knowlton, and W. C. Glazener. "Effects of Dam-Newborn Fawn Behavior on Capture and Mortality." *Journal of Wildlife Management,* vol. 36, no. 3, p. 897, 1972.

Wiggers, Ernie P., and Samuel Beason. "Characterization of Sympatric or Adjacent Habitats of Two Deer Species in West Texas." *Journal of Wildlife Management,* vol. 50, no. 1, p. 129–134, 1986.

Wildlife and Fisheries Commission. "How Much Did that Deer Weigh on the Hoof?" *Bulletin,* Wildlife and Fisheries Commission, 1976.

Winter, Ruth. *The Smell Book: Scents, Sex and Society.* Philadelphia and New York: J. B. Lippincott Company, 1976.

Wishart, William D., "Frequency of Antlered White-tailed Does in Camp Wainwright, Alberta." *Journal of Wildlife Management,* vol. 49, no. 2, p. 386–388, 1985.

———. "Hybrids of White-tailed and Mule Deer in Alberta." *Journal of Mammalogy,* vol. 61, no. 4, p. 716–720.

Wobeser, G., and W. Runge. "Rumen Overload and Rumenitis in White-tailed Deer." *Journal of Wildlife Management,* vol. 39, no. 3, p. 596, 1975.

Wootters, John. "Buck Sense." *Petersen's Hunting Magazine,* p. 20–21, June 1981.

Young, Stanley P., and Edward A. Goldman. *The Puma, Mysterious American Cat.* Washington, D.C.: American Wildlife Institute, 1946.

Zagata, Michael, and Aaron N. Moen. "Antler Shedding by White-tailed Deer in Midwest." *Journal of Mammalogy,* vol. 55, no. 3, p. 656, 1974.

Zagata, Michael, and Arnold O. Haugen. "Influence of Light and Weather on Observability of Iowa Deer." *Journal of Wildlife Management,* vol. 38, no. 2, p. 220, 1974.

INDEX

intelligence and, 225
maternal, 251
Intelligence
coyote, 397
hunting pressure and, 343
instinct and, 225
subspecies variations, 348-349
Interbreeding
coyote, 398-399
deer, 337-340
Interdigital glands
alarm signals, 230
described, 116-117
Intestines, 171-172
Inyo mule deer, 35

Jackson, Lawrence W., 334
Jacobsen, Nadine, 257
Jacobson's organ, 315
Jaskaniec, Mike, 184
Jaw, 130. *See also* Teeth
Jones, Boyd, 157, 161
Jordan, James, 27
Jumping, 56-59

Kacsinko, Frank, 91
Kammermeyer, K.E., 306
Kansas whitetail deer, 32
Kehoe, Edward F., 481, 482
Keratin
hoof, 41
horns, 68
Key deer. *See* Florida Key whitetail
deer
Killing shots, 350-351
King, Bill, 335
Kolenosky, G.B., 403

Lacrymal glands. *See* Preorbital
(lacrymal) glands
Lactation
milk content, 261-262
nursing, 254-256
nutrition and, 249-250, 251
season and, 303
See also Nursing
Lactones, 117
Landscape damage protection,
442-444
Lang, Lincoln, 440
Laramie, Harry, 107
Lauckhart, J. Burton, 402
Laux, Charles, 415-416

Laws. *See* Game laws; Wildlife
management
Leabo, Henry, 372
Lead poisoning, 433
Learning
hunting pressure and, 343
migration routes and, 304
See also Instinct; Intelligence
Leaves, 302
Legs
gait and, 49-60
lie down and rise behavior, 62-64
snow and, 372
see also Gait
Lenander, Carl J., Jr., 159, 160
Leopold, A. Starker, 155
Licking behavior, 124
Licking branch, 205-206, 208,
309, 311
Licking stick, 211-212
Lie down behavior, 62-63
Life span, 133-138
hunting pressure and, 133
records for, 134-138
sex differences and, 133-134
teeth and, 129, 137
See also Age level
Lightning, 271
Light (warning), 442
Lingle, Ben, 137
Linnaeus, Carolus, 21-22
Linsdale, Jean M., 400, 406
Lipogenesis, 295
Lipomas, 286, 287
Liver, 380
Liver flukes, 282
Livetrapping, 370-371, 468-470
Locked antlers, 322, 324-325
Longevity. *See* Life span
Loomis, Forrest, 149
Lord, Rexford, 131
Lorenz, Konrad, 215
Louse flies, 277-278
Loveless, Charles M., 110, 126
Lumber companies, 445
Lung shots, 351, 353, 356
Lungworms, 282
Lyme disease, 279-281
Lynx, 116, 405

MacNamara, Les, 284
Madson, John, 316
Mallard duck, 339

Ozoga, John, 138, 375

Palmation, 95
Palmer, Cynthia, 415
Palmer, R.W., 134
Papillomas, 286-287, 425
Parasites, 275-284
    external, 275-281
    internal, 281-284
Patha, Henry, 253
Patten, Barney, 183
Paunch, 391-392. *See also* Digestive
    system
Paunch shots, 355
Pause scrapes, 205. *See also* Rubs;
    Scrapes
Peabody, W.C., 338-339
Pearlation, 74
Pears, 301-302
Pearson, Tom, 405
Pedicel, 69-70, 73, 77, 103, 104
Penis
    breeding activity, 332
    masturbation, 312-313
Periosteum, 70, 73
Permanone repellent, 281
Perrige, Dwain, 160
Peruke, 97
Peter, Arnold, 157
Peterson, Pete, 183
Pheromones
    rutting behavior, 315, 329
    sudoriferous glands, 114-115
    tarsal glands, 117
Photoperiodism
    antler development and, 70, 76-77,
        99
    antler shedding and, 103
    bird migrations and, 294
    fetal development and, 234
    rutting season and, 103, 336
Pigs, 300
Pimlott, D.H., 131
Pineal gland, 73, 76
Pituitary gland, 251
    antler calcification and, 99
    antler shedding and, 103
    fetal development and, 234
    testicles and, 289
Placenta, 244, 248
Poachers and poaching, 40, 471-478
    danger of, 478
    ethics and, 471-472

extent of, 477-478
increase in, 471
law enforcement problems,
    473-475
professional, 472
profile of, 475-476
statistics on, 474
trophy deer, 472-473, 477
Poisons, 398, 419
Politics, 479-484
Populations. *See* Deer populations;
    Overpopulation
Porath, Wayne, 160
Potter, Warren S., 156
Power lines, 438, 456-457
Powless, John, 406
Pratt, Jerome J., 153, 338
Predation
    artificial feeding and, 462
    bears, 406
    birthing and, 244, 245
    bobcat, 405
    communication and, 216
    cougar, 401-403
    coyote, 397-401
    dogs, 388, 407-416
    eagles, 406-407
    elimination of, 419
    fawns, 260, 264
    feeding behavior and, 164
    fox, 406
    game management and, 397-419,
        428, 490-491
    lynx, 405
    prey-predator relationship,
        416-419
    tail signals, 215
    wolves, 403-405
Preorbital (lacrymal) glands
    described, 114-116
    scrapes and, 206, 311
    subspecies variation, 340
Presevationists, 387-389, 426-428,
    490-491
Primary scrapes. *See* Scrapes
Professional hunting, 15-16. *See also*
    Hunting; Hunting pressure;
    Poachers and poaching
Prolactin, 251
Pronghorn antelope, 68-69, 106, 213
Pruitt, William, Jr., 203-204
Pruitt, William O., 204
Psychic communication, 223-224